Category Theory
———— and ————
Applications

A Textbook for Beginners

Category Theory

— and —

Applications

A Textbook for Beginners

Marco Grandis
Università di Genova, Italy

W World Scientific

NEW JERSEY · LONDON · SINGAPORE · BEIJING · SHANGHAI · HONG KONG · TAIPEI · CHENNAI · TOKYO

Published by

World Scientific Publishing Co. Pte. Ltd.

5 Toh Tuck Link, Singapore 596224

USA office: 27 Warren Street, Suite 401-402, Hackensack, NJ 07601

UK office: 57 Shelton Street, Covent Garden, London WC2H 9HE

Library of Congress Cataloging-in-Publication Data
Names: Grandis, Marco, author.
Title: Category theory and applications : a textbook for beginners / by Marco
   Grandis (Università di Genova, Italy).
Description: New Jersey : World Scientific, 2018. | Includes bibliographical
   references and index.
Identifiers: LCCN 2017042912 | ISBN 9789813231061 (hardcover : alk. paper)
Subjects: LCSH: Categories (Mathematics) | Mathematical analysis. | Morphisms
   (Mathematics) | Homology theory.
Classification: LCC QA278 .G6985 2018 | DDC 512/.6--dc23
LC record available at https://lccn.loc.gov/2017042912

British Library Cataloguing-in-Publication Data
A catalogue record for this book is available from the British Library.

For any available supplementary material, please visit
http://www.worldscientific.com/worldscibooks/10.1142/10737#t=suppl

Printed in Singapore

To

Maria Teresa

Preface

Category Theory now permeates most of Mathematics, large parts of theoretical Computer Science and parts of theoretical Physics. Its unifying power brings together different branches, and leads to a deeper understanding of their roots.

This book is addressed to students and researchers of these fields and can be used as a text for a first course in Category Theory. It covers its basic tools, like universal properties, limits, adjoint functors and monads. These are presented in a concrete way, starting from examples and exercises taken from elementary Algebra, Lattice Theory and Topology, then developing the theory together with new exercises and applications.

A reader should have some elementary knowledge of these three subjects, or at least two of them, in order to be able to follow the main examples, appreciate the unifying power of the categorical approach, and discover the subterranean links brought to light and formalised by this perspective.

Applications of Category Theory form a vast and differentiated domain. This book cannot give a global picture of them, but wants to present the basic applications and a choice of more advanced ones, based on the interests of the author. References are given for applications in many other fields.

Contents

Introduction

0.1 Categories

Category Theory originated in an article of Eilenberg and Mac Lane [EiM], published in 1945. It has now developed into a branch of mathematics, with its own internal dynamics. Here we want to present its elementary part, and its strong links with the origins in Algebra and Topology.

Many mathematical theories deal with a certain kind of mathematical objects, like groups, or ordered sets, or topological spaces, etc. Each kind has its own privileged mappings, that preserve the structure in some sense, like homomorphisms of groups, or order-preserving mappings, or continuous mappings.

We have thus the 'category' Gp of groups and their homomorphisms, Ord of ordered sets and monotone mappings, Top of topological spaces and continuous mappings, and so on. More elementarily, we have the category Set of sets and their mappings.

In all these instances, the privileged mappings are called morphisms or arrows of the category; an arrow from the object X to the object Y is written as $f\colon X \to Y$.

Two consecutive morphisms, say $f\colon X \to Y$ and $g\colon Y \to Z$, can be composed giving a morphism $gf\colon X \to Z$; this partial composition law is 'as regular as it can be', which means that it is associative (when legitimate) and has a partial identity $\mathrm{id}X\colon X \to X$ (written also as 1_X) for every object, which acts as a unit for every legitimate composition.

'Concrete categories' are often associated with mathematical structures, in this way; but we shall see that categories are not limited to these instances, by far.

1

0.2 Universal properties

In these categories (and many 'similar' ones) we have cartesian products, constructed by forming the cartesian product of the underlying sets and putting on it the 'natural' structure of the kind we are considering, be it of algebraic character, or an ordering, or a topology, or something else.

All these procedures can be unified, so that we can better understand what we are doing: we have a family $(X_i)_{i \in I}$ of objects of a category, indexed by a set I, and we want to find an object X equipped with a family of morphisms $p_i \colon X \to X_i$ $(i \in I)$, called *cartesian projections*, which satisfies the following *universal property*:

- for every object Y and every family of morphisms $f_i \colon Y \to X_i$ $(i \in I)$ in the given category

$$
\begin{array}{ccc}
Y & \dashrightarrow^{f} & X \\
 & {}_{f_i}\searrow & \downarrow{}^{p_i} \\
 & & X_i
\end{array}
\tag{0.1}
$$

there exists precisely one morphism $f \colon Y \to X$ such that $p_i f = f_i$, for all $i \in I$.

We shall see that this property determines the solution *up to isomorphism*, i.e. an invertible morphism of the given category. In fact, the proof is quite easy and some hints can be useful as of now: given two solutions $(X, (p_i))$, $(Y, (q_i))$, we can determine two morphisms $f \colon X \to Y$ and $g \colon Y \to X$ and prove that their composites coincide with the identities of X and Y.

(Let us remark, incidentally, that the product topology is much less obvious than the other product structures we have mentioned; the universal property tells us that it is indeed the 'right' choice.)

These facts bring to light a crucial aspect: a categorical definition (as the previous one) *is based on morphisms and their composition*, while the objects only step in as domains and codomains of arrows. If we want to understand what unifies the product of a family (X_i) of sets, or groups, or ordered sets, or topological spaces we should forget the nature of the objects, and think of the family of cartesian projections $p_i \colon X \to X_i$, together with the previous property. Then – in each category we are interested in – we come back to the objects in order to prove that a solution exists (and also to fix a particular solution, when this can be useful).

From a structural point of view, a category only 'knows' its objects by their morphisms and composition.

This is even more evident if we think of another procedure, which in category theory is called a 'sum', or 'coproduct'. We start again from a family $(X_i)_{i \in I}$ of objects; its (categorical) *sum* is an object X equipped with a family of morphisms $u_i \colon X_i \to X$ $(i \in I)$, called *injections*, which satisfies the following universal property:

- for every object Y and every family of morphisms $f_i \colon X_i \to Y$ $(i \in I)$ in the given category

$$
\begin{array}{ccc}
X & \overset{f}{\dashrightarrow} & Y \\
{\scriptstyle u_i}\big\uparrow & \nearrow{\scriptstyle f_i} & \\
X_i & &
\end{array}
\tag{0.2}
$$

there exists precisely one morphism $f \colon X \to Y$ such that $f u_i = f_i$, for all $i \in I$.

Again, the solution is determined up to isomorphism; its existence depends on the category.

The sum of a family of objects is easy to construct in the category Set of sets, by a disjoint union. We have similar solutions in Ord and Top. But in Gp the categorical sum of a family of groups is called the *free product* of the family; its construction is rather complex, and its underlying set has little to do with the disjoint union of the sets underlying the groups. The categorical approach highlights the fact that we are 'solving the same problem' and – finally – makes clear what we are doing.

It is also important to note that the categorical definitions of product and coproduct are *dual* to each other: each of them is obtained from the other by reversing all arrows and compositions. This only makes sense – in general – within category theory, because the dual of a given category, formed by reversing its arrows and partial composition law, is a formal construction: the dual of a category of structured sets and mappings is *not* a category of the same kind (even though, *in certain cases*, it may essentially be, as a result of some duality theorem).

0.3 Diagrams in a category

In a category the objects and morphisms that we are considering are often represented by vertices and arrows in a *diagram*, as above in (0.1) and (0.2), to make evident their relationship and which compositions are legitimate.

As an important property, satisfied in the previous cases, we say that a diagram in a category is *commutative* if:

(i) whenever we have two 'paths' of consecutive arrows, from a certain object to another, the two composed morphisms are the same,

(ii) whenever we have a 'loop' of consecutive arrows, from an object to itself, then the composed morphism is the identity of that object.

Let us consider some other instances

$$
\begin{array}{ccc}
X \xrightarrow{\ f\ } Y & A \xrightarrow{\ f\ } B & \\[-2pt]
_{h}\searrow\ \downarrow{g} & ^{h}\downarrow\ \searrow^{d}\ \downarrow^{k} & X \underset{v}{\overset{u}{\rightleftarrows}} Y \\[-2pt]
\qquad Z & C \xrightarrow{\ g\ } D &
\end{array}
\qquad (0.3)
$$

The first diagram above is commutative if and only if $gf = h$. For the second, commutativity means that $kf = d = gh$. For the third, it means that $vu = \mathrm{id}X$ and $uv = \mathrm{id}Y$; let us note that in this case the morphisms u and v are inverse to each other, and each of them is an isomorphism of our category.

Formal definitions of these notions, diagrams and commutative diagrams, will be given in 1.4.9.

0.4 Functors

It becomes now possible to view on the same level, so to say, mathematical theories of different branches, and formalise their links. A well-behaved mapping between categories is called a *functor*.

Among the simplest examples there are the *forgetful functors*, that forget the structure (or part of it), like:

$$\mathsf{Gp} \to \mathsf{Set}, \qquad\qquad \mathsf{Ord} \to \mathsf{Set}, \qquad\qquad \mathsf{Top} \to \mathsf{Set}, \qquad (0.4)$$

For instance, $U\colon \mathsf{Gp} \to \mathsf{Set}$ takes a group G to its underlying set $U(G)$, and a homomorphism $f\colon G \to G'$ to its underlying mapping $U(f)\colon U(G) \to U(G')$. The whole procedure is well-behaved, in the sense that it preserves composition and identities.

These functors are so obvious that they are often overlooked, in mathematics; but here it will be important to keep trace of them. In particular, we shall see that they often determine other functors 'backwards', which are much less obvious: like the *free-group* functor $F\colon \mathsf{Set} \to \mathsf{Gp}$, 'left adjoint' to $U\colon \mathsf{Gp} \to \mathsf{Set}$ (of which more will be said below).

In another range of examples, a reader with some knowledge of Algebraic Topology will know that the core of this discipline is constructing *functors from a category of topological spaces to a category of algebraic structures*, and using them to reduce topological problems to simpler algebraic ones. We have thus the sequence of *singular homology functors*, with values in the category of abelian groups

$$H_n\colon \mathsf{Top} \to \mathsf{Ab} \qquad\qquad (n \geqslant 0), \qquad\qquad (0.5)$$

and the fundamental group functor $\pi_1\colon \mathsf{Top}_\bullet \to \mathsf{Gp}$, defined on the category of pointed topological spaces and pointed continuous mappings.

0.5 Natural transformations and adjunctions

The third basic element of category theory is a *natural transformation* $\varphi\colon F \to G$ between two functors $F, G\colon \mathsf{C} \to \mathsf{D}$ with the same domain C and the same codomain D. We also write $\varphi\colon F \to G\colon \mathsf{C} \to \mathsf{D}$.

This simply amounts to a family of morphisms of D, indexed by the objects X of C

$$\varphi_X\colon F(X) \to G(X), \qquad\qquad (0.6)$$

under a condition of 'naturality' which will be made explicit in the text.

Here we just give an example, based on groups and free groups (but the reader can replace groups with semigroups, or abelian groups, or R-modules, or any algebraic structure defined by 'equational axioms'). We have mentioned in the previous point the forgetful functor $U\colon \mathsf{Gp} \to \mathsf{Set}$ and the free-group functor $F\colon \mathsf{Set} \to \mathsf{Gp}$, that turns a set X into the free group generated by X.

The insertion of X in $F(X)$, as its basis, is a canonical mapping in Set

$$\eta_X\colon X \to U(F(X)). \qquad\qquad (0.7)$$

All of them give a natural transformation

$$\eta\colon \mathrm{id} \to UF\colon \mathsf{Set} \to \mathsf{Set}, \qquad\qquad (0.8)$$

where id is the identity functor of the category Set (turning objects and arrows into themselves) and $UF\colon \mathsf{Set} \to \mathsf{Set}$ is the composed functor (turning each set into the underlying set of the free group on it). This natural transformation is essential in linking the functors U, F and making the functor $F\colon \mathsf{Set} \to \mathsf{Gp}$ *left adjoint to* $U\colon \mathsf{Gp} \to \mathsf{Set}$.

The link is represented by the *universal property* of the insertion of the basis, namely the fact that every mapping $f\colon X \to U(G)$ with values in a group G (more precisely in its underlying set) can be uniquely extended to a homomorphism $g\colon F(X) \to G$. Formally:

- for every morphism $f\colon X \to U(G)$ in Set, there exists precisely one morphism $g\colon F(X) \to G$ in Gp such that $U(g).\eta_X = f$.

One can reformulate this link in an equivalent presentation of the adjunction, which may be more familiar to the reader: there is a (natural) bijective correspondence

$$\varphi_{XG} \colon \mathsf{Gp}(F(X), G) \to \mathsf{Set}(X, U(G)),$$
$$\varphi_{XG}(g) = U(g).\eta_X, \tag{0.9}$$

between the set of group-homomorphisms $F(X) \to G$ and the set of mappings $X \to U(G)$.

We shall see many constructions of free algebraic structures, or more generally of left (or right) adjoints of given functors (in particular in Chapters 3, 4 and 6). Many of them are 'real constructions', which give a good idea of the backward procedure; others are so complicated that one can doubt of their constructive character. In such cases, one can prefer to prove the existence of the adjoint, by the Adjoint Functor Theorem (in Section 3.5) or some other general statement: then the result is determined up to isomorphism, and its universal property allows its use. Much in the same way as we can define a real function as the solution of a certain differential equation with initial data, as soon as we know that the solution exists and is unique.

0.6 A brief outline

The first chapter deals with the basic tools: categories, functors and natural transformations. Ordered sets can be seen as categories of a special kind; adjunctions in this particular case are well known, as 'covariant Galois connections'; studying them (in Section 1.7) will also serve as an elementary introduction to general adjunctions.

Chapter 2 introduces the limits in a category (including products and the classical projective limits) and their dual notion, the colimits (including sums and the classical inductive limits); universal arrows with respect to a functor are a general formulation of universal properties (in Section 2.7).

Chapter 3 studies the crucial notion of adjunction: many important constructions in mathematics can be described as an adjoint functor of some obvious functor: from free algebraic structures to Stone-Čech compactification, metric completion etc. We also explore the related notion of monad and its algebras, which investigates when a category can be thought of as a category of 'algebras' over another category, in a very wide sense.

The next three chapters explore applications of category theory in Algebra, Topology and Algebraic Topology, Homological Algebra.

Finally, Chapter 7 is a brief introduction to higher dimensional category theory, with some of its applications. The starting point is the fact that 'small' categories, functors and natural transformations – with all their composition laws – form a 'two-dimensional category'. Another aspect of the power of category theory is that it is able to 'study itself'.

0.7 Classes of categories

It is a common feature of Mathematics to look for the natural framework where certain properties should be studied: for instance, the properties of ordinary polynomials with real coefficients are usually examined in general polynomial rings, or in more general algebras. Besides yielding more general results, the natural framework gives a deeper comprehension of what we are studying.

Category theory makes a further step in this sense. For instance, categories of modules are certainly important, but – since Buchsbaum's *Appendix* [Bu2] and Grothendieck's paper [Gt] in the 1950's – a consistent part of Homological Algebra finds its natural framework in *abelian categories* and their generalisations (see Chapter 6). Similarly, the categories of structured sets can be viewed as particular *concrete categories* (defined in 1.4.8), and the categories of equational algebras as particular *monadic categories* (see Sections 3.6 and 4.4).

In other words, when studying certain 'classes' of categories, we may (or perhaps had better) look for a *structural definition* including this class, rather than some general way of constructing the important examples we want to study.

This leads again to considering categories of (small) categories, and to higher dimensional category theory.

0.8 Our approach

Notions will be presented in a concrete way, starting from examples taken from elementary mathematical theories. Then their theory is developed, with new examples and many exercises; the latter are generally endowed with a solution, or a partial solution, or suitable hints. Three chapters are devoted to applications.

We hope that a beginner can get, from these examples and applications, a concrete grasp of a theory which might risk of being quite abstract.

Many examples and applications are standard. But some of them may come out as unexpected, and hopefully intriguing, like those devoted to distributive lattices in 1.2.7, or to chains of adjunctions in 1.7.7 and 3.2.8, or to networks in 1.8.9.

The author's comments on some possibly unclear or controversial points are expressed:

- in 1.1.5, on the relationship between mathematical structures and categories,

- in 1.1.6 and 3.6.1, on varieties of algebras and artificial exclusions,

- in 1.5.9, 3.3.7 and 3.6.6, on favouring notions invariant up to categorical equivalence,

- in the introduction to Chapter 7, on the relationship between (strict and weak) 2-categories and double categories,

- in 7.3.3, about the 'naive view' of enrichment of a category over a monoidal category.

The foundational setting we use is based on standard set theory, assuming the existence of Grothendieck universes to formalise some necessary 'smallness' conditions. This aspect, presented in 1.1.3, will be mostly left as understood.

0.9 Literature

For further study of general category theory there are excellent texts, like Mac Lane [M4], Borceux [Bo1, Bo2, Bo3], Adámek, Herrlich and Strecker [AHS], Freyd and Scedrov [FrS]. References for higher dimensional category theory can be found in Chapter 7.

Sheaf categories and toposes, which are not treated here, are presented in Mac Lane–Moerdijk [MaM] and [Bo3], with extensive references to more advanced texts like Johnstone's [Jo3, Jo4]. A peculiar, conceptual introduction to category theory can be found in Lawvere and Schanuel [LwS].

As already warned, the range of applications of category theory is much wider than what will be seen here, and can be presented in an elementary text.

For instance, we only explore a few categorical properties of topological vector spaces and Banach spaces, for which the reader is referred to Semadeni's book [Se]. The applications in Homological Algebra examined here follow a particular approach, discussed in Chapter 6, where other approaches are cited. Some old and new applications of categories to the theory of networks are briefly presented in 1.8.9. Morita equivalence is only mentioned, in 1.5.6.

We do not examine the deep relationship among category theory, Logic and theoretical Computer Science, which is explored in texts like Makkai and Reyes [MkR], Lambek and Scott [LaS], Barr and Wells [BarW], Crole

[Cr], nor the growing influence of categories and higher dimensional categories in parts of theoretical Physics, for which one can see Baez and Lauda [BaL].

Differential Geometry is studied in a 'synthetic' way in Kock [Ko], making formal use of infinitesimals. A new book by Bunge, Gago and San Luis [BunGS] extends this subject to Synthetic Differential Topology. Different forms of Galois Theory are explored in Borceux and Janelidze [BoJ]. A categorical view of Set Theory can be found in Lawvere and Rosebrugh [LwR].

The relationship of category theory with Algebraic Geometry is perhaps too complex to be simply referred to.

0.10 Notation

The symbol $\subset$ denotes *weak* inclusion.

As usual, the symbols $\mathbb{N}$, $\mathbb{Z}$, $\mathbb{Q}$, $\mathbb{R}$, $\mathbb{C}$ denote the sets of natural, integral, rational, real or complex numbers; $\mathbb{N}^*$ is the subset of the positive integers. Open and semi-open real intervals are denoted as $]a, b[$, $[a, b[$, etc. This notation, loosely taken from Bourbaki, has the advantage of distinguishing the interval $]a, b[$ from the pair (a, b).

A singleton set is often written as $\{*\}$. The equivalence class of an element x, with respect to an assigned equivalence relation, is generally written as $[x]$. A bullet in a diagram stands for an arbitrary object.

Categories, 2-categories and bicategories are generally denoted as A, B...; strict or weak double categories as $\mathbb{A}$, $\mathbb{B}$...

A section, subsection or part marked with * deals with some topic out of the main line of this book, and is often addressed to readers having some knowledge of the subject.

0.11 Acknowledgements

Many points have been discussed with my colleagues and friends: in particular with Bob Paré and George Janelidze, during a long collaboration.

1

Categories, functors and natural transformations

Categories were introduced by Eilenberg and Mac Lane [EiM] in 1945, together with the other basic terms of category theory.

1.1 Categories

We start by considering concrete categories, associated with mathematical structures. But categories are not restricted to these instances, and the theory must be developed in a general way.

Given a mathematical discipline, it may not be obvious which category or categories are best suited for its study. This questionable point is discussed in 1.1.5, 1.1.6.

1.1.1 Some examples

Loosely speaking, before giving a precise definition, a category C consists of *objects* and *morphisms* together with a (partial) *composition law*: given two 'consecutive' morphisms $f\colon X \to Y$ and $g\colon Y \to Z$ we have a composed morphism $gf\colon X \to Z$. This partial operation is associative (whenever composition is legitimate) and every object X has an *identity*, written as $\mathrm{id}X\colon X \to X$ or 1_X, which acts as a unit for legitimate compositions.

The prime example is the category Set *of sets* (and mappings), where:
- an object is a set,
- the morphisms $f\colon X \to Y$ between two given sets X and Y are the (set-theoretical) mappings from X to Y,
- the composition law is the usual composition of mappings, where $(gf)(x) = g(f(x))$.

The following categories of structured sets and structure-preserving mappings (with the usual composition) will often be used and analysed:

- the category Top *of topological spaces* (and continuous mappings),

- the category Hsd *of Hausdorff spaces* (and continuous mappings),

- the category Gp *of groups* (and their homomorphisms),

- the category Ab *of abelian groups* (and homomorphisms),

- the category Mon *of monoids*, i.e. unitary semigroups (and homomorphisms),

- the category Abm *of abelian monoids* (and homomorphisms),

- the category Rng *of rings*, understood to be associative and unitary (and homomorphisms),

- the category CRng *of commutative rings* (and homomorphisms),

- the category R Mod *of left modules* on a fixed unitary ring R (and homomorphisms),

- the category K Vct ($= K$ Mod) *of vector spaces* on a commutative field K (and homomorphisms),

- the category RAlg *of unitary algebras* on a fixed commutative unitary ring R (and homomorphisms),

- the category Ord of *ordered sets* (and monotone mappings),

- the category pOrd of *preordered sets* (and monotone mappings),

- the category Set. *of pointed sets* (and pointed mappings),

- the category Top. *of pointed topological spaces* (and pointed continuous mappings),

- the category Ban of *Banach spaces and continuous linear mappings*.

- the category Ban$_1$ of *Banach spaces and linear weak contractions* (with norm $\leqslant 1$).

A homomorphism of a 'unitary structure', like a monoid or a unitary ring, is always assumed to preserve units.

For Set. we recall that a *pointed set* is a pair (X, x_0) consisting of a set X and a *base*-element $x_0 \in X$, while a pointed mapping $f: (X, x_0) \to (Y, y_0)$ is a mapping $f: X \to Y$ such that $f(x_0) = y_0$.

Similarly, a *pointed topological space* (X, x_0) is a space with a base-point, and a pointed map $f: (X, x_0) \to (Y, y_0)$ is a continuous mapping from X to Y such that $f(x_0) = y_0$. The reader may know that the category Top. is important in Algebraic Topology: for instance, the fundamental group $\pi_1(X, x_0)$ is defined for a pointed topological space.

For the categories Ban and Ban$_1$ it is understood that we have chosen *either* the real *or* the complex field; when using both one can write $\mathbb{R}$Ban and $\mathbb{C}$Ban.

When a category is named after its objects alone (e.g. the 'category of groups'), this means that the morphisms are understood to be the obvious ones (in this case the homomorphisms of groups), with the obvious composition law. Different categories with the same objects are given different names, like Ban and Ban$_1$ above.

1.1.2 Definition

A *category* C consists of the following data:

(a) a set ObC, whose elements are called *objects* of C,

(b) for every pair X, Y of objects, a set $C(X, Y)$ (called a *hom-set*) whose elements are called *morphisms* (or *maps*, or *arrows*) of C from X to Y and denoted as $f: X \to Y$,

(c) for every triple X, Y, Z of objects of C, a mapping of *composition*

$$C(X, Y) \times C(Y, Z) \to C(X, Z), \qquad (f, g) \mapsto gf,$$

where gf is also written as $g.f$.

These data must satisfy the following axioms.

(i) *Associativity.* Given three consecutive arrows, $f: X \to Y$, $g: Y \to Z$ and $h: Z \to W$, one has: $h(gf) = (hg)f$.

(ii) *Identities.* Given an object X, there exists an *endo*map $e: X \to X$ which acts as an identity whenever composition makes sense; in other words if $f: X' \to X$ and $g: X \to X''$, one has: $ef = f$ and $ge = g$. One shows, in the usual way, that e is determined by X; it is called the *identity* of X and written as 1_X or id$X$.

We generally assume that the following condition is also satisfied:

(iii) *Separation.* For all X, X', Y, Y' objects of C, if $C(X, Y) \cap C(X', Y') \neq \emptyset$ then $X = X'$ and $Y = Y'$.

Therefore a map $f: X \to Y$ has a well-determined *domain* and *codomain*

$$\text{Dom}\,(f) = X, \qquad \text{Cod}\,(f) = Y.$$

Concretely, when constructing a category, one can forget about condition (iii), since one can always satisfy it by *redefining* a morphism $\hat{f}: X \to Y$ as a triple (X, Y, f) where f is a morphism from X to Y in the original sense (possibly not satisfying the Separation axiom).

MorC denotes the set of all the morphisms of C, i.e. the disjoint union of its hom-sets. Two morphisms f, g are said to be *parallel* when they have the same domain and the same codomain.

If C is a category, the *opposite* (or *dual*) category, written as C^{op}, has the same objects as C, reversed arrows and reversed composition

$$C^{op}(X, Y) = C(Y, X),$$
$$g * f = fg, \qquad \mathrm{id}^{op}(X) = \mathrm{id}X. \tag{1.1}$$

Every topic of category theory has a dual instance, which comes from the opposite category (or categories). A dual notion is generally distinguished by the prefix 'co-'.

A set X can be viewed as a *discrete* category: its objects are the elements of X, and the only arrows are their (formal) identities; here $X^{op} = X$.

As usual in category theory, the term *graph* will be used to denote a simplified structure, with objects (or *vertices*) and morphisms (or arrows) $f \colon x \to y$, but no assigned composition nor identities. (This is called a *directed multigraph* in graph theory, or also a *quiver*). A *morphism of graphs* preserves objects, arrows, domain and codomain. Every category has an *underlying graph*.

1.1.3 Small and large categories

We shall not insist on set-theoretical foundations. Yet some caution is necessary, to avoid speaking of 'the set of all sets', or requiring of a category properties of completeness that are 'too large for its size' (as we shall see in 2.2.3).

We assume the existence of a (Grothendieck) *universe* $\mathcal{U}$, which is fixed throughout; its axioms – recalled below, in 1.1.7 – say that we can perform inside it the usual operations of set theory. Its elements are called *small sets* (or $\mathcal{U}$-small sets, if necessary).

A category is understood to have objects and arrows belonging to this universe, and is said to be *small* if its set of morphisms belongs to $\mathcal{U}$, *large* if it does not (and is just a subset of $\mathcal{U}$). As a consequence, in a small category the set of objects (which is in bijective correspondence with the set of identities) also belongs to $\mathcal{U}$. A category C is said to *have small hom-sets* if all its sets $C(X, Y)$ are small; in this case C is small if and only if its set of objects is.

The categories of structured sets that we consider are generally large $\mathcal{U}$-categories, like the category Set of small sets (and mappings), or Top of small topological spaces (and continuous mappings), or Ab of small abelian groups (and homomorphisms); *in such cases, the term 'small' (referred to these structured sets) will be generally understood*, and we speak – as usual – of the 'category of sets', and so on.

*In fact one often needs a hierarchy of universes. For instance, Cat will denote the category of small categories and functors, introduced in 1.4.1. In order to view the (large) categories Set, Top, Ab, etc. in a similar structure we should assume the existence of a second universe $\mathcal{V}$, with $\mathcal{U} \in \mathcal{V}$, and use the category $Cat_{\mathcal{V}}$ (also written as CAT) of $\mathcal{V}$-small categories. In a more complex situation one may need a longer chain of universes. Most of the time *these aspects will be left understood.**

1.1.4 Isomorphisms and groupoids

In a category C a morphism $f: X \to Y$ is said to be *invertible*, or an *isomorphism*, if it has an inverse, i.e. a morphism $g: Y \to X$ such that $gf = 1_X$ and $fg = 1_Y$. The latter is uniquely determined; it is called the *inverse* of f and written as f^{-1}.

In the categories listed in 1.1.1 this definition gives the usual isomorphisms of the various structures – called 'homeomorphisms' in the case of topological spaces.

The *isomorphism relation* $X \cong Y$ between objects of C (meaning that there exists an isomorphism $X \to Y$) is an equivalence relation.

A *groupoid* is a category where every map is invertible; it is interesting to recall that this structure was introduced before categories, by H. Brandt in 1927 [Bra].

The fundamental groupoid of a space X is an important structure, that contains all the fundamental groups $\pi_1(X, x)$ for $x \in X$. It will be reviewed in 5.2.9.

1.1.5 A digression on mathematical structures and categories

When studying a mathematical structure with the help of category theory, it is crucial to choose the 'right' kind of structure and the 'right' kind of morphisms, so that the result is sufficiently general and 'natural' to have good properties with respect to the goals of our study – even if we are interested in more particular situations.

(a) A first point to be verified is that the isomorphisms of the category (i.e. its invertible arrows) preserve the structure we are interested in, or we risk of studying something different from our purpose.

As a trivial example, the category T of topological spaces and *all* mappings between them has little to do with topology: an isomorphism of T is any bijection between topological spaces. Indeed T is 'equivalent' to the category of sets (as we shall see in 1.5.5), and is a 'deformed' way of looking at the latter.

Less trivially, the category M of metric spaces and continuous mappings misses crucial properties of metric spaces, since its invertible morphisms need not preserve completeness: e.g. the real line is homeomorphic to any non-empty open interval. In fact, M is equivalent to the category of *metrisable topological spaces* and continuous mappings (by 1.5.5, again), and should be replaced with the latter.

A 'reasonable' category of metric spaces should be based on *Lipschitz* maps, or – more particularly – on weak contractions, so that its isomorphisms (bi-Lipschitz or isometric bijections, respectively) do preserve metric properties, like being complete or bounded (see 5.1.7).

(b) Other points will become clearer below. For instance, the category Top of topological spaces and continuous mappings is a classical framework for studying Topology. Among its good properties there is the fact that all 'categorical products' and 'categorical sums' (studied in Section 2.1, but already sketched in the Introduction) exist, and are computed *as in* Set, then equipped with a suitable topology determined by the structural maps.

(More generally, this is true of all limits and colimits, and – as we shall see – is a consequence of the fact that the forgetful functor Top → Set has a left *and* a right adjoint, corresponding to discrete and chaotic topologies).

Hausdorff spaces are certainly important, but it is often better to view them *in* Top, as their category Hsd is less well behaved: colimits exist, but are not computed as in Set, and the simplest way to compute them – generally – is to take the colimit in Top and 'make it Hausdorff' (see 5.1.4(b)).

*(c) Many category theorists would agree with Mac Lane, saying that even Top is not sufficiently good ([M4], Section VII.8), because it is not a cartesian closed category (see 5.1.1), and prefer – for instance – the category of compactly generated Hausdorff spaces (see 2.6.3(d)). However, researchers interested in Homotopy Theory and Algebraic Topology might be satisfied with the fact that the standard interval (with its cartesian powers) is exponentiable in Top, as we shall exploit in Section 5.2.

We shall also hint, in 5.1.8, at another approach called 'pointless topology', which is based on the category of locales and is favoured in topos theory.

(d) Finally we remark that artificial exclusions 'most of the time' give categories of poor properties, like the category of *non-abelian groups*, or *non-empty semigroups*. The latter case needs some further comment.

1.1.6 *Variety of algebras and* horror vacui

In Universal Algebra, a 'variety of algebras' includes all the algebraic structures of a given signature (i.e. with a certain family of operations, of assigned arity), that satisfy a given set of equational axioms (or universally
quantified identities): e.g. all groups, or all rings; but not all fields, because multiplicative inverses only exist for non-zero elements, and cannot
be given by a 'general' unary operation satisfying some *universal* identities.

Here a *variety of algebras* will mean a category of objects defined in this
way, with their homomorphisms (as made precise in Section 4.3). *We do
not follow the convention* that the underlying set should be non empty; a
convention which has unlucky consequences for any theory without 0-ary
operations, like semigroups: for instance two subalgebras of an algebra may
not have a meet (as subalgebras).

This convention is rather usual in Universal Algebra (cf. Grätzer [Gr1]),
but is not followed in Cohn's book [Coh]. (Of course, generally speaking,
results can be easily translated from one setting to the other.)

*For a reader with some knowledge of limits and colimits, dealt with in
Chapter 2, we can add that a variety of algebras (in the present sense) has
all limits and colimits, while the category of non-empty semigroups lacks
certain limits (like equalisers and pullbacks) and certain colimits (as an
initial object), precisely because we have artificially taken out a solution.*

Other comments in this sense can be found in 3.6.1; but there is also a
comment *in the opposite sense*, in 5.3.9, for categories used in an 'auxiliary
way'.

1.1.7 *Grothendieck universes*

For the interested reader, we recall the definition of a *universe* as given
in [M4], Section I.6. It is a set $\mathcal{U}$ satisfying the following (redundant)
properties:

(i) $x \in u \in \mathcal{U}$ implies $x \in \mathcal{U}$,

(ii) $u, v \in \mathcal{U}$ implies that $\{u, v\}$, (u, v) and $u \times v$ belong to $\mathcal{U}$,

(iii) $x \in \mathcal{U}$ implies that $\mathcal{P}x$ and $\bigcup x$ belong to $\mathcal{U}$,

(iv) the set $\mathbb{N}$ of finite ordinals belongs to $\mathcal{U}$,

(v) if $f \colon x \to y$ is a surjective mapping with $x \in \mathcal{U}$ and $y \subset \mathcal{U}$, then $y \in \mathcal{U}$.

Here $\mathcal{P}x$ is the set of subsets of x and $\bigcup x = \{y \mid y \in z \text{ for some } z \in x\}$.

1.2 Monoids and preordered sets as categories

Monoids and preordered sets can be viewed as categories of two simple kinds, providing intuition and models for some aspects of category theory.

We also review here some basic facts about the theory of lattices, to be used later on. A reader interested in this beautiful domain will find pleasure in browsing, or studying, the classical texts of Birkhoff and Grätzer [Bi, Gr2].

1.2.1 Monoids and categories

As we have seen, monoids (i.e. semigroups with unit) and their homomorphisms form a category, which we write as Mon. But we deal here with a different aspect.

A single monoid M can (and will often) be viewed as a category with one formal object $*$. The morphisms $x: * \to *$ are the elements of M, composed by the multiplication xy of the monoid, with identity $\mathrm{id}(*) = 1$, the unit of the monoid. If M is a group, the associated category is a groupoid.

On the other hand, in every category C, the *endomorphisms* $X \to X$ of any object form a (possibly large) monoid, under the composition law

$$\mathrm{End}(X) = \mathsf{C}(X, X), \tag{1.2}$$

and the invertible ones form the group $\mathrm{Aut}(X)$ of *automorphisms* of X.

In this way a monoid is essentially the same as a category on a single object, while a category can be thought to be a 'multi-object generalisation' of a monoid. Groups and groupoids have a similar relationship.

The theory of regular, orthodox and inverse semigroups (see [ClP, Ho, Law]) has a strong interplay with the categories of relations and their applications in Homological Algebra, which is investigated in [G9].

1.2.2 Preordered and ordered sets

We shall use the following terminology for orderings. A *preorder* relation $x \prec x'$ is reflexive and transitive. An *order* relation, generally written as $x \leqslant x'$, is also anti-symmetric: if $x \leqslant x' \leqslant x$ then $x = x'$.

The category of *ordered sets and increasing mappings* (the order preserving ones, also called *monotone*) will be written as Ord, while we write as pOrd the category of *preordered sets and monotone mappings*.

An order relation is said to be *total* if for all x, x' we have $x \leqslant x'$ or $x' \leqslant x$.

(An ordered set is often called a '*partially* ordered set', abbreviated to 'poset', to stress that the ordering is not assumed to be total. Accordingly, the reader can find the notation Pos for the category Ord.)

A preordered set X has an associated equivalence relation $x \sim x'$ defined by the conjunction: $x \prec x'$ and $x' \prec x$. The quotient $X/\sim$ has an induced order: $[x] \leqslant [x']$ if $x \prec x'$.

If X is a preordered set, X^{op} is the opposite one (with reversed preorder). If $a \in X$, the symbols $\downarrow a$ and $\uparrow a$ denote the downward or upward closed subsets of X generated by the element a

$$\downarrow a = \{x \in X \mid x \prec a\}, \qquad \uparrow a = \{x \in X \mid a \prec x\}. \tag{1.3}$$

It will be important to note that every hom-set $\mathsf{Ord}(X, Y)$ is canonically ordered by the *pointwise order* relation, defined as follows for $f, g \colon X \to Y$

$$f \leqslant g \quad \text{if for all } x \in X \text{ we have } f(x) \leqslant g(x) \text{ in } Y. \tag{1.4}$$

Similarly, every hom-set $\mathsf{pOrd}(X, Y)$ has a canonical preorder $f \prec g$.

In a preordered set X, the *infimum* (or *meet*) of a subset A, written as $\inf A$ or $\wedge A$, is defined as the greatest element of X smaller than all the elements of A, and is determined up to the associated equivalence relation (provided it exists). Dually, the *supremum* (or *join*) $\sup A = \vee A$ is the infimum of A in X^{op}. In an ordered set these results are uniquely determined – if extant.

1.2.3 Preorders and categories

A preordered set X will often be viewed as a category, where the objects are the elements of X and the set $X(x, x')$ contains precisely one arrow if $x \prec x'$, which can be written as $(x, x') \colon x \to x'$, and no arrow otherwise. Composition and units are (necessarily) as follows

$$(x', x'').(x, x') = (x, x''), \qquad \mathrm{id}(x) = (x, x).$$

In this way a preordered set is essentially the same as a category where each hom-sets has at most one element. All diagrams in these categories commute. Two elements x, x' are isomorphic objects if and only if $x \sim x'$.

In particular, each ordinal defines a category, written as $\mathbf{0}, \mathbf{1}, \mathbf{2}, \ldots$

- $\mathbf{0}$ is the empty category,

- $\mathbf{1}$ is the *singleton category*, i.e. the discrete category on one object,

- $\mathbf{2}$ is the *arrow category*, with two objects (0 and 1), and precisely one non-identity arrow, $0 \to 1$.

Here **2** should not be confused with the cardinal $2 = \{0, 1\}$, which is viewed as a discrete category.

1.2.4 Lattices

Classically a lattice is defined as an ordered set X such that every pair x, x' of elements has a *join* $x \vee x' = \sup\{x, x'\}$ (the least element of X greater than both) and a *meet* $x \wedge x' = \inf\{x, x'\}$ (the greatest element of X smaller than both).

Here we follow a different convention, usual in category theory: *lattice* will always mean an ordered set with *finite* joins and meets, which amounts to the existence of binary joins and meets *together with* the least element $0 = \vee \emptyset$ and the greatest element $1 = \wedge \emptyset$. (This structure is called a 'bounded lattice' in Lattice Theory.)

The bounds 0 and 1 (the empty join and the empty meet) are equal in the one-point lattice $\{*\}$, and only there.

Consistently with this terminology, a *lattice homomorphism* has to preserve finite joins and meets; a *sublattice* of a lattice X is closed under such operations (and has the same bounds as X). The category of lattices and homomorphisms will be written as Lth.

Occasionally we speak of a *quasi lattice* when we only assume the existence of binary joins and meets; a homomorphism of quasi lattices only has to preserve them. A *quasi sublattice* Y of a quasi lattice X is closed under binary joins and meets in X; when X is a lattice, Y may have different bounds, or lack some of them.

For instance, if X is a lattice and $a \in X$, the downward and upward closed subsets $\downarrow a$, $\uparrow a$ of X generated by a (see 1.2.2) are quasi sublattices of X, and lattices in their own right.

Let us note that the *free lattice* (see 2.7.3) generated by an element x *has three elements*: $0 < x < 1$, while the *free quasi lattice* L generated by x is just the singleton $\{x\}$: in fact every mapping $\{x\} \to X$ with values in a lattice (resp. in a quasi lattice) has a unique extension to a homomorphism $L \to X$ (resp. $\{x\} \to X$). In the same way, the free lattice L generated by a set S can be obtained from the corresponding free quasi-lattice M by adding a (new) minimum and a (new) maximum, even when M is *already* bounded – as above.

1.2.5 Exercises and complements

(a) *Lattices as algebras.* The reader may know, or be interested to prove, that a lattice can be equivalently presented as a set X equipped with two operations, $x \vee y$ and $x \wedge y$, called *join* and *meet*, that satisfy the following axioms:

(i) both operations are associative, commutative and idempotent:

$$x \vee (y \vee z) = (x \vee y) \vee z, \qquad x \wedge (y \wedge z) = (x \wedge y) \wedge z,$$
$$x \vee y = y \vee x, \qquad x \wedge y = y \wedge x,$$
$$x \vee x = x, \qquad x \wedge x = x,$$

(ii) each operation has a unit, written as 0 and 1, respectively

$$x \vee 0 = x = x \wedge 1,$$

(iii) the following *absorption laws* hold:

$$x \vee (x \wedge y) = x = x \wedge (x \vee y).$$

Given this presentation, one defines the ordering by letting $x \leqslant y$ if $x \vee y = y$, or equivalently $x \wedge y = x$. Again, one should be aware that $0 = 1$ is not excluded; we already know that in this case all the elements coincide. The reader will easily guess how the opposite lattice X^{op} is defined here.

(b) *Complete lattices.* The reader may know, or easily prove, that a pre-ordered set has all infima (of its subsets) if and only if it has all suprema. In this case – if it is an ordered set – it is called a complete lattice. This agrees with the notion of a 'complete category', as we shall see in 2.2.5(e).

Frames are particular complete lattices, related with topological spaces and 'pointless topology', and will be briefly introduced in 5.1.8.

1.2.6 Distributive and modular lattices

A lattice is said to be *distributive* if the meet operation distributes over the join operation, or equivalently if the join distributes over the meet. In fact, if we assume that meets distribute over joins, we have:

$$(x \vee y) \wedge (x \vee z) = ((x \vee y) \wedge x) \vee ((x \vee y) \wedge z)$$
$$= x \vee (x \wedge z) \vee (y \wedge z) = x \vee (y \wedge z).$$

A *boolean algebra* is a distributive (bounded) lattice where every element x has a (necessarily unique) *complement* x', defined by the properties:

$$x \wedge x' = 0, \qquad x \vee x' = 1.$$

The subsets of a set X form the classical boolean algebra $\mathcal{P}X$, which is a complete lattice.

The (complete) lattice Sub$A$ of subgroups of an abelian group (or sub-modules of a module) is not distributive, generally (see 1.2.7(d)); but one can easily check that it always satisfies a weaker, restricted form of distributivity, called modularity.

Namely, a lattice is said to be *modular* if it satisfies the following selfdual property (for all elements x, y, z)

(i)   if $x \leqslant z$ then $(x \vee y) \wedge z = x \vee (y \wedge z)$.

The category of modular (resp. distributive) lattices and their homomorphisms will be written as Mlh (resp. Dlh).

By Birkhoff's representation theorem ([Bi] III.5, Theorem 5) the free distributive lattice on n generators is finite and isomorphic to a lattice of subsets. The reader may also be interested to know that the free modular lattice on three elements is finite and (obviously!) not distributive (see [Bi], III.6, Fig. 10), while four generators already give an infinite free modular lattice (see the final Remark in [Bi], III.6).

1.2.7 *Exercises and complements* (Distributive lattices)

The goal of this point is to show that the (complete) lattice $X = \mathrm{Sub}(\mathbb{Z})$ of subgroups of the abelian group $\mathbb{Z}$ (or ideals of the ring $\mathbb{Z}$) is distributive – a fact that will have unexpected links with Homological Algebra, in Section 6.6.

The interested reader is invited to directly investigate the problem, before considering the layout given below. The first step is showing that X is anti-isomorphic to the divisibility lattice of natural numbers (a well-known, nearly obvious point). Then one proves that the latter is distributive, a (hopefully amusing) exercise based on our school knowledge – prime factorisation.

(a) The reader likely knows, or should prove, that each subgroup of $\mathbb{Z}$ is of the form $n\mathbb{Z}$, for a unique $n \in \mathbb{N}$ (and is an ideal of the ring of integers).

This gives us an isomorphism $X \to Y^{\mathrm{op}}$, where $Y = (\mathbb{N}, |)$ is the set of natural numbers ordered by the divisibility relation $m|n$. Therefore Y is a (complete) lattice as well, and we recognise its operations $m \vee n$ and $m \wedge n$ as fairly well-known. We also note that $1 = \min Y$ and $0 = \max Y$.

(b) Now, each $n \in \mathbb{N}^*$ has a unique decomposition $n = \prod_p p^{n_p}$ as a product of powers of non-invertible prime numbers p; of course the natural exponents n_p are quasi-null (i.e. all of them are 0, out of a finite number of prime indices p), so that the factorisation is essentially finite.

The reader will use this fact to prove that Y can be embedded in the cartesian product $\prod_p \overline{\mathbb{N}}$ of countably many copies of the set $\overline{\mathbb{N}} = \mathbb{N} \cup \{\infty\}$, with the natural order. One can now show that Y is distributive, and X as well.

(c) As a consequence, the lattice $\mathrm{Sub}(A)$ of any cyclic group A is also distributive.

(d) The reader can easily show that this property fails for the abelian group $\mathbb{Z}^2$, or for any non-trivial abelian group $A \oplus A$. (Again, this will be of use in Section 6.6.)

On the other hand, $\mathbb{Z}/2 \oplus \mathbb{Z}/3 \cong \mathbb{Z}/6$ has a distributive lattice of subgroups.

(e) A reader acquainted with *principal ideal domains* may like to rethink the whole thing.

1.3 Monomorphisms and epimorphisms

In a category, monomorphisms and epimorphisms (monos and epis for short) are defined by cancellation properties with respect to composition.

For categories of structured sets, they represent an 'approximation' to the injective and surjective mappings of the category.

1.3.1 Main definitions

In a category C a morphism $f \colon X \to Y$ is said to be a *monomorphism*, or *mono*, if it satisfies the following cancellation property: for every pair of maps $u, v \colon X' \to X$ (defined on an arbitrary object X') such that $fu = fv$, one has $u = v$ (see the left diagram below)

$$X' \underset{v}{\overset{u}{\rightrightarrows}} X \xrightarrow{f} Y \qquad\qquad X \xrightarrow{f} Y \underset{v}{\overset{u}{\rightrightarrows}} Y' \qquad\qquad (1.5)$$

Dually, the morphism $f \colon X \to Y$ is said to be an *epimorphism*, or *epi*, if it satisfies the dual cancellation property: for every pair of maps $u, v \colon Y \to Y'$ such that $uf = vf$, one has $u = v$ (see the right diagram above).

An arrow $\rightarrowtail$ will always denote a monomorphism, while $\twoheadrightarrow$ stands for an epimorphism.

Every isomorphism is mono and epi. A category is said to be *balanced* if the converse holds: every morphism which is mono and epi is invertible.

Suppose now that we have, in a category C, two maps $m \colon A \to X$ and $p \colon X \to A$ such that $pm = \mathrm{id}A$. It follows that m is a monomorphism (called a *section*, or a *split monomorphism*), while p is an epimorphism (called a *retraction*, or a *split epi*); A is said to be a *retract* of X.

A family of morphisms $f_i \colon X \to Y_i$ $(i \in I)$ with the same domain is said to be *jointly mono* if for every pair of maps $u, v \colon X' \to X$ such that $f_i u = f_i v$ (for all indices i) one has $u = v$. Dually a family $f_i \colon X_i \to Y$ is *jointly epi* if for all $u, v \colon Y \to Y'$ such that $u f_i = v f_i$ (for all i) one has $u = v$.

The general properties of monos, epis and retracts will be examined in 1.3.6. Related notions, like regular monos and epis, strong monos and epis, subobjects and quotients, will be seen in the next chapter.

1.3.2 Comments

In a category of structured sets and structure-preserving mappings, an injective mapping (of the category) is obviously a monomorphism, while a surjective one is an epimorphism. The converse may require a non-trivial proof, or even fail. This can only be understood by working out the examples below.

Interestingly, a divergence appears between monos and epis: the theory of categories is self-dual, but our frameworks of structured sets are not! When we classify monos in Set, this tells us everything about the epis of $\mathsf{Set}^{\mathrm{op}}$ but nothing about the epis of Set.

In fact, in all the examples below it will be easy to prove that the monomorphisms coincide with the injective morphisms. Later we shall see, in 2.7.4(d), some conditions that ensure this fact, and hold true in all the 'usual categories of structured sets'.

On the other hand, various problems occur with epimorphisms: classifying them in various categories of algebraic structures leads to difficult problems with no elementary solution (and no real need of it).

1.3.3 Exercises and complements, I

(a) The first point is to prove that in Set every mono is an injective mapping and every epi is surjective. We write down the proof, but a beginner should try to give an independent solution; this is quite easy for monos and slightly less for epis.

If $f \colon X \to Y$ is a monomorphism in Set, let us suppose that $f(x) = f(x')$, for $x, x' \in X$. We consider the mappings u, v

$$\{*\} \underset{v}{\overset{u}{\rightrightarrows}} X \xrightarrow{f} Y \qquad\qquad u(*) = x, \quad v(*) = x'. \tag{1.6}$$

Now we have $fu = fv$, whence $u = v$ and $x = x'$, which shows that f is injective. Note that the proof works by simulating an element of X with a map $\{*\} \to X$.

On the other hand, if $f\colon X \to Y$ is an epimorphism in Set, we define two mappings u, v with values in the set $\{0, 1\}$

$$X \xrightarrow{\ f\ } Y \underset{v}{\overset{u}{\rightrightarrows}} \{0, 1\} \qquad (1.7)$$

where u is the characteristic function of the subset $f(X) \subset Y$ (with $u(y) = 1$ if and only if $y \in f(X)$) and v is the constant map $v(y) = 1$. Then $uf = vf$, whence $u = v$ and $f(X) = Y$. (A different proof can be based on the set $Y \times \{0, 1\}$, the disjoint union of two copies of Y; or a quotient of the latter.)

Since the invertible morphisms in Set are obviously the same as the bijective mappings, we remark that the category Set is balanced.

(b) Now the reader should prove that also in Top and Ab monos and epis coincide with the injective and surjective mappings of the category, respectively.

For monos the proof is quite similar to the previous one, making use of the singleton in Top and of the group $\mathbb{Z}$ in Ab. Note that the latter allows us to simulate an element $x \in X$ by a homomorphism $u\colon \mathbb{Z} \to X$, sending the generator 1 to x.

To prove that an epi is surjective, one can use a two-point codiscrete space in Top and the quotient group $Y/f(X)$ in Ab. Note that the last point follows a different pattern: constructing arrows is fairly free in Set, somewhat less in Top, much less in categories of algebraic structures.

We conclude here that Ab is balanced, while Top is not: a bijective continuous mapping need not be invertible in Top, i.e. a homeomorphism.

(c) The reader should prove that, in the categories Mon of monoids and Rng of rings, the monomorphisms coincide again with the injective homomorphisms. Then one can easily show that the inclusion $\mathbb{N} \to \mathbb{Z}$ (of additive monoids) is mono and *epi* in Mon, which is not balanced. The same holds for the inclusion $\mathbb{Z} \to \mathbb{Q}$ in Rng.

In fact epimorphisms in Mon and Rng have no elementary characterisation and the 'regular epimorphisms' (namely the surjective homomorphisms) are more important, as we shall see in Section 4.4.

(d) In a preordered set, viewed as a category, all arrows are mono and epi. The category is balanced under a precise condition on the preordering.

1.3.4 Exercises and complements, II

(a) In the category Gp of groups all epimorphisms are surjective: a non-obvious fact, whose proof can be found in [M4], Section I.5, Exercise 5.

(One cannot use the same argument as in Ab, because the image of a homomorphism need not be normal in its codomain.)

(b) In the category Hsd of Hausdorff spaces, monomorphisms coincide with the injective continuous mappings, with the same proof as in Top. But our argument for the surjectivity of epimorphisms in Top is based on the codiscrete two-point space and does not work here.

Indeed the reader may know that two maps $u, v \colon Y \to Y'$ between Hausdorff spaces always coincide on a closed subset of their domain. It follows that *every map whose image is dense* (in its codomain) *is an epimorphism* in Hsd. We shall see in 2.3.6(c) that this condition is also necessary.

(c) In pOrd and Ord every mono is injective, with the same argument as in Set, based on the singleton.

To show that epis are surjective in pOrd one can use a two-point set with chaotic preorder. In Ord the conclusion is the same, but the proof is more complicated, if easy; for a monotone mapping $f \colon X \to Y$ and a point $y_0 \in Y \setminus f(X)$, one can construct an ordered set Y' where y_0 is duplicated.

(d) Prove that the monomorphisms of Ban and Ban_1 coincide with their injective morphisms.

Prove that the epimorphisms of each of these categories coincide with the morphisms whose image is dense, using the previous point (b) and the quotient of a Banach space modulo a closed subspace.

1.3.5 Exercises and complements (Retracts)

For the sake of simplicity, in these exercises we always replace a monomorphism $m \colon A \rightarrowtail X$, in a category of structured sets, by its image. (The reader can check that 'nothing is missed', in the examples below.)

(a) Prove that in Set a retract of a set $X \neq \emptyset$ is any non-empty subset.

(b) Prove that in Ab retracts coincide with direct summands. More precisely, a subgroup A of an abelian group X is a retract if and only if there exists a subgroup $B \subset X$ such that $A \cap B = \{0\}$ and $A + B = X$; then X is isomorphic to the direct sum $A \oplus B$.

For a reader acquainted with elementary Homological Algebra, this condition amounts to saying that the short exact sequence $A \rightarrowtail X \twoheadrightarrow X/A$ splits.

(c) There is no elementary characterisation of retracts in Top. To prove that a subspace $A \subset X$ is a retract one 'simply' has to construct a retraction; but proving that it is not a retract can be an important, non-trivial result.

For instance, saying that a subspace $\mathbb{R} \setminus \{a\}$ of the euclidean line $\mathbb{R}$ is not a retract is a way of stating the Intermediate Value Theorem. These problems are generally studied with the tools of Algebraic Topology: see 1.4.7.

1.3.6 General properties of monos and epis

The reader should (easily) prove the following elementary properties, that will be used throughout without reference.

We have two consecutive maps $f: X \to Y$ and $g: Y \to Z$ in a category. Property (a) is dual to (a*), and we only need to prove one of them.

(a) If f and g are both mono, gf is also mono; if gf is mono, f is also.

(a*) If f and g are both epi, gf is also epi; if gf is epi, g is also.

(b) If f and g are both split mono, gf is also; if gf is a split mono, f is also.

(b*) If f and g are both split epi, gf is also; if gf is a split epi, g is also.

(c) If f is a split mono and an epi, then it is invertible.

(c*) If f is a split epi and a mono, then it is invertible.

1.4 Functors

The structure-preserving mappings between categories are called 'functors'.

1.4.1 Functors and isomorphisms of categories

A (covariant) *functor* $F: \mathsf{C} \to \mathsf{D}$ consists of the following data:

(a) a mapping

$$F_0: \mathrm{Ob}\mathsf{C} \to \mathrm{Ob}\mathsf{D},$$

whose action is generally written as $X \mapsto F(X)$,

(b) for every pair of objects X, X' in C, a mapping

$$F_{XX'}: \mathsf{C}(X, X') \to \mathsf{D}(F(X), F(X')),$$

whose action is generally written as $f \mapsto F(f)$.

Composition and identities must be preserved. In other words:

(i) if f, g are consecutive maps of C then $F(gf) = F(g).F(f)$,

(ii) if X is an object of C then $F(\mathrm{id}X) = \mathrm{id}(F(X))$.

Given a second functor $G: \mathsf{D} \to \mathsf{E}$, one defines in the obvious way the *composed* functor $GF: \mathsf{C} \to \mathsf{E}$. This composition is associative and has identities: the *identity functor* of each category

$$\mathsf{idC}: \mathsf{C} \to \mathsf{C}, \qquad X \mapsto X, \quad f \mapsto f. \qquad (1.8)$$

An *isomorphism of categories* is a functor $F: \mathsf{C} \to \mathsf{D}$ which is invertible, i.e. admits an *inverse* $G: \mathsf{D} \to \mathsf{C}$; this means a functor such that $GF = \mathsf{idC}$ and $FG = \mathsf{idD}$. It is easy to verify that the functor F is an isomorphism if and only if all the mappings F_0 and $F_{XX'}$ considered above are bijective.

Being isomorphic categories is an equivalence relation, written as $\mathsf{C} \cong \mathsf{D}$.

A *contravariant functor* $F: \mathsf{C} \dashrightarrow \mathsf{D}$ is defined as a covariant functor $\mathsf{C}^{\mathrm{op}} \to \mathsf{D}$, and takes a morphism $f: X \to X'$ in C to a morphism $F(f): F(X') \to F(X)$ in D, preserving identities and reversing compositions: $F(gf) = F(f).F(g)$ (with respect to the compositions of C and D, of course). By a common abuse of terminology, it is often referred to as a *contravariant functor* $F: \mathsf{C}^{\mathrm{op}} \to \mathsf{D}$, a way of recalling that we are interested in C rather than in C^{op}.

Cat will denote the category of small categories and their functors. (Its 2-dimensional structure, including the natural transformations, will be examined in 1.5.3 and Section 7.1.)

1.4.2 Examples and comments

(a) A functor between two monoids, viewed as categories (see 1.2.1), is the same as a homomorphism of monoids.

(b) A functor between two preordered sets, viewed as categories (see 1.2.3), is the same as a monotone function, i.e. a preorder-preserving mapping.

(c) A functor $G \to K\,\mathsf{Vct}$ defined on a group (viewed as a category) amounts to a *representation* of G, in the usual sense of a homomorphism $G \to \mathrm{Aut}(X)$ with values in the group of automorphisms of a K-vector space.

(d) Categories linked by an obvious isomorphism are often perceived as 'the same thing'. For instance Ab is isomorphic to the category $\mathbb{Z}\,\mathsf{Mod}$ of modules on the ring of integers. The various equivalent ways of defining topological spaces give rise to isomorphic categories that are nearly never distinguished; the same holds for the two equivalent ways of defining lattices (recalled in 1.2.4, 1.2.5).

1.4.3 Subcategories

Let C be a category. A *subcategory* D is defined by assigning:

- a subset $\mathrm{ObD} \subset \mathrm{ObC}$, whose elements are called *objects of* D,

- for every pair of objects X, Y of D, a subset $D(X, Y) \subset C(X, Y)$, whose elements are called *morphisms of* D, from X to Y,

so that the following conditions hold:

(i) for every pair of consecutive morphisms of D their composite in C belongs to D,

(ii) for every object of D its identity in C belongs to D.

D, equipped with the induced composition law, is then a category. It comes with an *inclusion* functor $U: D \to C$, also written as $D \subset C$.

One says that D is a *full* subcategory of C if, for every pair of objects X, Y of D, we have $D(X, Y) = C(X, Y)$, so that D is determined by assigning its subset of objects. One says that D is a *wide* subcategory of C if it has the same objects, so that D is determined by assigning its subset of maps (closed under composition and containing all identities).

For instance, Ab is a full subcategory of Gp while Ban_1 is a wide subcategory of Ban. Of course a full and wide subcategory must be the total one.

1.4.4 Forgetful and amnestic functors

Forgetting structure, or part of it, yields various examples of functors between categories of structured sets, like the following obvious instances

$$Top \to Set, \qquad\qquad Ord \to Set,$$
$$Rng \to Ab \to Set, \qquad Rng \to Mon \to Set. \qquad (1.9)$$

These are called *forgetful* functors, and often denoted by the letter U, which refers to the *underlying* set, or *underlying* abelian group, and so on. It is also common to write $U(X)$ as $|X|$.

The term forgetful functor can also refer to the inclusion functor of a subcategory, which *forgets properties* rather than structure, like

$$Ab \to Gp \to Mon, \qquad\qquad Ord \to pOrd, \qquad (1.10)$$

or to functors that forget (a part of) both, like the functor $CRng \to Mon$ that sends a commutative ring to its underlying multiplicative monoid (forgetting the additive structure and the commutative property of the multiplication).

A functor $F: C \to D$ is said to be *amnestic* if, for every isomorphism f in C, $F(f)$ is an identity in D (if and) only if f is.

All the forgetful functors above are amnestic. Those dealing with categories of algebraic structures even *reflect identities*: for every map f in C, $F(f)$ is an identity in D (if and) only if f is.

1.4.5 Quotients of categories

A *congruence* $R = (R_{XY})$ in a category C consists of a family of equivalence relations R_{XY} in each set of morphisms $C(X, Y)$ that is consistent with composition:

(i) if $f \, R_{XY} \, f'$ and $g \, R_{YZ} \, g'$, then $gf \, R_{XZ} \, g'f'$.

The *quotient category* $D = C/R$ has the same objects of C and

$$D(X, Y) = C(X, Y)/R_{XY}.$$

In other words, a morphism $[f] \colon X \to Y$ in D is an equivalence class of morphisms $X \to Y$ in C. The composition is induced by that of C, which is legitimate because of condition (i):

$$[g].[f] = [gf]. \tag{1.11}$$

The *projection functor* $P \colon C \to C/R$ is the identity on objects and sends a morphism f to its equivalence class $[f]$.

In Top the homotopy relation $f \simeq f'$ is a congruence of categories; the quotient category

$$\mathsf{hoTop} = \mathsf{Top}/\simeq$$

is called the *homotopy category of topological spaces*, and is important in Algebraic Topology. Plainly, a continuous mapping $f \colon X \to Y$ is a homotopy equivalence (i.e. there exists a continuous mapping $g \colon Y \to X$ such that $gf \simeq \mathrm{id}X$ and $fg \simeq \mathrm{id}Y$) if and only if its homotopy class $[f]$ is an isomorphism of hoTop. All this will be examined in a more detailed way in Section 5.2.

Every category C has a finest congruence $f = g$, whose quotient is trivially isomorphic to C, and a coarsest congruence $f \sim g$, meaning that f and g are parallel arrows. In the second case, the quotient category is the (possibly large) set $\mathrm{Ob}(C)$ preordered by the relation $X \prec Y$ whenever there is some arrow $X \to Y$ in C; it will be called the *preordered set associated to* C (in a universal way, as we shall see in Exercise 3.1.4(f)), and denoted as $\mathrm{po}(C)$.

Let us note that the relation of isomorphism is wider in a quotient category, but may happen to coincide.

1.4.6 Products of categories

If C and D are categories, one defines the *product category* $C \times D$.

An object is a pair (X, Y) where X is in C and Y in D. A morphism is a pair of morphisms

$$(f,g)\colon (X,Y) \to (X',Y'), \tag{1.12}$$

for $f \in C(X,X')$ and $g \in D(Y,Y')$.

The composition of (f,g) with $(f',g')\colon (X',Y') \to (X'',Y'')$ is (obviously) component-wise:

$$(f',g').(f,g) = (f'f,g'g).$$

Similarly one defines the product $C = \Pi\, C_i$ of a family of categories indexed by a small set. It has (cartesian) *projection functors* $P_i\colon C \to C_i$.

A *functor in two variables* is an ordinary functor $F\colon C \times D \to E$ defined on the product of two categories. Fixing an object X_0 in C we have a functor $F(X_0,-)\colon D \to E$; and symmetrically.

If the category C has small hom-sets $C(X,Y)$ (see 1.1.3), there is a functor of morphisms, or *hom-functor*

$$\mathrm{Mor}\colon C^{\mathrm{op}} \times C \to \mathsf{Set},$$
$$(X,Y) \mapsto C(X,Y), \qquad (f,g) \mapsto g.-.f. \tag{1.13}$$

Here, for $f\colon X' \to X$ and $g\colon Y \to Y'$ in C, the mapping $g.-.f$ acts as:

$$g.-.f\colon C(X,Y) \to C(X',Y'), \qquad u \mapsto guf.$$

This quite important functor will reappear in various 'enriched' forms. It is described as 'a functor in two variables, contravariant in the first and covariant in the second'.

Fixing one of these variables one gets a 'representable functor' (covariant or contravariant on C), as we shall see in 1.6.4.

1.4.7 Faithful and full functors

For a functor $F\colon C \to D$ let us consider again the mappings (of sets):

$$F_{XX'}\colon C(X,X') \to D(F(X),F(X')), \qquad f \mapsto F(f). \tag{1.14}$$

F is said to be *faithful* if all these mappings are injective (for X,X' in C); F is said to be *full* if all of them are surjective.

An isomorphism of categories is always full and faithful. The inclusion functor $D \to C$ of a subcategory is always faithful; it is full if and only if D is a full subcategory of C. A projection functor $P\colon C \to C/R$ on a quotient category is always full and bijective on objects.

There are some (obvious) preservation and reflection properties of functors:

(a) every functor preserves commutative diagrams, isomorphisms, retracts, split monos and split epis (commutative diagrams were presented in Section 0.3 of the Introduction and are formalised below, in 1.4.9),

(b) a faithful functor *reflects* monos and epis (i.e. if $F(f)$ is mono or epi, then f is also) and commutative diagrams,

(c) a full and faithful functor reflects isomorphisms, split monos and split epis.

Applying point (a), the usual way of proving that a topological subspace $A \subset X$ is not a retract (in Top) is to find a functor $F\colon \mathsf{Top} \to \mathsf{Ab}$ such that the associated homomorphism $F(A) \to F(X)$ is not a split mono in Ab. In this way, any homology functor H_n shows that the n-dimensional sphere $\mathbb{S}^n$ is not a retract of the euclidean space $\mathbb{R}^{n+1}$.

1.4.8 Concrete categories

The notion of 'category of structured sets' is either vague or has different, complicated formalisations, which can hardly cover all the cases we may want.

As a formal and simple alternative, a *concrete category* is defined as a category A *equipped* with a faithful functor $U\colon \mathsf{A} \to \mathsf{Set}$, called its *forgetful functor*. As a consequence U *reflects monos and epis*, but need not preserve them. Concrete categories are extensively studied in [AHS].

It is not easy to find categories that can *not* be made concrete: even $\mathsf{Set}^{\mathrm{op}}$ can be, by the *contravariant functor of subsets*

$$\mathcal{P}^*\colon \mathsf{Set}^{\mathrm{op}} \to \mathsf{Set},$$
$$X \mapsto \mathcal{P}X, \qquad (f\colon X \to Y) \mapsto (f^*\colon \mathcal{P}Y \to \mathcal{P}X), \tag{1.15}$$

where $f^*(B) = f^{-1}(B)$ is the pre-image of $B \subset Y$.

As a consequence, if the category C is concrete via $U\colon \mathsf{C} \to \mathsf{Set}$, then C^{op} *can* be made concrete via the composed functor $\mathcal{P}^*U^{\mathrm{op}}\colon \mathsf{C}^{\mathrm{op}} \to \mathsf{Set}$ (letting apart whether this is of interest).

But it is important to note that the homotopy category hoTop (introduced in 1.4.5) *cannot* be made concrete, as was proved by Freyd [Fr3, Fr4].

More generally a category A equipped with a faithful functor $U\colon \mathsf{A} \to \mathsf{C}$ is said to be *concrete over* C. (We shall see in Lemma 1.5.7 how this notion is related with subcategories.)

1.4.9 Exercises and complements (Functors as diagrams)

Diagrams in a category have already been presented in Section 0.3 of the Introduction, in an informal way. We can now be more precise.

(a) Let S be a small category. A *diagram* in the category C, *of type* S (or *based on* S) will be any functor $X: S \to C$.

It can be written in *index notation*, with indices $i \in$ ObS and $a: i \to j$ in MorS:

$$X: S \to C, \qquad i \mapsto X_i, \quad a \mapsto (X_a: X_i \to X_j), \qquad (1.16)$$

or as a system $((X_i), (u_a))$, with $u_a = X_a$.

(b) The diagram $X: S \to C$ is said to be *commutative* if it factorises through the canonical projection $P: S \to \text{po}(S)$ with values in the preordered set po(S) defined in 1.4.5. The reader will easily see that this amounts to the conditions (i), (ii) of Section 0.3.

Therefore, every functor defined on a preordered set S is automatically a commutative diagram.

In particular, a *coherent system of isomorphisms* is a functor $X: S \to C$ where the preorder of S is an equivalence relation $i \sim j$ (i.e. a symmetric preorder). The system amounts thus to a family (X_i) of objects of C equipped with a family of (iso)morphisms $u_{ij}: X_i \to X_j$ for $i \sim j$, such that

$$u_{jk}.u_{ij} = u_{ik}, \quad u_{ii} = \text{id}(X_i) \qquad (\text{for } i \sim j \sim k \text{ in } S). \qquad (1.17)$$

Since the well-known Coherence Theorem of monoidal categories, by Mac Lane [M3], a 'coherence theorem' in category theory states that, under suitable hypotheses, all the isomorphisms of a certain kind form a coherent system. The extended version for bicategories is in [MaP].

(c) A functor $\mathbf{1} \to C$ amounts to an object of C, while a functor $\mathbf{2} \to C$ amounts to a morphism of C.

(d) Let $E = \{1, e\}$ be the idempotent monoid on two elements, with $e^2 = e$. A functor $F: E \to C$ amounts to an object X of C equipped with an idempotent endomorphism $e: X \to X$; or – more simply – to an idempotent endomorphism e (which determines its domain-codomain).

(Note that E is isomorphic to the join-semilattice $(\mathbf{2}, \vee)$, with unit 0 and $1 \vee 1 = 1$. But of course one should not confuse the ordinal category $\mathbf{2}$, which has two objects, with the monoid-category E, which has only one.)

(e) A functor $X: (\mathbb{N}, \leqslant) \to C$, defined on the ordered set of natural numbers (i.e. the ordinal ω), amounts to a sequence of consecutive morphisms of C

$$X_0 \longrightarrow X_1 \longrightarrow X_2 \longrightarrow \ \dots \ \longrightarrow X_n \longrightarrow \ \dots \qquad (1.18)$$

while a functor $X\colon (\mathbb{N}, +) \to \mathsf{C}$, defined on the additive monoid of natural numbers, amounts to an endomorphism $u_1\colon X_* \to X_*$ (and its powers, $u_n = (u_1)^n$, including $u_0 = \mathrm{id}(X_*)$).

(f) The product $\mathbf{2} \times \mathbf{2}$ has four objects and five non-identity arrows, as in the left diagram below

$$
\begin{array}{ccc}
00 \longrightarrow 10 & \qquad & X_{00} \longrightarrow X_{10} \\
\downarrow \searrow \ \downarrow & & \downarrow \quad = \quad \downarrow \\
01 \longrightarrow 11 & & X_{01} \longrightarrow X_{11}
\end{array}
\qquad (1.19)
$$

A functor $X\colon \mathbf{2} \times \mathbf{2} \to \mathsf{C}$ amounts to a commutative square in C, as in the right diagram above.

(g) In some texts a diagram in C is defined as a morphism of graphs $X\colon \Gamma \to \mathsf{C}$, where Γ is a small graph. This is not more general, because every graph generates a free category, in a simple way – by finite paths of consecutive arrows (see 3.7.3). On the contrary, this second definition is *less expressive*: for instance, commutative squares or idempotent endomorphisms cannot be formalised as arbitrary diagrams on a given graph.

1.5 Natural transformations and equivalence of categories

Given two *parallel* functors $F, G\colon \mathsf{C} \to \mathsf{D}$ (between the same categories) there can be 'higher arrows' $\varphi\colon F \to G$, called 'natural transformations' (and functorial isomorphisms, or natural isomorphisms, when they are invertible).

'Category', 'functor' and 'natural transformation' are the three basic terms of category theory, since the very beginning of the theory in [EiM].

It is interesting to note that only the last term is taken from the common language: one can say that Eilenberg and Mac Lane introduced categories and functors because they wanted to formalise the natural transformations that they were encountering in Algebra and Algebraic Topology (as Mac Lane remarks in [M4], at the end of Section I.6). Much in the same way as a general theory of *continuity* (a familiar term for a familiar notion) requires the introduction of *topological spaces* (denoted by a theoretical term).

A reader acquainted with basic homotopy theory can take advantage of a formal parallelism, where *spaces* correspond to *categories*, *continuous mappings* to *functors*, *homotopies* of mappings to *invertible natural transformations*, and *homotopy equivalence* of spaces to *equivalence* of categories. This analogy is even deeper in the domain of Directed Algebraic Topology, where *directed homotopies* need not be reversible and correspond to general *natural transformations*, see [G8].

1.5.1 Natural transformations

Given two functors F, G: C $\to$ D between the same categories, a *natural transformation* φ: $F \to G$ (or φ: $F \to G$: C $\to$ D) consists of the following data:

- for each object X of C a morphism φX: $FX \to GX$ in D (called the *component* of φ on X and also written as φ_X, or sometimes just φ),

so that, for every arrow f: $X \to X'$ in C, we have a commutative square in D (*naturality condition* of φ on f)

$$\begin{array}{ccc} FX & \xrightarrow{\varphi X} & GX \\ {\scriptstyle Ff}\downarrow & & \downarrow{\scriptstyle Gf} \\ FX' & \xrightarrow[\varphi X']{} & GX' \end{array} \qquad \varphi X'.F(f) = G(f).\varphi X. \qquad (1.20)$$

In particular, *the identity of a functor* F: C $\to$ D is the natural transformation $\mathrm{id}F$: $F \to F$, with components $(\mathrm{id}F)_X = \mathrm{id}(FX)$: $FX \to FX$.

A natural transformation φ: $F \to G$ often comes out of a 'canonical choice' of a family $(\varphi X: FX \to GX)_X$ of morphisms, *but these two aspects should not be confused.* There are canonical choices of such families that are not natural (see 1.5.2(d)), and natural transformations 'defined' through the axiom of choice (see 1.5.8).

Replacing the category C with a graph Γ (see 1.1.2), one can consider a *natural transformation* φ: $F \to G$: $\Gamma \to$ D between two morphisms of graphs defined on Γ, *with values in a category.*

1.5.2 Exercises and complements

(a) Let F: $R\,\mathsf{Mod} \to R\,\mathsf{Mod}$ be the identity functor, and let the scalar λ belong to the centre of a ring, i.e. the subring of the elements that commute with all the others.

Prove that there is a natural transformation λ: $F \to F$, whose component λ_A: $A \to A$ on the left module A is the multiplication by λ. Prove that every natural transformation $F \to F$ is of this type (working on its component on R, as a module), for a unique scalar λ.

This bijective correspondence shows that the category $R\,\mathsf{Mod}$ *determines* the centre of the ring R, in a structural way, and leads to considering the relation of Morita equivalence of rings (see 1.5.6(b)).

(b) Determine the natural endotransformations of the forgetful functors $R\,\mathsf{Mod} \to \mathsf{Set}$ and $R\,\mathsf{Mod} \to \mathsf{Ab}$.

(c) Determine the natural endotransformations of the *contravariant functor of subsets* $\mathcal{P}^* \colon \mathsf{Set}^{\mathrm{op}} \to \mathsf{Set}$ (defined in (1.15)).

(d) Determine the natural endotransformations of the *covariant functor of subsets*

$$\mathcal{P}_* \colon \mathsf{Set} \to \mathsf{Set},$$
$$X \mapsto \mathcal{P}X, \qquad (f \colon X \to Y) \mapsto (f_* \colon \mathcal{P}X \to \mathcal{P}Y), \tag{1.21}$$

where $f_*(A) = f(A)$ is the image of $A \subset X$. Show that the canonical choices $\varphi_X(A) = X$ and $\psi_X(A) = X \setminus A$ are *not* natural.

(e) A natural transformation $f \to g \colon X \to Y$ between increasing mappings of preordered sets amounts to the preordering $f \prec g$ introduced in 1.2.2, and is determined by f, g.

(f) Characterise the natural transformations $f \to g \colon X \to Y$ between homomorphisms of monoids.

1.5.3 Operations

Two natural transformations $\varphi \colon F \to G$ and $\psi \colon G \to H$ have a *vertical composition* $\psi\varphi \colon F \to H$ (also written as $\psi.\varphi$)

$$\mathsf{C} \xrightarrow[\;\;H\;\;]{\substack{\;F\; \\ \downarrow\varphi \\ \downarrow\psi}} \mathsf{D} \qquad\qquad (\psi\varphi)(X) = \psi X.\varphi X \colon FX \to HX. \tag{1.22}$$

Moreover there is a *whisker composition* of natural transformations with functors, or *reduced horizontal composition*, written as $K\varphi H$ (or $K{\circ}\varphi{\circ}H$, when useful to distinguish it)

$$\mathsf{C}' \xrightarrow{\;\;H\;\;} \mathsf{C} \xrightarrow[\;\;G\;\;]{\substack{\;F\; \\ \downarrow\varphi}} \mathsf{D} \xrightarrow{\;\;K\;\;} \mathsf{D}' \tag{1.23}$$

$$K\varphi H \colon KFH \to KGH, \qquad (K\varphi H)(X') = K(\varphi(HX')).$$

(The binary operations φH and $K\varphi$ are also called whisker compositions; they can be obtained from the ternary operation by inserting identity functors.)

This 2-dimensional structure of the framework of categories, where natural transformations play the role of 2-dimensional arrows between functors, will be further analysed in Section 7.1 (including the *full* horizontal composition of natural transformations).

An *isomorphism of functors*, or functorial isomorphism, or natural isomorphism, is a natural transformation $\varphi \colon F \to G \colon \mathsf{C} \to \mathsf{D}$ which is invertible with respect to vertical composition. It is easy to see that this happens if and only if all the components φX are invertible in D.

(The original term of 'natural equivalence', in [EiM], is now obsolete and will not be used, because it can lead to confusion with an equivalence of categories.)

Isomorphism of (parallel) functors is an equivalence relation, written as $F \cong G$.

1.5.4 Exercises

The reader should prove the following properties of the compositions of natural transformations, which will often be used in computations. All the proofs are trivial, except the last, which is easy.

(a)
$$\chi(\psi\varphi) = (\chi\psi)\varphi,$$
$$K'(K\varphi H)H' = (K'K)\varphi(HH') \qquad (associativities),$$

(b)
$$\varphi.\mathrm{id}F = \varphi = \mathrm{id}G.\varphi,$$
$$1_Y \circ \varphi \circ 1_X = \varphi, \qquad K \circ \mathrm{id}F \circ H = \mathrm{id}(KFH) \qquad (identities),$$

(c)
$$K(\psi\varphi)H = (K\psi H).(K\varphi H) \qquad (distributivity),$$

(d)
$$(\psi G).(H\varphi) = (K\varphi).(\psi F) \qquad (interchange),$$

The last, more precisely a 'reduced interchange property', is about two 'horizontally consecutive' natural transformations, as in the following diagram

$$\mathsf{C} \xrightarrow[\substack{G}]{\substack{F \\ \downarrow\varphi}} \mathsf{D} \xrightarrow[\substack{K}]{\substack{H \\ \downarrow\psi}} \mathsf{E} \qquad\qquad (1.24)$$

1.5.5 Equivalence of categories

Isomorphisms of categories have been defined in 1.4.1.

More generally, an *equivalence of categories* is a functor $F \colon \mathsf{C} \to \mathsf{D}$ which is invertible up to isomorphism of functors (defined in 1.5.3), i.e. there exists a functor $G \colon \mathsf{D} \to \mathsf{C}$ such that $GF \cong \mathrm{id}\mathsf{C}$ and $FG \cong \mathrm{id}\mathsf{D}$. The functor G can be called a *weak inverse* of F.

An *adjoint equivalence of categories* is a coherent version of this notion; namely it is a four-tuple $(F, G, \eta, \varepsilon)$ where

- $F\colon \mathsf{C} \to \mathsf{D}$ and $G\colon \mathsf{D} \to \mathsf{C}$ are functors,
- $\eta\colon \mathrm{id}\mathsf{C} \to GF$ and $\varepsilon\colon FG \to \mathrm{id}\mathsf{D}$ are functorial isomorphisms,
- $F\eta = (\varepsilon F)^{-1}\colon F \to FGF, \qquad \eta G = (G\varepsilon)^{-1}\colon G \to GFG.$

(The direction of η and ε is written above as in the more general case of an adjunction, where these transformations need not be invertible: see Section 3.1.)

The following conditions on a functor $F\colon \mathsf{C} \to \mathsf{D}$ are equivalent, forming a very useful *characterisation of the equivalence of categories*:

(i) F is an equivalence of categories,

(ii) F can be completed to an adjoint equivalence of categories $(F, G, \eta, \varepsilon)$,

(iii) F is faithful, full and *essentially surjective on objects*.

The last property means that: for every object Y of D there exists some object X in C such that $F(X)$ is isomorphic to Y in D. The proof of the equivalence of these three conditions requires the axiom of choice and is given below, in Theorem 1.5.8.

One says that the categories C, D are *equivalent*, written as $\mathsf{C} \simeq \mathsf{D}$, if there exists an equivalence of categories between them (or, equivalently, an *adjoint* equivalence of categories). This is indeed an equivalence relation, as is easily proved using condition (iii) above, or can be proved directly using the operations we have seen above (in 1.5.3).

The role of equivalences in category theory is fundamental. Like everything, it should not be viewed as 'absolute'. Comments on this point can be found in 1.5.9, 3.3.7 and 3.6.6.

1.5.6 Exercises and complements

(a) The category of finite sets (and mappings between them) is equivalent to its full subcategory of finite cardinals, which is small and therefore cannot be isomorphic to the former.

*(b) The reader may know that two rings R, S are said to be *Morita equivalent* if their categories of modules are equivalent [Mo, AnF]. This is a crucial notion in ring theory, that becomes trivial in the domain of commutative rings: in fact, if two rings are Morita equivalent one can prove that their centres are isomorphic (using what we have seen in 1.5.2); but – quite interestingly – commutative rings can be Morita equivalent to non-commutative ones, like their rings of square matrices. Thus, studying left modules on the ring $M_n(\mathbb{R})$ is equivalent to studying real vector spaces.*

(c) A category is said to be *skeletal* if it has no pair of distinct isomorphic objects. It is easy to show, by the previous characterisation of equivalences, that every category has a *skeleton*, i.e. an equivalent skeletal category. The latter can be obtained as a full subcategory, by *choosing* precisely one object in every class of isomorphic objects.

(d) We have described in (a) a skeleton of the category of finite sets that can be constructed without any choice, even though we do need the axiom of choice to prove that the inclusion of this skeleton has a weak inverse.

In a different way, a preordered set X has a natural skeleton *formed by a quotient*, the associated ordered set $X/\!\!\sim$ (see 1.2.2); but again we need the axiom of choice to prove that the projection $X \to X/\!\!\sim$ has a weak inverse.

(e) The reader can easily prove that two categories are equivalent if and only if they have isomorphic skeletons; loosely speaking, this says that an equivalence of categories amounts to multiplying or deleting isomorphic copies of objects, even though there can be no canonical way of doing this.

1.5.7. *Lemma* (Modifying functors and categories)

(a) *Starting from a functor* $F \colon \mathsf{C} \to \mathsf{D}$ *and an arbitrary family of isomorphisms* $\varphi_X \colon F(X) \to G(X)$ *in* D *(for X in* C*), there is precisely one way of extending these data to a functor* $G \colon \mathsf{C} \to \mathsf{D}$ *and a natural transformation* $\varphi \colon F \to G$ *(obviously invertible).*

(b) *Given a faithful functor* $U \colon \mathsf{A} \to \mathsf{C}$, *one can embed* A *as a subcategory of a suitable category* C' *equivalent to* C.

(c) *If the faithful functor* U *is essentially surjective on the objects, one can make this embedding wide (and even the identity on objects).*

Note. As an example, the forgetful functor $U \colon \mathsf{Top} \to \mathsf{Set}$ is surjective on objects, and Top is a wide subcategory of the category $\mathsf{T} \simeq \mathsf{Set}$ introduced in 1.1.5(a).

Proof (a) We define G on a morphism $f \colon X \to Y$ of C as

$$G(f) = \varphi_Y . F(f) . (\varphi_X)^{-1} \colon G(X) \to G(Y).$$

Besides preserving identities and composition, this is necessary and sufficient to make φ natural.

(b) We construct C' with objects C_A where A belongs to A and $U(A) = C$, and other (distinct) objects C_* where C belongs to C. Then we let $\mathsf{C}'(C_X, D_Y) = \mathsf{C}(C, D)$, where X (resp. Y) is either an object of A or the symbol $*$. The composition is that of C.

Now the functor $U' \colon \mathsf{A} \to \mathsf{C}'$ that takes A to $U(A)_A$ and f to $U(f)$ is injective on objects and arrows, and identifies A to a subcategory of C'. On the other hand, the obvious full embedding $\mathsf{C} \to \mathsf{C}'$ taking C to C_* is an equivalence of categories, since ever object C_A is isomorphic to C_*.

(c) If U is essentially surjective on the objects, we can omit the objects of type C_* in the construction of C', and rewrite the object C_A as A so that $\mathrm{Ob}\mathsf{C}' = \mathrm{Ob}\mathsf{A}$. $\qquad\square$

1.5.8 *Theorem* (Characterisation of equivalences)

The conditions (i)–(iii) of 1.5.5 are equivalent.

Proof Part of the proof is a consequence of general properties of adjunctions, but it is useful to have now this important theorem, much before Chapter 3.

(i) $\Rightarrow$ (iii). We assume that $F \colon \mathsf{C} \to \mathsf{D}$ has a quasi inverse $G \colon \mathsf{D} \to \mathsf{C}$, and we write $\eta \colon 1 \to GF$ a functorial isomorphism. Then F is faithful: given two parallel morphisms $f, f' \colon X \to X'$, we have two commutative squares with horizontal isomorphisms

$$\begin{array}{ccc} X & \xrightarrow{\ \eta X\ } & GFX \\ {\scriptstyle f}\big\downarrow\big\downarrow{\scriptstyle f'} & {\scriptstyle GFf}\big\downarrow\big\downarrow{\scriptstyle GFf'} & \\ X' & \xrightarrow[\ \eta X'\ ]{} & GFX' \end{array} \qquad (1.25)$$

so that $Ff = Ff'$ implies $f = f'$. Similarly, G is faithful (because $FG \cong 1$).

To show that F is full, we take a morphism $g \colon FX \to FX'$ and define $f = (\eta X')^{-1}.Gg.\eta X \colon X \to X'$. Now we have the parallel morphisms $g, Ff \colon FX \to FX'$ in D. Both $G(g)$ and $G(Ff)$ make the following square commute in C

$$\begin{array}{ccc} X & \xrightarrow{\ \eta X\ } & GFX \\ {\scriptstyle f}\big\downarrow & {\scriptstyle Gg}\big\downarrow{\scriptstyle GFf} & \\ X' & \xrightarrow[\ \eta X'\ ]{} & GFX' \end{array} \qquad (1.26)$$

Since the horizontal arrows are isomorphisms, $G(g) = G(Ff)$; since G is faithful, $g = F(f)$.

Finally it is obvious that F is essentially surjective on objects, because each object Y of D is isomorphic to $F(GY)$.

(iii) $\Rightarrow$ (ii). Assuming that F is faithful, full and essentially surjective on objects, we 'construct' with the axiom of choice a quasi inverse G.

For every object Y of D we choose an object $G(Y)$ in C *and* an isomorphism $\varepsilon Y \colon F(G(Y)) \to Y$ in D. For every morphism $g \colon Y \to Y'$ in D, we define $G(g) \colon GY \to GY'$ as the unique morphism of C such that

$$F(G(g)) = (\varepsilon Y')^{-1}.g.\varepsilon Y \colon F(GY) \to F(GY'),$$

which makes the following square commutative in D

$$
\begin{array}{ccc}
FGY & \xrightarrow{\ \varepsilon Y\ } & Y \\
{\scriptstyle FGg}\downarrow & & \downarrow{\scriptstyle g} \\
FGY' & \xrightarrow[\ \varepsilon Y'\ ]{} & Y'
\end{array}
\qquad g.\varepsilon Y = \varepsilon Y'.FG(g). \qquad (1.27)
$$

It is easy to verify that G is a functor. For instance, for a composed morphism $g'g \colon Y \to Y' \to Y''$, we have:

$$F(G(g'g)) = (\varepsilon Y'')^{-1}.g'g.\varepsilon Y = ((\varepsilon Y'')^{-1}.g'.\varepsilon Y').((\varepsilon Y')^{-1}.g.\varepsilon Y)$$
$$= FG(g').FG(g) = F(Gg'.Gg),$$

and $G(g'g) = Gg'.Gg$, because F is faithful.

The family $\varepsilon = (\varepsilon Y)_Y$ of isomorphisms is a natural transformation $\varepsilon \colon FG \to \mathrm{id}\mathsf{D}$, by (1.27), whence a functorial isomorphism.

Moreover, for every X in C there is precisely one morphism $\eta X \colon X \to GFX$ such that

$$F(\eta X) = (\varepsilon FX)^{-1} \colon FX \to FG(FX),$$

and ηX is invertible (by 1.4.7(c)).

Using again the faithfulness of F, one shows that $\eta = (\eta X)_X$ is natural, whence a functorial isomorphism.

We already have one coherence condition: $F\eta = (\varepsilon F)^{-1}$. For the other, let us first note that the naturality of $\varepsilon \colon FG \to 1 \colon \mathsf{D} \to \mathsf{D}$ on its own (invertible) component $\varepsilon Y \colon FGY \to Y$

$$
\begin{array}{ccc}
FGFGY & \xrightarrow{\ \varepsilon FGY\ } & FGY \\
{\scriptstyle FG\varepsilon Y}\downarrow & & \downarrow{\scriptstyle \varepsilon Y} \\
FGY & \xrightarrow[\ \varepsilon Y\ ]{} & Y
\end{array}
\qquad \varepsilon Y.\varepsilon(FGY) = \varepsilon Y.FG(\varepsilon Y), \qquad (1.28)
$$

gives $\varepsilon(FGY) = FG(\varepsilon Y)$. Therefore:

$$F(\eta GY) = (\varepsilon FGY)^{-1} = FG(\varepsilon Y)^{-1} = F(G\varepsilon Y)^{-1},$$

and the faithfulness of F gives $\eta GY = (G\varepsilon Y)^{-1}$, for all Y in D.

(ii) $\Rightarrow$ (i). Obvious. $\qquad\qquad\qquad\qquad\qquad\qquad\qquad\qquad\qquad\qquad\quad \square$

1.5.9 A first digression on categorical equivalence

There is now a tendency in category theory to prefer notions invariant up to equivalence of categories (let us say an 'E-notion', for short), rather than just invariant up to isomorphism of categories (let us say an 'I-notion'), or not even invariant in this sense (let us say an 'S-notion', for 'strict'). Non E-notions have even been described as 'evil'.

In the author's opinion invariance up to categorical equivalence is certainly an important property. Yet, it should be put in balance with other aspects, like simplicity, effectiveness and adherence to well-founded practice in Mathematics.

(a) The product $C \times D$ of categories has been defined in a strict way (in 1.4.6). Replacing this notion by a weaker one, invariant up to equivalence, would be particularly strange and ineffective: we no longer have a functor $Cat^2 \to Cat$. A reader familiar with the cartesian closedness of Cat (to be seen in 3.4.7(a)) may wonder what this may mean, with respect to an E-notion of product.

(b) A *faithful functor* can be viewed as an 'E-replacement' of the S-notion of *subcategory* (as shown by Lemma 1.5.7). We can agree that a faithful functor is a more important notion; nevertheless the notion of subcategory keeps at least its practical importance, for instance when we want to quickly describe a new structure as a subcategory of a known one.

(c) Some instances where an I-notion can be preferable to the corresponding E-notion will be seen in 3.3.7 (for comma categories), in 3.6.6 (for monadicity, following Mac Lane) and in 7.1.5 (for flexible limits, following [BKPS]).

1.6 Categories of functors, representable functors

This brief section is about categories $Cat(S, C)$ of functors $S \to C$, which includes the well-known case of simplicial sets. Representable functors are also introduced here.

S is always a small category.

1.6.1 Categories of functors

For any category C (possibly large) we write as $Cat(S, C)$, or C^S, the category whose objects are the functors $F \colon S \to C$ and whose morphisms are the natural transformations $\varphi \colon F \to G$, with vertical composition. These objects can be viewed as diagrams in C, based on S, as in 1.4.9.

In particular the arrow category **2**, with two objects (0 and 1) and precisely one non-identity arrow, $0 \to 1$ (see 1.2.3) gives the *category of morphisms* C^2 of C, where a map $f = (f', f''): x \to y$ is a commutative square of C. These are composed *by pasting* commutative squares, as in the right diagram below

$$
\begin{array}{ccc}
X' & \xrightarrow{f'} & Y' \\
{\scriptstyle x}\downarrow & & \downarrow{\scriptstyle y} \\
X'' & \xrightarrow[g']{} & Y''
\end{array}
\qquad
\begin{array}{ccccc}
X' & \xrightarrow{f'} & Y' & \xrightarrow{f''} & Z' \\
{\scriptstyle x}\downarrow & & \downarrow{\scriptstyle y} & & \downarrow{\scriptstyle z} \\
X'' & \xrightarrow[g']{} & Y'' & \xrightarrow[g'']{} & Z''
\end{array}
\qquad (1.29)
$$

The identity of x is: $\mathrm{id}(x) = (\mathrm{id}X', \mathrm{id}X''): x \to x$.

1.6.2 Exercises and complements

(a) Prove that a natural transformation $\varphi: F \to G: C \to D$ can be viewed as a functor $C \to D^2$, or equivalently as a functor $C \times \mathbf{2} \to D$.

(b) Analyse the category

$$
G\mathsf{Set} = \mathsf{Cat}(G, \mathsf{Set}), \qquad (1.30)
$$

where G is a group (viewed as a category). An object, called a *G-set*, is a set equipped with an action of G, and an arrow, called a *G-morphism*, is a G-equivariant mapping – notions that the reader may already know or can deduce from (1.30). Similarly we have the category $G\mathsf{Top} = \mathsf{Cat}(G, \mathsf{Top})$ of *G-spaces*. All this can be extended to monoids.

(c) With a basic knowledge of Homological Algebra, one can describe the category $\mathsf{Ch}^+\mathsf{Ab}$ of positive cochain complexes of abelian groups as a full subcategory of the category $\mathsf{Cat}(\mathbb{N}, \mathsf{Ab})$ that we have seen in 1.4.9(e), and give a similar description of the category $\mathsf{Ch}_+\mathsf{Ab}$ of positive chain complexes (to be examined in 5.3.7).

1.6.3 Categories of presheaves

A functor $S^{\mathrm{op}} \to C$, defined on the opposite category S^{op}, is also called a *presheaf* of C on the (small) category S. They form the *presheaf category* $\mathsf{Psh}(S, C) = C^{S^{\mathrm{op}}}$.

We are particularly interested in the category $\mathsf{Psh}(S, \mathsf{Set}) = \mathsf{Set}^{S^{\mathrm{op}}}$ of *presheaves of sets* on S.

The category S is canonically embedded in the latter, by the *Yoneda embedding*

$$
y: S \to \mathsf{Set}^{S^{\mathrm{op}}}, \qquad y(i) = S(-, i): S^{\mathrm{op}} \to \mathsf{Set}, \qquad (1.31)
$$

which sends every object i to the corresponding presheaf $y(i)$, *represented by i* (according to the definition below, in 1.6.4).

Taking as S the category Δ of finite positive ordinals (and increasing maps), one gets the category $\mathrm{Smp}(\mathsf{C}) = \mathsf{C}^{\Delta^{\mathrm{op}}}$ of *simplicial objects* in C, and – in particular – the well-known category SmpSet of *simplicial sets*, that will be studied in Section 5.3. Here the Yoneda embedding sends the positive ordinal $\{0, ..., n\}$ to the simplicial set Δ^n, freely generated by one simplex of dimension n.

Simplicial groups, in the category SmpGp, are also important.

Cubical sets, another category of presheaves of interest in Topology, will be studied in Section 5.4.

A reader interested in categories of *sheaves* is referred to [MaM, Bo3].

1.6.4 Representable functors

Assume now that C has small hom-sets. A functor $F\colon \mathsf{C} \to \mathsf{Set}$ is said to be *representable* if it is isomorphic to a functor

$$\mathsf{C}(X_0, -)\colon \mathsf{C} \to \mathsf{Set}, \qquad X \mapsto \mathsf{C}(X_0, X),$$
$$(f\colon X \to Y) \mapsto (f.-\colon \mathsf{C}(X_0, X) \to \mathsf{C}(X_0, Y)), \tag{1.32}$$

for some object X_0 in C; this object is said *to represent F*.

Then the Yoneda Lemma (in 1.6.6) describes the natural transformations $F \to G$, for every functor $G\colon \mathsf{C} \to \mathsf{Set}$; it also proves that the object which represents a functor is determined up to isomorphism (see 1.6.7).

A contravariant functor $F\colon \mathsf{C} \dashrightarrow \mathsf{Set}$ is in fact a covariant functor $F\colon \mathsf{C}^{\mathrm{op}} \to \mathsf{Set}$; therefore it is *representable* if it is isomorphic to a functor

$$\mathsf{C}(-, X_0)\colon \mathsf{C}^{\mathrm{op}} \to \mathsf{Set}, \qquad X \mapsto \mathsf{C}(X, X_0),$$
$$(f\colon X \to Y) \mapsto (-.f\colon \mathsf{C}(Y, X_0) \to \mathsf{C}(X, X_0)), \tag{1.33}$$

for some object X_0 in C.

1.6.5 Exercises and complements

(a) Show that the forgetful functors

$$\mathsf{Top} \to \mathsf{Set}, \qquad R\,\mathsf{Mod} \to \mathsf{Set}, \qquad \mathsf{Rng} \to \mathsf{Set},$$

are representable.

(b) By definition, the canonical forgetful functors of Ban and Ban_1 are represented by the scalar field K ($\mathbb{R}$ or $\mathbb{C}$):

$$U = \mathsf{Ban}(K, -)\colon \mathsf{Ban} \to \mathsf{Set}, \qquad U(X) = |X|,$$
$$B_1 = \mathsf{Ban}_1(K, -)\colon \mathsf{Ban}_1 \to \mathsf{Set}, \quad B_1(X) = \{x \in X \mid ||x|| \leqslant 1\}, \tag{1.34}$$

and give, respectively, the underlying set $|X|$ and the unit ball $B_1(X)$ of a Banach space (up to functorial isomorphism).

(c) Show that the contravariant functor of subsets $\mathcal{P}^*\colon \mathsf{Set}^{\mathrm{op}} \to \mathsf{Set}$ (defined in (1.15)) is representable, making use of the characteristic function of a subset.

(d) A representable covariant functor preserves monomorphisms. A representable contravariant functor takes epimorphisms of its domain to injective mappings.

(e) The reader can review the exercises 1.5.2(b) and (c) at the light of the Yoneda Lemma, below.

1.6.6 Yoneda Lemma

Let $F, G\colon \mathsf{C} \to \mathsf{Set}$ be two functors, with $F = \mathsf{C}(X_0, -)$. The canonical mapping

$$y\colon \mathrm{Nat}(F, G) \to G(X_0), \qquad \varphi \mapsto \varphi_{X_0}(\mathrm{id}X_0) \in G(X_0), \tag{1.35}$$

from the set of natural transformations $\varphi\colon F \to G$ to the set $G(X_0)$ is a bijection.

   Note. This result can be extended to a representable functor $F \cong \mathsf{C}(X_0, -)$, making use of this isomorphism.

Proof We construct a backward mapping

$$y'\colon G(X_0) \to \mathrm{Nat}(F, G),$$
$$(y'(x))_X\colon \mathsf{C}(X_0, X) \to GX, \quad (f\colon X_0 \to X) \mapsto (Gf)(x), \tag{1.36}$$

where the family $\psi = y'(x)$ is indeed a natural transformation $F \to G$. In fact, for $g\colon X \to Y$ in C

$$
\begin{array}{ccc}
\mathsf{C}(X_0, X) & \xrightarrow{\psi X} & GX \\
{\scriptstyle g.-}\downarrow & & \downarrow{\scriptstyle Gg} \\
\mathsf{C}(X_0, Y) & \xrightarrow[\psi Y]{} & GY
\end{array}
\tag{1.37}
$$

$$(G(g).\psi X)(f) = G(g)((Gf)(x)) = G(gf)(x) = \psi_Y(gf).$$

Then we have to show that the mappings y, y' are inverse to each other.

First, a natural transformation $\varphi \colon F \to G$ is taken to the element $x = y(\varphi) = \varphi_{X_0}(\mathrm{id}X_0) \in G(X_0)$, and the latter to the natural transformation $y'(x) \colon F \to G$. The component $(y'(x))_X$ takes the morphism $f \colon X_0 \to X$ to $(Gf)(x)$, and it is sufficient to prove that $\varphi_X(f) = (Gf)(x)$; this follows from the naturality of φ on f

$$
\begin{array}{ccc}
\mathsf{C}(X_0, X_0) & \xrightarrow{\varphi X_0} & GX_0 \\
{\scriptstyle f.-}\downarrow & & \downarrow{\scriptstyle Gf} \\
\mathsf{C}(X_0, X) & \xrightarrow[\varphi X]{} & GX
\end{array}
\tag{1.38}
$$

$$\varphi_X(f) = \varphi_X(f.\mathrm{id}X_0) = (Gf)(\varphi_{X_0}(\mathrm{id}X_0)) = (Gf)(x).$$

The other way round, an element $x \in G(X_0)$ is taken to the natural transformation $\varphi = y'(x) \colon F \to G$ and then to

$$y(\varphi) = \varphi_{X_0}(\mathrm{id}X_0) = (G(\mathrm{id}X_0))(x) = x.$$

$\square$

1.6.7 Exercises

(a) Prove that, if the functor $F \colon \mathsf{C} \to \mathsf{Set}$ is represented by two objects, these are isomorphic in C.

(b) More generally, prove that a natural transformation $\varphi \colon F_0 \to F_1$ between two representable functors $F_i = \mathsf{C}(X_i, -)$ is invertible if and only if $y(\varphi) \in F_1(X_0) = \mathsf{C}(X_1, X_0)$ is an isomorphism.

1.7 A category of Galois connections

After the category Ord of ordered sets and monotone mappings, we now study the category AdjOrd of ordered sets and (covariant) Galois connections between them – or adjunctions between ordered sets. Motivations of this notion can be found in the comments and examples below.

All this will also serve as an introduction to general adjunctions, dealt with in Section 3.1.

The section ends with the *contravariant* form of Galois connections, the classical one.

1.7.1 Definition

Given a pair X, Y of ordered sets, a (covariant) *Galois connection* (f, g) between them can be presented in the following equivalent ways.

(i) We assign two increasing mappings $f\colon X \to Y$ and $g\colon Y \to X$ such that:

$$f(x) \leqslant y \text{ in } Y \quad \Leftrightarrow \quad x \leqslant g(y) \text{ in } X. \tag{1.39}$$

(ii) We assign an increasing mapping $g\colon Y \to X$ such that for every $x \in X$ there exists in Y:

$$f(x) = \min\{y \in Y \mid x \leqslant g(y)\}. \tag{1.40}$$

(ii*) We assign an increasing mapping $f\colon X \to Y$ such that for every $y \in Y$ there exists in X:

$$g(y) = \max\{x \in X \mid f(x) \leqslant y\}. \tag{1.41}$$

(iii) We assign two increasing mappings $f\colon X \to Y$ and $g\colon Y \to X$ such that

$$\mathrm{id}X \leqslant gf, \qquad\qquad fg \leqslant \mathrm{id}Y. \tag{1.42}$$

By these formulas g determines f (called its *left adjoint*) and f determines g (its *right adjoint*). One writes $f \dashv g$ (as in the general notation of adjoints in category theory), and the connection (f, g) is also called an *adjunction* between ordered sets.

The relations (1.42) imply that

$$f = fgf, \qquad\qquad g = gfg. \tag{1.43}$$

1.7.2 Exercises and complements

(a) Prove that the four conditions above are indeed equivalent.

(b) Of course an isomorphism of ordered sets is, at the same time, left and right adjoint to its inverse. Conversely, if $f \dashv g \dashv f$ then f and g are isomorphisms, inverse to each other.

(c) More generally, an increasing mapping $f\colon X \to Y$ between ordered sets *may* have adjoints $h \dashv f \dashv g$ (one or both), which can be viewed as 'best approximations', of different kinds, to an inverse which may not exist

$$fh \geqslant 1, \quad hf \leqslant 1, \qquad\qquad gf \geqslant 1, \quad fg \leqslant 1. \tag{1.44}$$

For instance, the embedding of ordered sets $f: \mathbb{Z} \to \mathbb{R}$ has a well-known right adjoint, the integral-part function, or *floor* function

$$[-]: \mathbb{R} \to \mathbb{Z}, \qquad [x] = \max\{k \in \mathbb{Z} \mid k \leqslant x\}. \tag{1.45}$$

The left adjoint also exists: it is the *ceiling* function

$$\min\{k \in \mathbb{Z} \mid k \geqslant x\} = -[-x],$$

linked (here) to the right adjoint $[-]$ by the anti-isomorphism $x \mapsto (-x)$ of the real and integral lines.

In this example the left adjoint is the 'lowest upper inverse', while the right adjoint is the 'highest lower inverse'; but this comes out of the injectivity of f, which forces $hf = 1$ and $gf = 1$ in (1.44), leaving the inequalities $fh \geqslant 1$ and $fg \leqslant 1$. The general situation is more complex.

(d) The embedding $\mathbb{Q} \to \mathbb{R}$ has no left nor right adjoint: an irrational number has no distinguished rational approximation.

(e) Prove that two mappings $f: X \rightleftarrows Y : g$ between preordered sets, that satisfy condition (1.39), are necessarily monotone.

1.7.3 Properties

Let us come back to a general Galois connection $f \dashv g$ between ordered sets X, Y.

The mapping f *preserves all the existing joins* (also infinite), while g *preserves all the existing meets*. In fact, if $x = \vee x_i$ in X then $f(x_i) \leqslant f(x)$ (for all indices i). Supposing that $f(x_i) \leqslant y$ in Y (for all i), it follows that $x_i \leqslant g(y)$ (for all i); but then $x \leqslant g(y)$ and $f(x) \leqslant y$.

From the relations $f = fgf$ and $g = gfg$ it follows that:

(a) $gf = \mathrm{id} \iff f$ is injective $\iff f$ is a split mono $\iff g$ is surjective
$\iff g$ is a split epi $\iff f$ reflects the order relation,

(a*) $fg = \mathrm{id} \iff f$ is surjective $\iff f$ is a split epi $\iff g$ is injective
$\iff g$ is a split mono $\iff g$ reflects the order relation.

Moreover the connection restricts to an isomorphism (of ordered sets) between the sets of *closed elements* of X and Y, defined as follows

$$\begin{aligned}
\mathrm{cl}(X) &= g(Y) = \{x \in X \mid x = gf(x)\}, \\
\mathrm{cl}(Y) &= f(X) = \{y \in Y \mid y = fg(y)\}.
\end{aligned} \tag{1.46}$$

1.7.4 A category of adjunctions

An adjunction $f \dashv g$ will also be written as an arrow $(f, g): X \to Y$, often dot-marked and conventionally directed as the left adjoint $f: X \to Y$. Such arrows have an obvious composition

$$(f', g').(f, g) = (f'f, gg') \qquad \text{(for } (f', g') \colon Y \to Z), \qquad (1.47)$$

and form the category AdjOrd *of ordered sets and Galois connections.*

Each hom-set $\mathrm{AdjOrd}(X, Y)$ is canonically ordered: for two adjunctions $(f, g), (f', g') \colon X \to Y$ we let $(f, g) \leqslant (f', g')$ if the following equivalent conditions hold

$$f \leqslant f', \qquad\qquad g' \leqslant g, \qquad\qquad (1.48)$$

since $f \leqslant f'$ gives $g' \leqslant gfg' \leqslant gf'g' \leqslant g$.

The relationship between AdjOrd and the usual category Ord (on the same objects) will become clearer in Chapter 7, where we shall amalgamate them to form the double category AdjOrd, with horizontal arrows in Ord, vertical arrows in AdjOrd and convenient double cells (see 7.2.7). Other similar double categories of lattices will also be of interest (see 7.4.2).

1.7.5 Direct and inverse images of subsets

The transfer of subobjects along morphisms is an important feature, that we examine here in Set and will be developed in Chapter 6 in other frameworks related with Homological Algebra.

Every set A has an ordered set $\mathrm{Sub}A = \mathcal{P}A$ of subsets, which actually is a complete boolean algebra (as recalled in 1.2.6). A mapping $f \colon A \to B$ gives two increasing mappings, of *direct* and *inverse image*, or image and preimage

$$f_* \colon \mathrm{Sub}A \;\rightleftarrows\; \mathrm{Sub}B \colon f^*,$$
$$f_*(X) = f(X), \quad f^*(Y) = f^{-1}(Y) \qquad (X \subset A,\ Y \subset B). \qquad (1.49)$$

These mappings form a Galois connection $f_* \dashv f^*$, since

$$X \subset f^* f_*(X), \qquad f_* f^*(Y) = Y \cap f(A) \subset Y, \qquad (1.50)$$

which implies, in particular, that f_* preserves unions and f^* preserves intersections. (In fact f^* is a homomorphism of complete boolean algebras and also has a right adjoint, related to the universal quantifier, while f_* is related to the existential one; we shall not use these facts.)

All this defines a *transfer functor* for subobjects of Set

$$\mathrm{Sub} \colon \mathsf{Set} \to \mathsf{AdjOrd}, \qquad\qquad (1.51)$$

with values in the category of ordered sets and Galois connections defined above – a sort of 'amalgamation' of the covariant and contravariant functors of subsets, considered in 1.5.2.

1.7.6 The graph of a Galois connection

The *graph* of the adjunction $(f, g) \colon X \to Y$ will be the following ordered set

$$
\begin{aligned}
G(f, g) &= \{(x, y) \in X \times Y \mid f(x) \leqslant y\} \\
&= \{(x, y) \in X \times Y \mid x \leqslant g(y)\},
\end{aligned}
\tag{1.52}
$$

with the order induced by the cartesian product $X \times Y$ (that is a categorical product *in* Ord).

The given adjunction has a canonical factorisation in two adjunctions

$$
X \mathrel{\underset{g'}{\overset{f'}{\rightleftarrows}}} G(f, g) \mathrel{\underset{g''}{\overset{f''}{\rightleftarrows}}} Y
\tag{1.53}
$$

$$
\begin{aligned}
f'(x) &= (x, f(x)), & g'(x, y) &= x, \\
f''(x, y) &= y, & g''(y) &= (g(y), y).
\end{aligned}
$$

This forms a *natural weak factorisation system* in AdjOrd, in the sense of [GT] (see 2.5.5). Note that the first morphism is a monomorphism of AdjOrd (in fact its 'covariant part' f' is an injective mapping, because $g'f' = \mathrm{id}$) and the second is an epimorphism (in fact f'' is surjective, because $f''g'' = \mathrm{id}$).

1.7.7 Exercises and complements (Chains of adjunctions)

Non-trivial *chains of Galois connections* between two ordered sets X, Y

$$
\ldots f_{-2} \dashv f_{-1} \dashv f_0 \dashv f_1 \dashv f_2 \ldots \qquad (f_{2i} \colon X \rightleftarrows Y \colon f_{2i+1}),
\tag{1.54}
$$

seem not to be frequent, 'in nature'.

Yet, as an interesting, unexpected fact, a Galois connection $\mathbb{Z} \rightleftarrows \mathbb{Z}$ on the integral (ordered) line *always* produces an unbounded chains of adjunctions.

The reader is invited to investigate this intriguing situation, construct examples and study variations, where $\mathbb{Z}$ is replaced with other ordered sets, like $\mathbb{N}$ or the real line. Then the results can be compared with the following outline – which has probably the fault of being too explicit.

(a) The following conditions on an increasing function $f \colon \mathbb{Z} \to \mathbb{Z}$ are equivalent:

(i) $f(\mathbb{Z})$ is lower and upper unbounded,

(ii) f has a right adjoint,

(ii*) f has a left adjoint.

In fact, by form (ii) of Definition 1.7.1, f has a right adjoint if and only if every subset $S(y) = \{x \in X \mid f(x) \leqslant y\}$ is non empty and upper bounded, which is equivalent to condition (i); the latter is self-dual.

(b) It follows that each adjunction $f_0 \dashv f_1$ on $\mathbb{Z}$ can be extended to a unique chain of Galois connections, as in (1.54).

(c) As a *trivial case*, if one of these mappings is invertible, all of them are and we get a chain of adjunctions of period 2

$$ \ldots f \dashv f^{-1} \dashv f \dashv f^{-1} \ldots \tag{1.55} $$

Note that a bijective increasing mapping $f : \mathbb{Z} \to \mathbb{Z}$ is necessarily a translation $f(x) = x + k$. Therefore the only chain of period 1 is that of $f = \mathrm{id}\mathbb{Z}$. (But the ordered set $\mathbb{Z}^2$ has the symmetry $f(x, y) = (y, x)$ with $f \dashv f$.)

(d) In the chain (1.54), f_0 is injective if and only if f_1 is surjective (by 1.7.3(a)). *In such a case* all f_{2i} are split mono, all f_{2i+1} are split epi, and the adjunctions are expressed by the following conditions (for $i \in \mathbb{Z}$)

$$ f_{2i+1}f_{2i} = \mathrm{id} = f_{2i-1}f_{2i}, \qquad f_{2i}f_{2i+1} \leqslant \mathrm{id}, \qquad f_{2i}f_{2i-1} \geqslant \mathrm{id}. \tag{1.56} $$

Moreover $f_{2i} \leqslant f_{2i+2}$ and $f_{2i-1} \geqslant f_{2i+1}$, because:

$$ f_{2i} \leqslant f_{2i+2}f_{2i+1}f_{2i} = f_{2i+2}, \qquad f_{2i-1} = f_{2i+1}f_{2i}f_{2i-1} \geqslant f_{2i+1}. $$

Once we exclude the trivial case (1.55), the sequence (f_{2i}) is strictly increasing while (f_{2i+1}) is strictly decreasing: in fact we know that the relation $f \dashv g \dashv f$ gives $g = f^{-1}$. Since no f_{2i} can be surjective, all mappings f_i are different.

The reader can compute the chain of adjunctions produced by the injective mapping $f_0(x) = 2x$.

(e) There is an interesting case of this kind, where all d_i are injective and all e_i are surjective

$$ d_i, e_i : \mathbb{Z} \to \mathbb{Z}, \qquad \ldots d_{i+1} \dashv e_i \dashv d_i \dashv e_{i-1} \dashv d_{i-1} \ldots \tag{1.57} $$

$$ d_i(x) = x \quad \text{if} \quad x < i, \qquad\qquad d_i(x) = x + 1 \quad \text{if} \quad x \geqslant i, $$

$$ e_i(x) = x \quad \text{if} \quad x \leqslant i, \qquad\qquad e_i(x) = x - 1 \quad \text{if} \quad x > i, $$

$$ e_i d_i = \mathrm{id} = e_{i-1}d_i, \qquad\qquad d_{i+1}e_i \leqslant \mathrm{id}, \qquad\qquad d_i e_i \geqslant \mathrm{id}, $$

$$ d_i > d_{i+1}, \qquad\qquad e_i < e_{i+1}. $$

These mappings (or better their restrictions to finite ordinals) are related with the faces and degeneracies of simplicial sets (see 5.3.6(a)).

(f) On the other hand, if we start from an increasing mapping $f : \mathbb{Z} \to \mathbb{Z}$ that satisfies condition (i) and is neither injective nor surjective, we get a chain (1.54) of adjunctions, with $f_0 = f$, where no f_i is injective or surjective.

For instance, from the chain (1.57) one can readily compute the chain of Galois connections of the mapping $f = d_i e_i$, by composing adjunctions (as in (1.7.4)).

(g) Replacing $\mathbb{Z}$ with the real line $\mathbb{R}$ *one cannot have non-trivial chains of adjunctions of length higher than* 3. In fact in an adjunction $f \dashv g$ of increasing functions $\mathbb{R} \to \mathbb{R}$, the left adjoint f is lower and upper unbounded (as above) and preserves all sup (hence it is left-continuous) while the right adjoint g is lower and upper unbounded and preserves all inf (hence it is right-continuous).

Therefore, in a chain $f \dashv g \dashv h$, the function g is lower and upper unbounded and continuous. If g is strictly increasing, then it is invertible and we fall in the trivial case of period 2. Otherwise g is constant on some non trivial interval $[a, b]$, but then both f and h are not continuous at the point $g([a, b])$ and the chain $f \dashv g \dashv h$ cannot be extended.

(h) Other variations on the ordered sets X, Y of diagram (1.54) are left to the reader; of course, they need not be the same. Chains of adjunctions between categories (that are not orders) will be examined in 3.2.8.

1.7.8 Contravariant Galois connections

A *contravariant Galois connection* between the ordered sets X, Y is a pair of *decreasing* (or anti-monotone) mappings $f\colon X \to Y$ and $g\colon Y \to X$ such that

$$ gf \geqslant \mathrm{id}X, \qquad\qquad fg \geqslant \mathrm{id}Y. \qquad\qquad (1.58) $$

This *symmetric notion* is equivalent to a covariant Galois connection:

- replacing Y with Y^{op} we have two monotone functions $f \dashv g$,

- replacing X with X^{op} we have two monotone functions $g \dashv f$.

Classically, the term 'Galois connection' used to refer to the contravariant setting, which covers important concrete cases – some of which are recalled below. This terminology is still used in [M4]. Presently, the covariant setting is often privileged because only in this case composition is possible and we have a category.

1.7.9 Exercises and complements

(a) We are given an arbitrary subset $\perp \subset X \times Y$ of a cartesian product of sets, which we view as an 'orthogonality relation': $x \perp y$ means that $(x, y) \in \perp$.

There is an associated contravariant Galois connection

$$R: \mathcal{P}X \rightleftarrows \mathcal{P}Y : L,$$

$$R(A) = A\perp = \{y \in Y \mid \text{for all } x \in A, \ x \perp y\}, \qquad (1.59)$$

$$L(B) = \perp B = \{x \in X \mid \text{for all } y \in B, \ x \perp y\},$$

that establishes a bijective correspondence between the closed subsets of X and those of Y

$$\mathrm{cl}(X) = \{A \subset X \mid A = \perp(A\perp)\} = \{\perp B \mid B \subset Y\},$$
$$\mathrm{cl}(Y) = \{B \subset Y \mid B = (\perp B)\perp\} = \{\perp A \mid A \subset X\}. \qquad (1.60)$$

(b) For a commutative unitary ring R, let us take the affine space $X = R^n$ and the polynomial ring $Y = R[x_1, ..., x_n]$ (for some $n \geqslant 1$). Every polynomial $p \in Y$ can be evaluated at any point $x \in X$, giving $p(x) \in R$. We write $x \perp p$ to mean that $p(x) = 0$. The associated contravariant Galois connection is well known in Algebraic Geometry, and sends:

- a subset $A \subset R^n$ to the ideal $R(A)$ of all polynomial which annihilate at every point of A,

- a subset $B \subset R[x_1, ..., x_n]$ to the *affine algebraic set* $L(B)$ of all points of R^n annihilated by every polynomial of B.

*The reader may know that, if R is an algebraically closed field, Hilbert's *Nullstellensatz* characterises the closed subsets of $R[x_1, ..., x_n]$ as the radical ideals of this ring. On the other hand, the closed subsets of R^n are its affine algebraic sets, by definition.*

(c) An orthogonality (endo)relation $f \perp g$ between morphisms of a category will be introduced in 2.4.2.

1.8 Elementary categories of relations and partial mappings

We now review the elementary construction of some categories of relations and partial mappings, for sets, or abelian groups, etc.

Various extensions of relations (to regular categories, abelian categories, etc.) will be examined in Chapters 4 and 6; in the latter we also show how categories of relations are important in Homological Algebra.

Relations and partial mappings will often (but not necessarily) be written as dot-marked arrows $X \rightarrow Y$, to distinguish them from ordinary mappings.

Categories of partial mappings will be denoted by calligraphic letters like $\mathcal{S}, \mathcal{I}, \mathcal{C}, \ldots$

1.8.1 Categories and involutions

We begin by fixing our terminology for involutive endofunctors, starting from two examples related with ordered sets.

(i) The category Ord, of ordered sets and increasing mappings, has an involutive *covariant* endofunctor:

$$X \mapsto X^{\mathrm{op}}, \qquad (f \colon X \to Y) \mapsto (f^{\mathrm{op}} \colon X^{\mathrm{op}} \to Y^{\mathrm{op}}), \qquad (1.61)$$

where X^{op} has reversed order and f^{op} is the 'same' mapping as f.

(ii) The category AdjOrd, of ordered sets and Galois connections (introduced in 1.7.4), has an involutive *contravariant* endofunctor:

$$X \mapsto X^{\mathrm{op}}, \qquad (u \colon X \to Y) \mapsto (u^{\mathrm{op}} \colon Y^{\mathrm{op}} \to X^{\mathrm{op}}), \qquad (1.62)$$

where $u^{\mathrm{op}} = (g^{\mathrm{op}}, f^{\mathrm{op}})$, if $u = (f, g)$ with $f \dashv g$.

This situation will become clearer in the domain of double categories: the two procedures (i) and (ii) – which have the same action on the objects – can be combined to give a *vertical involution* of the double category AdjOrd, horizontally covariant (on increasing mappings) and vertically contravariant (on adjunctions); see 7.2.8.

We shall use the following terminology.

(a) A *category with reversor* will be a category C equipped with an involutive *covariant* endofunctor $R \colon \mathsf{C} \to \mathsf{C}$, often denoted as in (1.61), and said to be *trivial* when $R = \mathrm{idC}$.

Obvious non-trivial examples are Ord and Cat. The name of 'reversor' is taken from Directed Algebraic Topology [G8], where the categories of 'directed spaces' in some sense (like ordered topological spaces, 'd-spaces', and small categories) have such an endofunctor, that turns a directed structure into the opposite one. In the trivial case, when $R = \mathrm{idC}$, we go back to classical, non-directed 'spaces'.

(b) A *category with involution* will be a category C equipped with an involutive *contravariant* endofunctor $J \colon \mathsf{C} \dashrightarrow \mathsf{C}$, often denoted as in (1.61). It follows that C is selfdual, i.e. isomorphic to C^{op}.

AdjOrd is an obvious example; the same holds for AdjCat, that will be introduced in 3.2.1.

(c) More particularly, an *involutive category* will be a category C equipped with a (contravariant) involution *which is the identity on objects*. In other words every morphism $f\colon X \to Y$ has an associated *opposite morphism* $f^{\sharp}\colon Y \to X$ so that

$$f^{\sharp\sharp} = f, \qquad (gf)^{\sharp} = f^{\sharp}g^{\sharp}, \qquad (1_X)^{\sharp} = 1_X. \qquad (1.63)$$

Classical examples are the categories of relations introduced below.

The involution is said to be *regular* (in the sense of von Neumann) if $ff^{\sharp}f = f$ (for all maps f), as happens with relations of groups or modules. In this case all the semigroups of endomorphisms $C(X, X)$ are regular, in the well-known sense of semigroup theory. The reader is warned that this has nothing to do with the notion of 'regular category', that will be seen in 4.2.1.

1.8.2 Ordered categories

In this book an *ordered category* is a category C equipped with an order relation $f \leqslant g$ between *parallel morphisms*, which is consistent with composition: for $f, g\colon X \to Y$ and $f', g'\colon Y \to Z$

(i) if $f \leqslant g$ and $f' \leqslant g'$ then $f'f \leqslant g'g$.

(This notion is related with 'enrichment', see 7.3.3; a different 'internal' notion is recalled in 7.3.6.)

A functor $F\colon C \to D$ between ordered categories is *increasing*, or *monotone*, if it preserves the order. Any subcategory of an ordered category inherits an ordering.

For instance the category Ord of ordered sets is canonically ordered by the *pointwise order* $f \leqslant g$, defined in (1.4). This ordering is inherited by the subcategories Lth, Mlh and Dlh of lattices, modular lattices and distributive lattices (see 1.2.4–1.2.6).

Also AdjOrd has a canonical order, defined in (1.7.4).

1.8.3 Relations of sets

The category Rel Set of *sets and relations* (or *correspondences*) is well known. A relation $u\colon X \to Y$, often denoted by a dot-marked arrow, is a subset of the cartesian product $X \times Y$. It can be viewed as a 'partially defined, multi-valued mapping', that sends an element $x \in X$ to the subset $\{y \in Y \mid (x, y) \in u\}$ of Y.

The identity relations and the composite of u with $v\colon Y \to Z$ are

$$\mathrm{id}X = \{(x, x) \mid x \in X\},$$
$$vu = \{(x, z) \in X \times Z \mid \exists y \in Y : (x, y) \in u \text{ and } (y, z) \in v\}. \qquad (1.64)$$

It is easy to verify that Rel Set is a category. It is actually an *involutive ordered category*, in the following sense. First it is an involutive category as defined in 1.8.1: the *opposite relation* $u^\sharp \colon Y \to X$ is obtained by reversing pairs. Second, it is an ordered category, in the sense of 1.8.2: given two parallel relations $u, u' \colon X \to Y$ we say that $u \leqslant u'$ if $u \subset u'$ as subsets of $X \times Y$. Third, these structures are consistent: if $u \leqslant v$ then $u^\sharp \leqslant v^\sharp$ (and conversely, as a consequence).

The involution is *not* regular (in the sense of 1.8.1): see 1.8.4(c).

The category Set is embedded in the category of relations Rel Set, identifying a mapping $f \colon X \to Y$ with its graph $\{(x, y) \in X \times Y \mid y = f(x)\}$; the condition $f \leqslant g$ for parallel mappings amounts to $f = g$.

1.8.4 Exercises and complements

(a) It may help to draw here a synopsis of future investigations of Rel Set, as endowed with important structures – some of them perhaps unexpected:

(i) as a semiadditive category with arbitrary biproducts, in 2.2.6(b) and 6.4.5(d),

(ii) as a symmetric monoidal closed category (with a 'peculiar' internal hom), in 3.2.8(d) and 3.4.7(g),

(iii) as a Kleisli category for a monad on Set, in 3.7.2(c),

(iv) as a locally ordered 2-category in 4.5.1 (or a locally preordered bicategory, in a modified form Rel′Set; see 4.5.5).

The reader is warned that the points (i)–(iii) have no counterpart in many other categories of relations – for instance in Rel Ab, introduced below.

The interpretation of Rel Set as a category of partially defined, multivalued mappings will be better understood at the light of point (iii). Bicommutative squares in Rel Set are analysed in 2.3.8.

(b) Prove the associativity of the composition of relations of sets.

(c) Show that the involution is not regular, by a suitable endorelation of the set $\{0, 1\}$.

(d) A relation $f \colon X \to Y$ of sets is a mapping if and only if it is everywhere defined and single-valued. These conditions can be characterised as follows

$$f^\sharp f \geqslant \mathrm{id} X, \qquad\qquad f f^\sharp \leqslant \mathrm{id} Y. \tag{1.65}$$

(This means that $f \dashv f^\sharp$ in the ordered category Rel Set, viewed as a 2-category: see 7.1.2.)

1.8.5 *Relations of abelian groups and modules*

Abelian groups have a similar category of relations, also called *additive relations* or *additive correspondences* (see [M1, M2, Hi]).

A *relation of abelian groups* $u \colon X \to Y$ is a subgroup of the cartesian product $X \times Y$. Composition, order and involution are defined as above, giving the involutive ordered category Rel Ab.

Ab is embedded in the former as the wide subcategory of everywhere defined, single-valued relations.

It is important to note that the involution of Rel Ab is regular, i.e.

$$uu^{\sharp}u = u, \qquad \text{for all } u \text{ in Rel Ab.} \tag{1.66}$$

In fact the inclusion $u \leqslant uu^{\sharp}u$ is obvious (and also holds for relations of sets). The other way round, if (x, y), (x', y) and (x', y') are in u, we have

$$(x, y') = (x, y) - (x', y) + (x', y') \in u. \tag{1.67}$$

More generally for any (unitary) ring R we have the involutive ordered category Rel $(R \operatorname{Mod})$ of left R-modules, where a relation is a submodule of the cartesian product.

We also have the involutive ordered category Rel Gp of groups, where a relation is a subgroup of the cartesian product. The involution is still regular, with the same proof as in (1.67), since the commutativity of addition has not been used.

1.8.6 *Partial mappings*

A partial mapping (of sets) $f \colon X \to Y$ is a mapping Def $f \to Y$ defined on a subset of X, its subset of *definition*. Equivalently, it is a single-valued relation.

For a partial mapping $g \colon Y \to Z$, the composite $gf \colon X \to Z$ is obvious (as in the well-known case of partially defined functions $\mathbb{R} \to \mathbb{R}$):

$$\operatorname{Def}(gf) = f^{-1}(\operatorname{Def} g), \qquad (gf)(x) = g(f(x)), \tag{1.68}$$

for $x \in \operatorname{Def}(gf)$.

We have thus a wide subcategory $\mathcal{S} \subset \operatorname{Rel}(\operatorname{Set})$ that contains Set. In the induced order of $\mathcal{S}$, $f' \leqslant f$ means that f' is a restriction of f to a smaller subset of X.

The category $\mathcal{S}$ *should not be viewed as essentially different* from the usual categories of structured sets and 'total' mappings, since *it is equivalent* to the category Set. of pointed sets, via the functor

$$R \colon \operatorname{Set}_{\bullet} \to \mathcal{S}, \qquad (X, x_0) \mapsto X \setminus \{x_0\}, \tag{1.69}$$

that sends a pointed mapping $f: (X, x_0) \to (Y, y_0)$ to its restriction

$$R(f): X \setminus \{x_0\} \to Y \setminus \{y_0\},$$

$$\text{Def}\,(R(f)) = f^{-1}(Y \setminus \{y_0\}).$$

In fact this functor is plainly faithful, full and surjective on objects (see 1.5.5).

The quasi inverse adjoint $S: \mathcal{S} \to \mathsf{Set}_\bullet$ is obtained by choosing, for every set X, a pointed set $S(X) = (X \cup \{x_0\}, x_0)$ where $x_0 \notin X$ and defining S on partial mappings in the obvious (and unique) way that gives $RS = 1$. Note that SR is only isomorphic to the identity.

(Set theory can give a canonical – if 'confusing' – choice for S: one can take $x_0 = X$, since $X \notin X$.)

This equivalence will be used in 2.2.5(d) to transfer to $\mathcal{S}$ categorical constructions that are easily computed and well known in $\mathsf{Set}_\bullet$, like categorical and tensor products.

1.8.7 Partial bijections

More particularly a *partial bijection* (or partial isomorphism) of sets $f: X \to Y$ is a bijection between a subset $\text{Def}\, f$ of X and a subset $\text{Val}\, f$ of Y, or equivalently a single-valued, injective relation.

Partial bijections form a wide subcategory

$$\mathcal{I} \subset \mathcal{S} \subset \text{Rel}\,(\mathsf{Set}). \tag{1.70}$$

It is equivalent to the wide subcategory of $\mathsf{Set}_\bullet$ formed by the pointed mappings $f: (X, x_0) \to (Y, y_0)$ that are injective out of the preimage $f^{-1}\{y_0\}$.

$\mathcal{I}$ is used as a basic tool for graphic representations of spectral sequences, in [G9]. We shall see in 6.3.3(d) that it is a pointed category with good exactness properties, but lacks products and sums.

1.8.8 *Partial continuous mappings

In various cases, one should consider partial morphisms defined on subobjects of a particular kind.

For topological spaces, for instance, partial continuous mappings should be 'reasonably' *defined on subspaces* (i.e. 'regular subobjects', as we shall see); but in fact the important structures of partial continuous mappings are even more restricted.

While studying topological manifolds, it can be useful to use the category

$\mathcal{C}$ of spaces and partial continuous mappings *defined on open subspaces*, together with its subcategory of spaces and partial homeomorphisms *between open subspaces* (see [G2]).

The interested reader can verify that the category $\mathcal{C}$ is equivalent to a full subcategory $\mathsf{Top}'_\bullet$ of the category $\mathsf{Top}_\bullet$ of pointed topological spaces, consisting of those objects (X, x_0) where the base point x_0 is closed and adherent to every other point (i.e. the only neighbourhood of x_0 is X).

The category $\mathcal{D}$ of topological spaces and partial continuous mappings *defined on closed subspaces* is also of some interest. It is equivalent to another full subcategory $\mathsf{Top}''_\bullet$ of $\mathsf{Top}_\bullet$, consisting of those objects (X, x_0) where the base point x_0 is open – i.e. an isolated point.

1.8.9 *Corelations and networks*

The involutive ordered category $\mathsf{C} = \mathsf{Cor}\,\mathsf{Set}$ *of sets and corelations* can be constructed in a dual way with respect to $\mathsf{Rel}\,\mathsf{Set}$, letting $\mathsf{C}(X, Y)$ be the ordered set of the quotients of the sum $X + Y$; more explicitly, this is the set of the equivalence relations on the disjoint union of X and Y.

The reader can easily find out how corelations compose, by using the equivalence relation generated by a relation. (The pictures below tell everything about that!)

Early in the history of categories, since the late 1960's, the full subcategory $\mathcal{T}$ *of finite sets and corelations* was used in the theory of networks, by G. Darbo [Da1, Da2] and some of his collaborators like F. Parodi and G. Testa. The subject was apparently ahead of its time, and later forgotten; but recently J. Baez has independently reinvented it, with his students, in a similar perspective [BaP, Fo].

In fact, a corelation $\varphi\colon \alpha \to \beta$ between finite sets (called a 'transducer' in [Da1] and related articles) can be viewed as an electrical circuit made of perfectly conductive wires, connecting the 'terminals' belonging to α to those belonging to β, as in the following picture

$\alpha \qquad\qquad \beta$

(1.71)

Transducers compose as if we were connecting wires

$$\tag{1.72}$$

In [Da1] a 'universe of devices' is defined as a pair $(\mathcal{D}, \sigma)$, formed of a functor and a natural transformation

$$\mathcal{D}: \mathcal{T} \to \mathsf{Set},$$
$$\sigma_{\alpha\beta}: \mathcal{D}\alpha \times \mathcal{D}\beta \to \mathcal{D}(\alpha + \beta),\tag{1.73}$$

satisfying some natural axioms.

(It would now be called a unitary lax monoidal functor, from the symmetric monoidal category $(\mathcal{T}, +, \emptyset)$ to the cartesian monoidal category of sets, according to definitions that we shall see in Section 3.4 and 7.2.4.)

In a model of the theory, $\mathcal{D}\alpha$ represents the set of devices with terminals belonging to the finite set α. A corelation $\varphi: \alpha \nrightarrow \beta$ takes a device $A \in \mathcal{D}\alpha$ to a device $\varphi_*(A) \in \mathcal{D}\beta$, as suggested in the diagram below

$$\tag{1.74}$$

Finally, the mapping $\sigma_{\alpha\beta}$ takes a pair (A, B) of devices to a device $A + B$ which, in a concrete model, is obtained by putting A and B in a box and letting out their terminals

$$\alpha + \beta$$

(1.75)

The theory is quite general, and not limited at all to electrical devices.

2
Limits and colimits

Limits and colimits in a category C are defined by a universal property, which determines a solution up to isomorphism. The general notion of a limit (introduced in Section 2.2) includes cartesian products, equalisers and pullbacks (analysed in Sections 2.1 and 2.3). Dually, colimits comprise sums, coequalisers, pushouts and the classical 'inductive limits' (analysed in Sections 2.1, 2.3 and 2.6). The classical 'projective limits' are also included in general limits, but – in the usual categories of structured sets – are not worthy of a separate study.

The construction of limits is dealt with in Theorem 2.2.4, and their description as universal arrows or Kan extensions in 2.7.2 and 2.8.5. Their relationship with adjoints will be examined in 3.2.2 and 3.2.4, with comma categories in 3.3.5.

2.1 Basic limits and colimits

We begin by defining *products* and *equalisers* in a category C, and computing them in various concrete categories. Then we do the same for the dual notions, *sums* and *coequalisers*. We end this section with a brief introduction to biproducts and additive categories.

The general definition of limits and colimits is deferred to the next section, where we shall see that all limits and colimits can be constructed with the basic ones, dealt with here (Theorem 2.2.4).

Various examples are presented, as exercises. The reader will find useful to complete the missing details, and work out similar results for other categories of structured sets.

Other results and computations of basic limits and colimits will be given later: for Ban and Ban$_1$ in 2.2.6; for ordered sets in 3.2.3(b); for varieties of algebras in Chapter 4; for various categories of topological or metric spaces in Chapter 5; for abelian categories and their generalisations in Chapter 6; etc.

2.1.1 Products

The simplest case of a limit, in a category C, is the *product* of a family $(X_i)_{i \in I}$ of objects (indexed by a *small set I*).

This is defined as an object X equipped with a family of morphisms $p_i \colon X \to X_i$ ($i \in I$), called (cartesian) *projections*, which satisfies the following universal property

$$Y \overset{f}{\dashrightarrow} X$$
$$f_i \searrow \quad \downarrow p_i$$
$$X_i$$

(2.1)

(i) for every object Y and every family of morphisms $f_i \colon Y \to X_i$ there exists a unique morphism $f \colon Y \to X$ such that, for all $i \in I$, $p_i f = f$.

The morphisms f_i are called the *components* of f, and the latter is often written as (f_i) (even though it is not the same as the family of its components, of course).

The product of a family of objects need not exist. If it does, it is determined up to a unique *coherent* isomorphism, in the sense that if also Y is a product of the family $(X_i)_{i \in I}$ with projections $q_i \colon Y \to X_i$, then the unique morphism $f \colon X \to Y$ which commutes with all projections (i.e. $q_i f = p_i$, for all indices i) is invertible. This follows easily from the fact that there is also a unique morphism $g \colon Y \to X$ such that $p_i g = q_i$; moreover $gf = \mathrm{id}X$ (because $p_i(gf) = p_i(\mathrm{id}X)$, for all i) and $fg = \mathrm{id}Y$.

Therefore one speaks of *the* product of the family (X_i), denoted as $\prod_i X_i$.

We say that a category C *has products* (resp. *finite products*) if every family of objects indexed by a small set (resp. by a finite set) has a product in C.

In particular the product of the empty family of objects $\emptyset \to \mathrm{Ob}\mathsf{C}$ means an object X (equipped with no projections) such that for every object Y (equipped with no maps) there is a unique morphism $f \colon Y \to X$ (satisfying no condition). The solution is called the *terminal* object of C; again, it need not exist, but is determined up to a unique isomorphism. It can be written as $\top$.

2.1.2 *Exercises and complements* (Products)

(a) In Set, Top, Ab, Ord all products exist and are the usual cartesian ones (constructed on the product of the underlying sets). The terminal object is the singleton, with the corresponding structure.

Let us recall that a topological product X with projections $p_i \colon X \to X_i$ has the coarsest topology making all projections continuous. Similarly, in Ord, the product has the coarsest order relation making all projections monotone: $(x_i) \leqslant (y_i)$ if, for all indices i, $x_i \leqslant y_i$ in X_i.

(b) Products in the categories Set. and Top. are also obvious

$$\Pi_i(X_i, \overline{x}_i) = (\Pi_i X_i, (\overline{x}_i)). \tag{2.2}$$

(c) In the category X associated to a preordered set, the categorical product of a family of points $x_i \in X$ amounts to their inf (the greatest lower bound), and the terminal object amounts to the greatest element of X (when such elements exist; notice that they are determined up to the equivalence relation associated to our preorder, and uniquely determined for an ordered set).

(d) Products in Cat have been considered in 1.4.6; the terminal object is the singleton category **1**.

(e) It is easy to prove that a category has finite products if and only if it has binary products $X_1 \times X_2$ and a terminal object.

2.1.3 *Equalisers and regular monomorphisms*

Another basic limit is the *equaliser* of a pair $f, g \colon X \to Y$ of parallel maps of C. This is an object E with a map $m \colon E \to X$ such that $fm = gm$ and the following universal property holds:

$$E \xrightarrow{m} X \underset{g}{\overset{f}{\rightrightarrows}} Y \qquad m = \mathrm{eq}(f, g), \tag{2.3}$$

(ii) every map $h \colon Z \to X$ such that $fh = gh$ factorises uniquely through m (i.e. there exists a unique map $w \colon Z \to E$ such that $mw = h$).

The equaliser morphism is necessarily a monomorphism, and is called a *regular monomorphism*. It is easy to see that a regular mono which is epi is an isomorphism.

Products and equalisers are particular instances of the limit of a functor, that will be introduced in 2.2.1.

2.1.4 *Exercises and complements* (Equalisers)

(a) In Set the natural solution for the equaliser of a pair of mappings $f, g \colon X \to Y$ is the subset

$$E = \{x \in X \mid f(x) = g(x)\}, \tag{2.4}$$

with the inclusion $m \colon E \subset X$.

(b) In Top, Ab, Ord, Set. and Top. (and many other concrete categories) we construct the equaliser in the same way, putting on E the structure induced by X as a subspace, or a subgroup, etc.

(c) In Cat, the equaliser of two functors $F, G \colon C \to D$ is the subcategory of the objects and arrows of C on which F and G coincide.

(d) In any category, every split monomorphism is regular.

(e) The reader will note that in Top a monomorphism $m \colon X' \to X$ is any injective map (see 1.3.3(b)), while a regular mono must be the inclusion of a subspace $E \subset X$, up to homeomorphism; this means a *topological embedding* $m \colon X' \to X$, i.e. an injective continuous mapping where X' has the coarsest topology making m continuous: the open subsets of X' are the preimages of those of X. (The converse is also true: every embedding of a subspace is a regular mono, see 2.4.5(c).)

(f) We have a similar situation in Ord: a monomorphism is any injective monotone mapping, and is a regular mono if and only if it *reflects* the ordering (inducing an isomorphism between the domain and an ordered subset of the codomain).

2.1.5 *Sums*

As already mentioned in Section 1.1, every notion of category theory has a dual notion.

The *sum*, or *coproduct*, of a family $(X_i)_{i \in I}$ of objects of C is dual to its product. Explicitly, it is an object X equipped with a family of morphisms $u_i \colon X_i \to X$ $(i \in I)$, called *injections*, which satisfy the following universal property:

$$\begin{array}{ccc} X & \dashrightarrow^{f} & Y \\ {\scriptstyle u_i}\big\uparrow & \nearrow_{f_i} & \\ X_i & & \end{array} \tag{2.5}$$

(i*) for every object Y and every family of morphisms $f_i \colon X_i \to Y$, there exists a unique morphism $f \colon X \to Y$ such that, for all $i \in I$, $f u_i = f_i$.

The map f will be written as $[f_i]$, by its *co-components*.

Again, if the sum of the family (X_i) exists, it is determined up to a unique coherent isomorphism, and denoted as $\Sigma_i X_i$, or $X_1 + ... + X_n$ in a finite case.

The sum of the empty family is the *initial* object $\perp$: this means that every object X has precisely one map $\perp \to X$.

2.1.6 Exercises and complements (Sums)

(a) Sums in Set are realised as 'disjoint unions'. There is a rather natural way of doing this

$$X = \Sigma_i X_i = \bigcup_i X_i \times \{i\}, \qquad u_i(x) = (x, i) \quad \text{(for } x \in X_i). \tag{2.6}$$

The initial object is the empty set.

(b) In Top we do the same, equipping the set X with the finest topology that makes all the injections continuous; in other words, the open subsets of X are arbitrary unions of open subsets of the summands. The initial object is the empty space.

Similarly, in Ord we endow the set X with the finest ordering that makes all the injections monotone: $(x, i) \leqslant (y, j)$ if and only if $i = j$ and $x \leqslant y$ in X_i.

(c) In Abm, Ab and R Mod categorical sums are realised as 'direct sums'

$$\bigoplus A_i = \{(a_i) \in \textstyle\prod_i A_i \mid a_i = 0 \text{ except for a finite set of indices}\}, \tag{2.7}$$

where the family (a_i) is said to be *quasi-null*. The initial object is the singleton.

(d) A sum $\Sigma_i(X_i, \overline{x}_i)$ in Set. or Top. can be constructed as a pointed subset or subspace of the cartesian product

$$\{(x_i) \in \textstyle\prod_i X_i \mid x_i = \overline{x}_i \text{ for all indices } i \text{ except one at most}\}. \tag{2.8}$$

For instance, $(\mathbb{R}, 0) + (\mathbb{R}, 0)$ can be realised as the subset (or subspace) of the cartesian plane formed of the two cartesian axes. (Equivalently, the sum can be realised as a quotient of the unpointed sum $\Sigma_i X_i$, by identifying all the base points $\overline{x}_i$.)

(e) In the category X associated to a preordered set, the categorical sum of a family of points $x_i \in X$ amounts to its sup (the least upper bound), while the initial object amounts to the least element of X.

(f) In Cat a sum ΣC_i of categories is their obvious disjoint union. The initial object is the empty category **0**.

*(g) In Gp categorical sums are classically known as 'free products'. In CRng, finite sums are realised as tensor products over $\mathbb{Z}$ (see 4.1.4(b)), and a general sum as a colimit of finite tensor products.

2.1.7 Coequalisers and regular epimorphisms

The *coequaliser* of a pair $f, g\colon X \to Y$ of parallel maps of C is a map $p\colon Y \to C$ such that $pf = pg$ and

$$X \underset{g}{\overset{f}{\rightrightarrows}} Y \overset{p}{\to} C \qquad p = \mathrm{coeq}(f, g), \tag{2.9}$$

(ii*) every map $h\colon Y \to Z$ such that $hf = hg$ factorises uniquely through p (i.e. there exists a unique map $w\colon C \to Z$ such that $wp = h$).

A coequaliser morphism is necessarily an epimorphism, and is called a *regular epimorphism*. A regular epi which is mono is an isomorphism.

Sums and coequalisers are particular instances of the *co*limit of a functor, that will be introduced in 2.2.1.

2.1.8 Exercises and complements (Coequalisers)

(a) In Set the natural solution for the coequaliser of two mappings $f, g\colon X \to Y$ is the projection

$$p\colon Y \to Y/R, \tag{2.10}$$

on the quotient modulo the equivalence relation of Y spanned by the pairs $(f(x), g(x)) \in Y^2$, for $x \in X$.

(b) In Top we do the same, putting on Y/R the quotient topology, namely the finest that makes the mapping p continuous. In pOrd we do the same, putting on Y/R the induced preordering (the finest that makes p monotone). In Ord we first compute the preorder Y/R in pOrd, and then take the associated ordered set; this fact will be extended to all colimits, in 3.2.3(b).

(c) In Ab we take the quotient Y/H, modulo the subgroup $H = \{f(x) - g(x) \mid x \in X\}$, which amounts to the quotient Y/R' modulo the congruence of abelian groups spanned by the previous equivalence relation R. The reader will adapt this construction to Gp.

(d) Coequalisers in Set. and Top. are (easy and) left to the reader.

(e) In Top an epimorphism $p\colon Y \to Y'$ is any surjective map (see 1.3.3(b)), while a regular epi must be the projection on a quotient space $p: Y \to Y/R$, up to homeomorphism; this means a *topological projection* $p: Y \to Y'$, i.e. a surjective continuous mapping where Y' has the finest topology making p continuous: a subset of Y is open if and only if its preimage is open in Y. (The converse is also true: every topological projection is a regular epi, see 2.4.5(c).)

*(f) The coequaliser in Cat of two functors $F, G: \mathsf{C} \to \mathsf{D}$ is the quotient of D modulo the *generalised congruence* generated by this pair; the latter, as defined in the paper [BBP], also involves equivalent objects. One can avoid giving a 'construction' of the coequaliser category (necessarily complicated) and just prove its existence by the Adjoint Functor Theorem 3.5.2.

2.1.9 Zero objects, biproducts and additive categories

We end this section by briefly introducing some important notions, that will be more completely analysed in Section 6.4.

First, a *zero object* of a category, often written as 0, is both initial and terminal. This exists in Abm, Ab, R Mod, Gp, Set$_\bullet$, Top$_\bullet$,... but not in Set, Top, Cat,... A category with zero object is said to be *pointed*; then each pair of objects A, B has a zero-morphism $0_{AB}: A \to B$ (also written as 0), which is given by the composite $A \to 0 \to B$.

As a stronger fact, the reader may have remarked that, in Abm, Ab and R Mod, a *finite* product ΠA_i and the corresponding finite sum ΣA_i are realised as the same object $A = \oplus A_i$, which satisfies:

- the property of the product, by a family of projections $p_i: A \to A_i$,
- the property of the sum, by a family of injections $u_i: A_i \to A$,
- the equations $p_i u_i = \operatorname{id} A_i$ and $p_j u_i = 0: A_i \to A_j$ (for $i \neq j$).

Such an object (in a pointed category) is called a *biproduct*. The empty biproduct is the zero object.

Other examples, including Ban and Rel Set (but not Ban$_1$), will be seen in 2.2.6.

As a related notion, a *preadditive*, or $\mathbb{Z}$-*linear*, category is a category C where every hom-set $\mathsf{C}(A, B)$ is equipped with a structure of abelian group, generally written as $f + g$, so that composition is additive in each variable (i.e. bilinear over $\mathbb{Z}$)

$$(f + g)h = fh + gh, \qquad k(f + g) = kf + kg, \qquad (2.11)$$

where $h: A' \to A$ and $k: B \to B'$. A preadditive category on a single object is 'the same' as a (unitary) ring.

An *additive category* C is a preadditive category with finite products, or equivalently with finite sums, which are then biproducts. The equivalence will be proved in Theorem 6.4.3, where we also see that, *in this case*, the sum of parallel maps is determined by the categorical structure.

Typical examples are Ab, R Mod, Ban (while Abm and Rel Set are just $\mathbb{N}$-linear and semiadditive: see 6.4.3, 6.4.5).

2.2 General limits and completeness

This sections contains the main definitions and basic results about limits and colimits in a category C. The main theorem proves that all limits can be constructed from products and equalisers.

2.2.1 General limits and colimits

Let S be a small category and $X\colon S \to C$ a functor, written in index notation (as in 1.4.9(a)), for $i \in S = \mathrm{Ob}S$ and $a\colon i \to j$ in S:

$$X\colon S \to C, \qquad i \mapsto X_i, \quad a \mapsto (X_a\colon X_i \to X_j). \qquad (2.12)$$

A *cone* for the functor X is an object A of C (the *vertex* of the cone) equipped with a family of maps $(f_i\colon A \to X_i)_{i\in S}$ in C such that all the following triangles commute

$$
\begin{array}{lll}
A \xdashrightarrow{\ f_i\ } X_i & \qquad & X_a.f_i = f_j \\[4pt]
\quad{}_{f_j}\searrow \quad \downarrow{\scriptstyle X_a} & & \\[4pt]
\qquad\quad X_j & & \text{(for } a\colon i \to j \text{ in S).}
\end{array}
\qquad (2.13)
$$

The *limit* of $X\colon S \to C$ is a universal cone $(L, (u_i\colon L \to X_i)_{i\in S})$. This means a cone of X such that every cone $(A, (f_i\colon A \to X_i)_{i\in S})$ factorises uniquely through the former; in other words, there is a unique map $f\colon A \to L$ such that, for all $i \in S$, $u_i f = f_i$. The solution need not exist; when it does, it is determined up to a unique coherent isomorphism and the limit object L is denoted as $\mathrm{Lim}(X)$.

The uniqueness part of the universal property amounts to saying that *the family $u_i\colon L \to X_i$ ($i \in S$) is jointly mono* (see 1.3.1). A cone of X that satisfies the existence part of the universal property is called a *weak limit*; of course such a solution is 'determined' in a (very) weak way.

Dually, a *cocone* $(A, (f_i\colon X_i \to A)_{i\in S})$ of X satisfies the condition $f_j.X_a = f_i$ for every $a\colon i \to j$ in S. The *colimit* of the functor X is a universal cocone $(L', (u_i\colon X_i \to L')_{i\in S})$: the universal property says now that for every cocone $(A, (f_i))$ of X there exists a unique map $f\colon L' \to A$ such that

$fu_i = f_i$ (for all i). The colimit object is denoted as $\mathrm{Colim}(X)$. The uniqueness part of the universal property means that the family $u_i\colon X_i \to L'$ is jointly epi. A cocone that satisfies the existence part of the universal property is called a *weak colimit*.

2.2.2 Particular cases and exercises

(a) The product ΠX_i of a family $(X_i)_{i\in S}$ of objects of C is the limit of the corresponding functor $X\colon \mathsf{S} \to \mathsf{C}$, defined on the *discrete* category whose objects are the elements $i \in S$ (and whose morphisms reduce to formal identities of such objects). The sum ΣX_i is the colimit of this functor X.

We recall that the family of cartesian projections (p_i) of a product is *jointly mono*. These projections are 'often' epi but *not necessarily*: for instance in Set this fails whenever some of the factors X_i are empty and others are not. However, one can easily show that, if C has a zero object (as defined in 2.1.9), all cartesian projections are split epi.

(b) The equaliser in C of a pair of parallel morphisms $f, g\colon X_0 \to X_1$ is the limit of the obvious functor defined on the category $0 \rightrightarrows 1$. The coequaliser is the colimit of this functor.

(c) Pullbacks and pushouts will be studied in the next section. Filtered colimits in Section 2.6.

(d) In Set every non-empty set is weakly terminal, while $\emptyset$ is the unique weakly initial object. In a pointed category, every object is weakly terminal and weakly initial.

(e) Weak pullbacks of sets are used in 2.3.8. A weakly initial object will be used in the proof of the Initial Object Theorem, in 3.5.1.

*(f) Weak limits and colimits are studied in [G4, G6]. They are important in Homotopy Theory. A reader familiar with the notion of homotopy pullback (resp. pushout) of topological spaces [Mat] will likely know, or easily prove, that it gives a weak pullback (resp. pushout) in the homotopy category hoTop.

2.2.3 Complete categories and the preservation of limits

A category C is said to be *complete* (resp. *finitely complete*) if it has a limit for every functor $\mathsf{S} \to \mathsf{C}$ defined on a *small* category (resp. a *finite* category).

More precisely one speaks – in the first case – of a *small-complete* category. Of course we cannot expect Set to have products indexed by a large set, or limits for all functors $\mathsf{S} \to \mathsf{Set}$ defined over a large category. In

other words, the basis S should be of a 'smaller size' than the category C
where we construct limits, or we force C to be just a preordered set (where
products are infima, and are not influenced by the size of the set of indices).

The question is settled by a neat result of P. Freyd: a small category C
has the limit of every functor defined on a small category (if and) only if
it is a preordered set with all infima (see 2.2.5(f)). This is precisely why,
*in an arbitrary category, the existence of all (small) limits does not imply
the existence of colimits*, as is the case of preordered sets (see 2.2.5(e)).

One says that a functor $F\colon C \to D$ *preserves the limit*

$$(L, (u_i\colon L \to X_i)_{i\in S})$$

of a functor $X\colon S \to C$ if the cone

$$(FL, (Fu_i\colon FL \to FX_i)_{i\in S})$$

is the limit of the composed functor $FX\colon S \to D$. One says that F *preserves
limits* if it preserves all the limits which exist in C. Analogously for the
preservation of products, equalisers, etc. Saying that a functor *does not
preserve limits* simply means that it does not preserve *some* of them.

Dually we have *cocomplete* categories and the property of *preservation
of colimits*.

2.2.4 *Theorem* (Construction and preservation of limits)

*A category is complete (resp. finitely complete) if and only if it has equalis-
ers and products (resp. finite products). Moreover, if C is complete (resp.
finitely complete), a functor $F\colon C \to D$ preserves all limits (resp. all finite limits) if and only if it preserves equalisers and products (resp. finite
products).*

*Dual results hold for colimits. In particular, a category is cocomplete
(resp. finitely cocomplete) if and only if it has coequalisers and sums (resp.
finite sums).*

Proof The reader can complete this outline. (Alternatively, a more detailed
argument can be found in [M4], Section V.2, Theorem 1.)

Let the functor $X\colon S \to C$ be written in index notation, as in 2.2.1, and
consider the small products

$\Pi_i X_i$ indexed by the objects $i \in \mathrm{Ob}S$, with projections p_i,

$\Pi_a X_{j(a)}$ indexed by the arrows a in S, with projections q_a.

$$(2.14)$$

where it is understood that $a\colon i(a) \to j(a)$ in S.

A cone $(A, (f_i))$ of X amounts to a map $f \colon A \to \Pi_i X_i$ that equalises the following two maps u, v (and gives back the original cone letting $f_i = p_i f$)

$$u, v \colon \textstyle\prod_i X_i \to \textstyle\prod_a X_{j(a)},$$

$$q_a u = p_{j(a)}, \quad q_a v = X_a \cdot p_{i(a)} \qquad \text{(for } a \text{ in S).} \tag{2.15}$$

Taking the equaliser $m \colon L \to \Pi_i X_i$ of the pair u, v, we have a cone $(L, (u_i))$ of X, with $u_i = p_i m \colon L \to X_i$ (for $i \in \mathrm{ObS}$). This cone is the limit.

The second part of the statement, on the preservation of limits, is an obvious consequence, once we assume that products and equalisers exist, so that all limits can be obtained as above. $\qquad\square$

2.2.5 Exercises and complements, I

(a) Taking into account the computations of Section 2.1 it follows that the categories Set, Top, Ab, Ord, Set. and Top. are complete and cocomplete. The forgetful functor Top $\to$ Set preserves limits and colimits, while Ab $\to$ Set only preserves limits; Ord $\to$ Set preserves limits and sums, but it does not preserve coequalisers (and general colimits, as a consequence).

(b) The reader will note that all limits in Top are constructed as in Set and equipped with the coarsest topology making all the structural mappings $u_i \colon L \to X_i$ continuous. Similarly, all colimits in Top are constructed as in Set, and equipped with the finest topology making all the structural mappings $u_i \colon X_i \to L'$ continuous.

(c) One can proceed in a similar way in the category pOrd of preordered sets. In Ord this procedure works for limits and sums; for coequalisers and general colimits it must be 'corrected', replacing the colimit in pOrd with the associated ordered set (see 3.2.3(b)).

(d) *The equivalence between* Set. *and* $\mathcal{S}$ (*in 1.8.6*) *proves that also the latter has all limits and colimits.* For instance, the product in $\mathcal{S}$ of a family (X_i) of sets is easily constructed 'along the equivalence' $R \colon \mathsf{Set.} \rightleftarrows \mathcal{S} \colon S$

$$R(\textstyle\prod_i S(X_i)) = (\textstyle\prod_i (X_i \cup \{\overline{x}_i\})) \setminus \{(\overline{x}_i)\}, \tag{2.16}$$

first adding a base point $\overline{x}_i \notin X_i$ to each factor, then taking their cartesian product in Set. and discarding the base point of the product. (Thus the product 1×1 in $\mathcal{S}$ has *three* elements, while 2×3 has *eleven*.) Sums in $\mathcal{S}$ can also be constructed in this way, but they simply amount to disjoint unions of sets.

(e) In a preordered set X, viewed as a category, equalisers and coequalisers are (obviously) trivial, while we have seen in Section 2.1 that products and sums amount to infima and suprema, respectively.

Therefore X is a complete category if and only if it has all infima, which is equivalent to the existence of all suprema (as recalled in 1.2.5): in other words, the category X is complete if and only if it is cocomplete. In the ordered case this means that X is a complete lattice.

Similarly, an ordered set is a (bounded) lattice if and only if, as a category, it is finitely complete and cocomplete.

*(f) Freyd's result recalled in 2.2.3 can be rewritten in this form: a small category C with all small products is a preordered set (with all infima).

The proof (also written in [M4], Section V.2, Proposition 3) is surprisingly simple. Assuming that C is not a preordered set, there is some hom-set $C(A, B)$ with at least two distinct arrows $f \neq g$. Take the (small) cartesian power $C = \prod_{j \in J} B$, where J is the cardinal of MorC; then the cardinal of $C(A, C) \cong \prod_{j \in J} C(A, B)$ is at least 2^J, a contradiction.

2.2.6 Exercises and complements, II

(a) It is interesting to consider limits and colimits in the categories Ban_1 and Ban (introduced in 1.1.1).

The reader can begin by showing that, in Ban_1, a binary product $X \times Y$ is the vector space $X \oplus Y$ with the l_∞-norm

$$||(x, y)||_\infty = ||x|| \vee ||y||,$$

while a binary sum $X + Y$ is the same vector space $X \oplus Y$ with the l_1-norm

$$||(x, y)||_1 = ||x|| + ||y||.$$

These objects are distinct in Ban_1, but they are isomorphic in Ban, where the *biproduct* $X \oplus Y$ (see 2.1.9) can be given any l_p-norm, for $1 \leqslant p \leqslant \infty$ (all of them being Lipschitz-equivalent).

After this, it is easy to show that both Ban_1 and Ban are finitely complete, constructing equalisers as closed linear subspaces with the restricted norm.

In fact Ban_1 is even complete and cocomplete, while Ban is finitely complete and cocomplete. The construction of coequalisers (as quotient linear spaces modulo closed linear subspaces, with the induced norm) is less elementary than the previous points above; the construction of arbitrary products and sums in Ban_1 is even more delicate. A reader interested in the non-elementary aspects of these categories is referred to the text of Semadeni [Se]. A fundamental result that simplifies these arguments is the fact that a normed linear space is complete if and only if every family of elements of summable norm is summable.

(b) Rel Set has arbitrary products and sums, that are *biproducts*, as above.

The reader will verify that an arbitrary sum $X = \Sigma X_i$ in Set is still a sum in Rel Set, with the same injections $u_i \colon X_i \to X$. Since Rel Set is an involutive category, X is also a product, with projections given by the opposite *relations* $u_i^\sharp \colon X \to X_i$. The zero object is $\emptyset$.

On the other hand, Rel Ab lacks products (and sums), including the terminal object. Both categories lack equalisers and coequalisers.

(c) If the category C is complete (resp. finitely complete), so is the category C^S (for any small category S), with limits computed pointwise on each object of S. Similarly for colimits.

2.2.7 Representable functors detect limits and colimits

Let C be a category with small hom-sets, and $X \colon S \to C$ a functor defined on a small category. The universal property of the limit can be rewritten as follows:

- a cone $(L, (u_i \colon L \to X_i))$ of X is a limit in C if and only if, for every A in C, the set $C(A, L)$ is the limit in Set of the sets $C(A, X_i)$, with limit-cone $u_i.- \colon C(A, L) \to C(A, X_i)$.

This rewriting tells us that:

(i) each representable covariant functor $F_A = C(A, -) \colon C \to$ Set preserves all (the existing) limits,

(ii) the family (F_A), indexed by the objects $A \in \mathrm{Ob}\,C$, reflects them.

Colimits in C are limits in C^{op}. Thus, a representable contravariant functor $C(-, A) \colon C^{\mathrm{op}} \to$ Set takes *colimits* of C to *limits* of sets; globally, their family reflects limits of Set into colimits of C.

For instance, for the contravariant functor of subsets $\mathrm{Set}^{\mathrm{op}} \to$ Set of 1.6.5(c), we (plainly) have: $\mathcal{P}(\Sigma_i X_i) \cong \Pi_i \mathcal{P}(X_i)$, since a subset $A \subset \Sigma_i X_i$ amounts to a family of subsets $A_i \subset X_i$.

2.2.8 Creating limits

There is another interesting relationship between functors and limits (or colimits); again, it does not assume their existence (as defined in [M4], Section V.1).

One says that a functor $F \colon C \to D$ *creates* limits for a functor $X \colon S \to C$ if:

(i) for every limit cone $(L', (v_i \colon L' \to FX_i)_{i \in S})$ of the composed functor

$FX\colon \mathsf{S} \to \mathsf{D}$, there is precisely one cone $(L, (u_i\colon L \to X_i)_{i \in S})$ of X in C taken by F to $(L', (v_i))$,

(ii) the cone $(L, (u_i))$ is a limit of X in C (preserved by F).

This 'nearly' implies that F preserves limits: the precise statement can be found in 2.2.9(b).

For instance the forgetful functor $U\colon \mathsf{Ab} \to \mathsf{Set}$ creates all small limits (which do exist). Indeed, a limit of abelian groups can be constructed by taking the limit L' of the underlying sets and putting on it the unique structure that makes the mappings v_i into homomorphisms. We shall see that the same happens for all varieties of algebras (in Section 4.4). On the other hand, these forgetful functors do not create general *colimits*: they do not even preserve them – with an important exception, filtered colimits, dealt with in Section 2.6.

On the other hand the forgetful functor $\mathsf{Top} \to \mathsf{Set}$ preserves all limits and colimits but does not create them: for instance the product topology is the *coarsest* topology on the product set that makes all cartesian projections continuous, and is not the only one having this outcome (generally): any topology intermediate between this topology and the discrete one would do.

Of course one can restrict the limits that we are considering to the S-based limits, for a given S, or for some class of small categories; e.g. finite limits, or products, etc.

2.2.9 Exercises and complements

Let $F\colon \mathsf{C} \to \mathsf{D}$ be a functor and S a small category.

(a) If F creates all S-based limits then it reflects them.

(b) If D has all S-based limits and F creates them, then C has all S-based limits and F preserves them.

(c) F creates unary products if and only if it reflects identities.

(d) If C has all S-based limits, and the functor F preserves them and reflects identities, then F creates these limits.

(e) The forgetful functor $\mathsf{Top} \to \mathsf{Set}$ does not reflect identities (but is amnestic, see 1.4.4).

(f) The reader likely knows that a differentiable manifold can be defined in two (slightly) different ways: as a topological space (satisfying some topological properties) equipped with a differentiable atlas, *or* equipped with an equivalence class of them. After defining the differentiable maps, these two approaches give equivalent categories of differentiable manifolds;

in each of them the forgetful functor with values in Top preserves finite products, but only in the second case it reflects identities, and creates finite products.

2.3 Pullbacks and pushouts

Pullbacks and pushouts are an important pair of limit and colimit, dual to each other. We work in an arbitrary category C.

2.3.1 Pullbacks

The *pullback* of a pair of morphisms (f, g) of C with the same codomain

$$f \colon X_1 \to X_0 \leftarrow X_2 \colon g$$

is the limit of the corresponding functor defined on the category $1 \to 0 \leftarrow 2$. This amounts to the usual definition: an object A equipped with two maps $u_i \colon A \to X_i$ $(i = 1, 2)$ which form a commutative square with f and g, in a universal way:

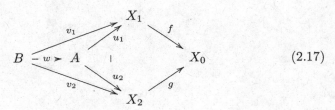 (2.17)

that is, $fu_1 = gu_2$, and for every triple (B, v_1, v_2) such that $fv_1 = gv_2$ there exists a unique map $w \colon B \to A$ such that $u_1 w = v_1$, $u_2 w = v_2$. Again, the pair (u_1, u_2) is jointly mono.

The pullback-object A is also called a *fibred product* over X_0 and written as $X_1 \times_{X_0} X_2$. The square diagram $fu_1 = gu_2$ in (2.17) is also called a *pullback*, or a *pullback diagram*; it can be put in evidence by some mark in the pullback corner, as above.

In the ordered set X the pullback of the diagram $x_1 \leqslant x_0 \geqslant x_2$ is the meet $x_1 \wedge x_2$. Saying that X has pullbacks means that every *upper-bounded* pair of elements has a meet.

In Set (resp. Top, Ab) the pullback-object can be realised as a subset (resp. subspace, subgroup) of the product $X_1 \times X_2$:

$$A = \{(x_1, x_2) \in X_1 \times X_2 \mid f(x_1) = g(x_2)\}. \tag{2.18}$$

2.3.2 Exercises and complements

(a) Generalising the last construction, prove that a category that has binary products and equalisers also has pullbacks: A is constructed as the equaliser of two maps $X_1 \times X_2 \to X_0$. Pullbacks in concrete categories are usually computed in this way.

(b) This result does not need the terminal object. When the latter exists, the product $X_1 \times X_2$ amounts to the pullback $X_1 \times_\top X_2$ over the terminal object.

(c) Note that the pullback of the symmetric pair $X \to \top \leftarrow X$ is the product $(X \times X, p, q)$, with *different* projections (generally). When a fixed choice of pullbacks is used, *one cannot assume that the choice is symmetric*; even though it certainly is, up to isomorphism.

2.3.3 Pushouts

Dually, the *pushout* of a pair (f, g) of morphisms with the same domain is the colimit of the obvious functor defined on the category $1 \leftarrow 0 \to 2$. This amounts to an object A equipped with two maps $u_i \colon X_i \to A$ $(i = 1, 2)$ which form a commutative square with f and g, in a universal way:

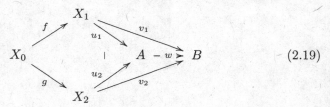

$$(2.19)$$

that is, $u_1 f = u_2 g$, and for every triple (B, v_1, v_2) such that $v_1 f = v_2 g$ there exists a unique map $w \colon A \to B$ such that $w u_1 = v_1$, $w u_2 = v_2$. The pushout square can also be highlighted by some mark in the appropriate corner.

A category that has binary sums and coequalisers also has pushouts: A is constructed as the coequaliser of the two obvious maps $X_0 \to X_i \to X_1 + X_2$ given by f, g and the injections.

The pushout-object A is also called a *pasting* over X_0 and written as $X_1 +_{X_0} X_2$. In Set this means a quotient of the sum $X_1 + X_2$ modulo the equivalence relation generated by all the pairs $(f(x), g(x))$ (viewed in the sum), for $x \in X_0$.

In the category Top of topological spaces, a pasting $X_1 +_{X_0} X_2$ is constructed as in Set and equipped with the quotient topology of the sum. It is a useful way of constructing new spaces. The reader can note that the usual process of verifying the continuity of a mapping $f \colon X \to Y$ on the subsets of a *closed* cover (X_1, X_2) of X amounts to saying that X is the pushout of X_1 and X_2 over $X_1 \cap X_2$. This fact can be easily generalised to a *finite closed cover* (or 'transferred' to an arbitrary open cover).

The well known Seifert–van Kampen Theorem exhibits the fundamental group of a space as a pushout of groups. R. Brown's version generalises this result, using fundamental groupoids [Bro1, Bro2].

2.3.4 Kernel and cokernel pairs

Given a morphism $f \colon A \to B$, the pullback $R \rightrightarrows A$ of the diagram $A \to B \leftarrow A$ is called the *kernel pair* of f

$$
\begin{array}{ccc}
A & \xrightarrow{\ f\ } & B \\
\uparrow{\scriptstyle u} & \nwarrow & \uparrow{\scriptstyle f} \\
R & \xrightarrow{\ v\ } & A
\end{array}
\tag{2.20}
$$

In Set it can be realised as $R = \{(x, x') \in A \times A \mid f(x) = f(x')\}$, and gives the equivalence relation R_f associated to f. Then, the coequaliser of the pair $u, v \colon R \to A$ is the (projection on) the quotient A/R_f. Kernel pairs and their coequalisers are basic tools of the theory of regular categories: see Chapter 4.

One can note that the kernel pair has little to do with the equaliser of f and f, which – in any category – is the identity of its domain.

The dual notion is called a *cokernel pair*.

2.3.5 Exercises and complements, I

The following properties of pullbacks and pushouts are important and can be easily verified.

(a) *Characterising monos and epis.* The morphism m is mono if and only if the left square below is a pullback, if and only if the kernel pair of m consists of two equal arrows

$$
\begin{array}{ccc}
\bullet & \xrightarrow{\ m\ } & \bullet \\
\uparrow{\scriptstyle 1} & \nwarrow & \uparrow{\scriptstyle m} \\
\bullet & \xrightarrow{\ 1\ } & \bullet
\end{array}
\qquad
\begin{array}{ccc}
\bullet & \xrightarrow{\ 1\ } & \bullet \\
\uparrow{\scriptstyle p} & \nwarrow & \uparrow{\scriptstyle 1} \\
\bullet & \xrightarrow{\ p\ } & \bullet
\end{array}
\tag{2.21}
$$

(a bullet stands for an arbitrary object). Dually the morphism p is epi if and only if the right square is a pushout, if and only if the cokernel pair of p consists of two equal arrows. Note that these characterisations do not require the existence of pullbacks and pushouts in our category.

A functor which preserves pullbacks (resp. pushouts) also preserves monomorphisms (resp. epimorphisms). This proves again that a representable functor preserves monomorphisms, as we have already seen in 1.6.5(d).

(b) *Pasting property.* If the two squares below are pullbacks, so is their 'pasting', i.e. the outer rectangle

$$(2.22)$$

(c) *Depasting property.* If in the commutative diagram above the outer rectangle is a pullback and the pair (u, v) is jointly mono then the left square is a pullback.

The hypothesis on the pair (u, v) is automatically satisfied in two common cases: when the right square is also a pullback, or one of the morphisms u, v is mono.

(d) Pushouts have dual properties with respect to (b) and (c). In an abelian category pullbacks and pushouts are characterised by exactness properties of an associated sequence of morphisms: see 6.4.8.

2.3.6 Exercises and complements, II

(a) In a category with equalisers and cokernel pairs, a monomorphism is regular if and only if it is the equaliser of its cokernel pair.

(b) A category with pullbacks and terminal object is finitely complete. Hint: an equaliser of two maps $A \rightrightarrows B$ can be constructed by suitable pullbacks, using also the diagonal $A \to A^2$.

(c) Epimorphisms in Hsd are precisely the maps whose image is dense in their codomain. Hints: we have already seen that this condition is sufficient (in 1.3.4(b)); conversely, one can consider the embedding $C \to Y$ of a closed subspace into a Hausdorff space, and show that its cokernel pair in Top also 'works' in Hsd.

*(d) The class of limits generated by the existence of pullbacks, in a category, is characterised in [Pa].

2.3.7 Lemma

In a category C we suppose that the following diagram is commutative. If the two lateral squares and the back square are pullbacks (with vertices P, Q and R), then the front square is also a pullback

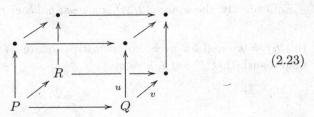

$$(2.23)$$

Proof It is a consequence of the pasting and depasting properties above, in 2.3.5. The pasting of the left and back squares is a pullback, and coincides with the pasting of the front and right squares; but the right square is a pullback, whence the front square is also. Also here one can drop the hypothesis that the right square be a pullback and replace it with: the pair (u, v) is jointly mono. $\qquad\qquad\qquad\qquad\qquad\square$

2.3.8 *Bicommutative squares and exact squares*

This point is about categories of relations of sets and modules, and will be used in Chapter 6.

(a) First we work in the category Set. It is easy to see that a square of mappings

$$\begin{array}{ccc} A & \xrightarrow{f} & X \\ {\scriptstyle g}\downarrow & & \downarrow{\scriptstyle h} \\ Y & \xrightarrow{k} & B \end{array} \qquad\qquad (2.24)$$

is commutative if and only if we have $gf^\sharp \leqslant k^\sharp h$ in Rel Set (or $fg^\sharp \leqslant h^\sharp k$, equivalently). In fact

$$hf = kg \quad\Rightarrow\quad gf^\sharp \leqslant k^\sharp kgf^\sharp = k^\sharp hff^\sharp \leqslant k^\sharp h,$$
$$gf^\sharp \leqslant k^\sharp h \quad\Rightarrow\quad kg \leqslant kgf^\sharp f \leqslant kk^\sharp hf \leqslant hf \quad\Rightarrow\quad kg = hf.$$

We say that the square (2.24) is *bicommutative* in Rel Set if it satisfies the stronger condition

(i) $gf^\sharp = k^\sharp h \colon X \to Y \qquad (\Leftrightarrow \quad fg^\sharp = h^\sharp k \colon Y \to X),$

i.e. if the original square (of mappings) commutes in Rel Set when we 're-verse' two parallel edges.

Plainly, condition (i) can be characterised as follows:

(ii) for $(x, y) \in X \times Y$, $h(x) = k(y)$ if and only if there exists some $a \in A$ such that $f(a) = x$, $g(a) = y$.

Equivalently, the square (2.24) is a *weak pullback* in Set (see 2.2.1), that is:

(iii) $hf = kg$ and for every commutative square $hf' = kg'$ there is *some* map u such that $f' = fu$, $g' = gu$

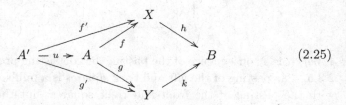

$$(2.25)$$

The sufficiency of (iii) follows from taking $A' = \{*\}$, its necessity from the axiom of choice.

(b) Similarly a square (2.24) of homomorphisms of $R\,\mathsf{Mod}$ (in particular of Ab) is said to be *exact* if it becomes *bicommutative* in the category of relations, i.e. if it satisfies (i). This is still equivalent to (ii), since the forgetful functor $\mathrm{Rel}\,(R\,\mathsf{Mod}) \to \mathrm{Rel}\,\mathsf{Set}$ is faithful; this condition can also be expressed by an exact sequence due to Hilton [Hi] (see 6.4.8). Characterisation (iii) holds here *if* we take $A' = R$ (or require A' to be a free module).

2.4 Subobjects and quotients, regular and strong

In a category, a subobject of an object A is defined as a 'selected' monomorphism $m \colon X \rightarrowtail A$, in a suitable equivalence class; dually, a quotient of A is a selected epimorphism $p \colon A \twoheadrightarrow X$.

Nevertheless, in many categories it is convenient to restrict the monos or epis we are considering, in order to get the 'good' notion of subobject or quotient: strong monomorphisms and strong epimorphisms (defined in 2.4.3) are important in categories of topological spaces, while strong epimorphisms and ordinary monos are important in varieties of algebras. This is discussed in 2.4.5.

In these cases, strong can be equivalently replaced with regular. But strong monos and epis have the advantage of being closed under composition, in every category.

Categories of structured sets, usually, have a natural choice for the 'good' notion of subobject and quotient.

2.4.1 Subobjects and quotients

Let A be an object of the category C. A subobject of A is defined as an equivalence class of monomorphisms, or better as a *selected representative* of such a class.

More precisely, given two monos m, n *with values in A*, we say that $m \prec n$ if there is a (uniquely determined) morphism u such that $m = nu$. We say that m and n *are equivalent*, or $m \sim n$, if $m \prec n \prec m$, which amounts to the existence of a (unique) isomorphism u such that $m = nu$.

In every class of equivalent monos with codomain A, precisely one is *selected* and called a *subobject* of A; in the class of isomorphisms we always choose the identity 1_A. The subobjects of A in C form the (possibly large) *ordered* set Sub$A$, with greatest element 1_A; here, the induced *order $m \prec n$* is also written as $m \leqslant n$.

Epimorphisms with a fixed domain A are dealt with in a dual way. Their preorder and equivalence relation are also written as $p \prec q$ (meaning that p factorises through q) and $p \sim q$. A *quotient* of A is a *selected* representative of an equivalence class of epimorphisms with domain A; they form the ordered set Quo$A$, with greatest element 1_A; again the induced order is also written as $p \leqslant q$.

The category C is said to be *well powered* (resp. *well copowered*) if all its sets Sub$A$ (resp. Quo$A$) are small, as is often the case with categories of (small) structured sets.

Duality of categories turns subobjects of C into quotients of C^{op}, *preserving* their order.

2.4.2 Orthogonal morphisms

The following relationship between morphisms of C will be useful. We say that f *is orthogonal to* g, and write $f \perp g$, if for all arrows u, v that make a commutative square $vf = gu$

$$
\begin{array}{ccc}
\bullet & \xrightarrow{\ f\ } & \bullet \\
u \downarrow & \nearrow^{w} & \downarrow v \\
\bullet & \xrightarrow[\ g\ ]{} & \bullet
\end{array}
\tag{2.26}
$$

there is precisely one morphism w such that $wf = u$ and $gw = v$.

Plainly, this relation is not symmetric. For a (generally large) subset $H \subset \mathrm{Mor}\,C$ we have the following *orthogonal classes*

$$
\perp H = \{ f \in \mathrm{Mor}\,C \mid \forall h \in H,\ f \perp h \},
$$
$$
H \perp = \{ f \in \mathrm{Mor}\,C \mid \forall h \in H,\ h \perp f \}.
\tag{2.27}
$$

Applying 1.7.9(a), the orthogonality relation gives a contravariant Galois connection

$$L\colon \mathcal{P}(\mathrm{Mor}\mathsf{C}) \rightleftarrows \mathcal{P}(\mathrm{Mor}\mathsf{C})\colon R,$$
$$L(H) = \bot H, \qquad R(H) = H\bot. \tag{2.28}$$

2.4.3 Regular and strong subobjects and quotients

Let us recall that an epimorphism $p\colon A \to X$ of C is said to be *regular* if it is the coequaliser of a pair of parallel morphisms with values in A (see 2.1.7).

As a more general property, an epimorphism $p\colon A \to X$ is said to be *strong* if it is orthogonal to every monomorphism: for every commutative square $mu = vp$ where m is mono, there exists a (unique) morphism w such that $wp = u$ and $mw = v$ (the last two conditions are actually equivalent)

$$\tag{2.29}$$

In fact, every regular epimorphism is strong, as proved below (in 2.4.4(a)). Plainly, both notions are invariant up to the equivalence relation $p \sim q$ (of epimorphisms defined on A). A quotient $p\colon A \to X$ is said to be *regular* (resp. *strong*) if it is, as an epimorphism.

Dually a monomorphism $m\colon Y \to B$ is said to be *strong* if it is orthogonal to every epimorphism: for every commutative square as above, where p is epi, there exists a morphism w such that $wp = u$ and $mw = v$. Every regular monomorphism is a strong mono. A subobject is said to be *regular* (resp. *strong*) if it is, as a monomorphism.

2.4.4 Exercises and complements, I

(a) It is easy to prove that every regular epimorphism is strong: in the diagram below (with a commutative square), let p be the coequaliser of f, g

$$\tag{2.30}$$

Then $uf = ug$ (cancelling the monomorphism m), and there exists a morphism w such that $wp = u$; cancelling p, we also have $mw = v$.

(b) Strong epis are closed under composition. If a composite gf is a strong epi, g also is. And dually.

(c) A strong epimorphism which is mono is an isomorphism. And dually.

2.4.5 Exercises and complements, II

(a) In Set every monomorphism (resp. epimorphism) is regular: the reader will show that every subset $E \subset X$ is the equaliser of its cokernel pair, and every quotient set X/R is the coequaliser of its kernel pair.

Therefore all subobjects are regular (and strong), and can be identified with subsets. All quotients are regular (and strong), and can be identified with set-theoretical quotients modulo an equivalence relation.

(b) In Ab and R Mod every monomorphism (resp. epimorphism) is regular (and even 'normal', as we shall see in 6.1.3). All subobjects are regular and can be identified with subgroups; all quotients are regular and can be identified with ordinary quotients modulo a subgroup.

(c) In Top every subspace $E \subset X$ is the equaliser of its cokernel pair, and every quotient space X/R is the coequaliser of its kernel pair.

Here every injective continuous mapping is a monomorphism, while the regular and strong subobjects coincide and amount to *inclusions of subspaces*: the 'general subobjects' are less important than the strong (or regular) ones. Similarly, every surjective continuous mapping is an epi, while the regular and strong quotients coincide and amount to projections on quotient spaces, in the usual sense.

(d) We shall see in Theorem 4.4.4 that, in a variety of algebras, regular epimorphisms coincide with the strong ones, and give the usual quotients of algebras, modulo a congruence. On the other hand, general subobjects amount to the usual subalgebras.

(e) In a variety of algebras, general epimorphisms on the one hand, regular and strong monomorphisms on the other, can be difficult to determine and – finally – of a limited interest. Knowing a few cases should be enough: for instance, all subgroups are regular subobjects in Gp (a non-trivial fact, see [AHS], Exercise 7H), while a subsemigroup need not be a regular subobject in the category of semigroups ([AHS], Exercise 7I).

2.4.6 Preimages of subobjects and images of quotients

The reader will likely find of interest to prove that, if the left square below is a pullback and n is a strong mono, or a regular mono, so is m.

$$(2.31)$$

The monomorphism $m\colon \bullet \rightarrowtail A$ is determined up to the equivalence relation described above; it is called the *preimage*, or *inverse image*, of n along f and written as $f^*(n)$. When we work with subobjects, $f^*(n)$ is strictly determined by their choice.

Dually, if the right square above is a pushout and p is epi, or a strong epi, or a regular epi, so is $q\colon B \to \bullet$. The latter is determined up to the equivalence relation of epimorphisms described above; it is called the *image* of p along f and written as $f_\circ(p)$.

We generally adopt the following conventions.

(i) *The unit constraint for pullbacks.* The pullback of an identity morphism along any morphism (always exists and) is chosen to be an identity: $f^*(1_B) = 1_A$.

(i*) *The unit constraint for pushouts.* The pushout of an identity morphism along any morphism (always exists and) is chosen to be an identity: $f_\circ(1_A) = 1_B$.

2.5 Factorisation systems

Many categories have a privileged factorisation system (E, M), e.g. the factorisation 'epi-mono' in Set and abelian categories, or 'regular epi - general mono' in any variety of algebras. Yet the general notion (in 2.5.2) does not require that the system be *proper*, i.e. that E be a set of epimorphisms and M a set of monomorphisms; there are important cases where this does not hold.

We briefly sketch this subject. The reader can find further information in [AHS, Bo1, FrK, CaJKP] and in the papers cited below. The (non elementary) relationship of factorisation systems with monads will be outlined in Section 7.6.

We work in a category C. The subsets of monomorphisms, epimorphisms and isomorphisms in MorC are written as Mono, Epi and Iso, respectively.

2.5.1 Examples

Two examples will be of help to introduce the subject.

(a) As an obvious, well known fact, every mapping $f \colon A \to B$ in Set has an *epi-mono factorisation* $f = me$, which is uniquely determined up to a unique 'central' isomorphism.

More precisely, given two factorisations $f = me = m'e'$ of this kind, there is a unique morphism w that makes the diagram below commutative, and it is an isomorphism

$$
\begin{array}{ccccc}
A & \xrightarrow{\ e\ } & X & \xrightarrow{\ m\ } & B \\[2pt]
\| & & \downarrow{\scriptstyle w} & & \| \\[2pt]
A & \xrightarrow[\ e'\ ]{} & X & \xrightarrow[\ m'\ ]{} & B
\end{array}
\qquad (2.32)
$$

This decomposition $f = me$ will also be called *the EM-factorisation* of f, where $E = \mathrm{Epi}$ and $M = \mathrm{Mono}$ are the (large) subsets of MorSet of epis and monos, respectively. The pair (E, M) is called a factorisation system for Set. It is a *proper* one – which in general just means that $E \subset \mathrm{Epi}$ and $M \subset \mathrm{Mono}$.

The reader likely knows various similar factorisations, in concrete categories; some of them will be recalled below.

(b) The category of morphisms C^2 of an arbitrary category C, described in 1.6.1, has a canonical factorisation which plays an important role in the theory of factorisations.

In fact every morphism $f = (f', f'') \colon x \to y$ can be factorised as in the right diagram below

$$
\begin{array}{ccc}
X' & \xrightarrow{\ f'\ } & Y' \\[2pt]
x\downarrow & & \downarrow y \\[2pt]
X'' & \xrightarrow[\ f''\ ]{} & Y''
\end{array}
\qquad\qquad
\begin{array}{ccccc}
X' & \xrightarrow{\ 1\ } & X' & \xrightarrow{\ f'\ } & Y' \\[2pt]
x\downarrow & & \downarrow{\scriptstyle \overline{f}} & & \downarrow y \\[2pt]
X'' & \xrightarrow[\ f''\ ]{} & Y'' & \xrightarrow[\ 1\ ]{} & Y''
\end{array}
\qquad (2.33)
$$

where $\overline{f} = vf' = f''u$ is the diagonal of the commutative square f.

If we let E be the set of morphisms (i, g) where i is an isomorphism, as in the left commutative square below, and M be the set of morphisms (f, j) where j is an isomorphism, as in the right square

$$
\begin{array}{ccc}
\bullet & \xrightarrow{\ i\ } & \bullet \\[2pt]
\downarrow & & \downarrow \\[2pt]
\bullet & \xrightarrow[\ g\ ]{} & \bullet
\end{array}
\qquad\qquad
\begin{array}{ccc}
\bullet & \xrightarrow{\ f\ } & \bullet \\[2pt]
\downarrow & & \downarrow \\[2pt]
\bullet & \xrightarrow[\ j\ ]{} & \bullet
\end{array}
\qquad (2.34)
$$

it is easy to see that the EM-factorisation considered above is unique up to isomorphism, in C^2. The reader will note that:

- this factorisation system (E, M) is not proper,

- restricting E to the set E_0 where i is an identity and M to the set M_0 where j is an identity, we even have a *strict factorisation system* (E_0, M_0), in the sense that the factorisation (2.33) is now strictly determined,

- such a restriction can also be made in case (a), taking – for instance – $E_0 = $ Epi and M the set of inclusions of subsets.

2.5.2 Factorisation systems

We now define a *factorisation system* for C as a pair (E, M) of (generally large) subsets of Mor$C$ such that:

(i) E and M are closed under composition and contain all the isomorphisms (forming two wide subcategories E, M of C),

(ii) every morphism f has a factorisation $f = me$, with $e \in E$ and $m \in M$,

(iii) (*uniqueness*) for any two factorisations $f = me = m'e'$ of this kind, there is a unique morphism w that makes the diagram (2.32) commute, and it is an isomorphism. (We speak of *the EM-factorisation $f = me$*.)

Axiom (iii) can be replaced with the following one (equivalent in the presence of the others):

(iii') (*orthogonality*) for every $e \in E$ and every $m \in M$, $e \perp m$ (see 2.4.2).

The equivalence is proved below, in 2.5.3(a).

The opposite category C^{op} inherits an *opposite* factorisation system, namely (M^{op}, E^{op}).

We say that (E, M) is a *proper*, or *epi-mono*, factorisation system if $E \subset$ Epi and $M \subset$ Mono. This need not be the case, as we have already seen in 2.5.1(b); moreover, every category admits two trivial factorisation systems: (Iso, Mor) and (Mor, Iso), which are not proper – in general.

2.5.3 Exercises and complements, I

A factorisation system (E, M) is given in C.

(a) (Here we only assume that (i) and (ii) hold.) Prove that (iii) $\Rightarrow$ (iii') using the left diagram below, and that (iii') $\Rightarrow$ (iii) using the right one

$$(2.35)$$

(b) $E \cap M$ coincides with the set of isomorphisms of C. The EM-factorisation $u = me \colon A \to B$ of an isomorphism u consists of two isomorphisms.

Hint, for the first statement: write $u \in E \cap M$ as $u = 1_B.u = u.1_A$ and apply uniqueness.

(c) Each of the sets E, M determines the other, as an orthogonal class in C (in the sense of 2.4.2)

- $E = \perp M = \{ f \in \operatorname{Mor} \mathsf{C} \mid \forall m \in M, \ f \perp m \}$,
- $M = E\perp = \{ f \in \operatorname{Mor} \mathsf{C} \mid \forall e \in E, \ e \perp f \}$.

Hint: It suffices to prove that every morphism $f \in \perp M$ belongs to E. The EM-factorisation $f = me$ gives the left commutative diagram below (because $f \perp m$), and the right square proves that w is an isomorphism

$$(2.36)$$

(d) For a composite gf in C, we have:
- if $gf \in E$ and $f \in E$, then $g \in E$,
- if $gf \in M$ and $g \in M$, then $f \in M$.

Hint: assuming that $gf \in E$ and $f \in E$, one can prove that $g \perp m$ (for every $m \in M$) by applying the hypotheses to the following commutative diagram

(e) The factorisation systems of C are ordered by the relation expressed by the following equivalent conditions

$$E \subset E', \qquad M' \subset M. \qquad (2.37)$$

In fact, because of the contravariant Galois connection (2.28), $E \subset E'$ implies $M = E^\perp \supset E'^\perp = M'$, and conversely. The trivial factorisation systems (Iso, Mor) and (Mor, Iso) on C are, respectively, the smallest and the largest.

(f) A *strict factorisation system* for C is a pair (E_0, M_0) of subsets of $\text{Mor}\mathsf{C}$ such that:

(i) E_0 and M_0 are closed under composition and contain all the identities,

(ii) every morphism f has a strictly unique factorisation $f = m_0 e_0$, with $e_0 \in E_0$ and $m_0 \in M_0$.

Note that this is *not* a factorisation system, because E_0 and M_0 do not contain all isomorphisms – generally. But there is a unique factorisation system (E, M) containing the former (or *spanned* by it), where $f = m_0 e_0$ is in E (resp. M) if and only if m_0 (resp. e_0) is an isomorphism. The only non obvious point – if not a difficult one – is the closure of E and M under composition. (One can see the proof in [G4], Section 2.)

2.5.4 Exercises and complements, II

(a) A category admits the pair $(\text{Epi}, \text{Mono})$ as a factorisation system if and only if

(i) all epimorphisms are strong epis and all monomorphisms are strong monos,

(ii) every morphism f has an epi-mono factorisation.

(b) The following categories have such a factorisation system: Set, $\mathsf{Set_\bullet}$, Ab, $\mathsf{R\,Mod}$, Gp. For the last case, one should recall that all epimorphisms in Gp are surjective (see 1.3.4(a)).

(c) We shall see in 4.2.3 that every regular category has a factorisation system $(\text{rEpi}, \text{Mono})$, where rEpi is the set of regular epis (coinciding with the strong ones). The examples of (b) are part of this case.

(d) In Top one can consider two main factorisation systems: $(\text{rEpi}, \text{Mono})$ and $(\text{Epi}, \text{rMono})$. One may prefer to use a 'ternary factorisation system' $(\text{rEpi}, \text{Iso}, \text{rMono})$, which generates each of the previous ones in the obvious way.

2.5.5 *Natural weak factorisation systems*

A *functorial weak factorisation system* in C is assigned by a functorial factorisation of its morphisms

$$X \xrightarrow{\lambda(f)} F(f) \xrightarrow{\rho(f)} Y \qquad\qquad f = \rho(f).\lambda(f), \qquad (2.38)$$

where $F \colon \mathsf{C}^2 \to \mathsf{C}$ is a functor and $\lambda \colon \mathrm{Dom} \to F$, $\rho \colon F \to \mathrm{Cod}$ are natural transformations.

Plainly, every factorisation system (E, M) in C produces such a structure, by *choosing* for every morphism $f \colon X \to Y$, a particular EM-factorisation $X \to F(f) \to Y$.

A *natural weak factorisation system* is a stronger notion that covers various situations, as the graph factorisation of Galois connections in 1.7.6 (and of general adjunctions in 3.3.6). It was introduced in [GT]; see also [Ga]. (Interestingly, these graph factorisations are *mono-epi*.)

2.6 Inductive limits and filtered colimits

The inductive limit of a diagram $X = ((X_i), (u_{ij}))$ is a classical notion, studied in many concrete categories of algebraic or topological character. In category theory it is a kind of *colimit*, and has a natural generalisation to a *filtered colimit*.

These colimits have important, peculiar properties: they are 'often' preserved by forgetful functors of concrete categories, and even created by the forgetful functors of equational algebras. Furthermore, they 'often' commute with finite limits.

The dual notion, classically called a *projective limit*, is less important in the usual categories of structured sets, where – generally speaking – *all* limits are already 'well-behaved'.

We keep this section at an elementary level, adequate to the present goals. More complete expositions can be found in [M4, Bo1].

2.6.1 *Inductive limits*

A *filtered* (ordered) *set I* will be a (small) ordered set where every finite subset has an upper bound. Or, equivalently, a non-empty ordered set such that each pair $i, j \in I$ has an upper bound $k \geqslant i, j$ in I. (This is also called a *directed* set.)

An *inductive limit* is the colimit of a functor $X \colon I \to \mathsf{C}$ defined on such a category. The functor X will be written in index notation (see 1.4.9(a)), as

$$X = ((X_i), (u_{ij})), \quad u_{ij} \colon X_i \to X_j, \qquad\qquad \text{for } i \leqslant j \text{ in } I,$$
$$u_{ii} = \mathrm{id} X_i, \qquad u_{jk}.u_{ij} = u_{ik}, \qquad\qquad \text{for } i \leqslant j \leqslant k \text{ in } I. \tag{2.39}$$

A cocone $(Y, (f_i))$ of X is thus an object Y with a family of morphisms f_i in C such that

$$f_i \colon X_i \to Y, \qquad f_j u_{ij} = f_i, \qquad\qquad \text{for } i \leqslant j \text{ in } I. \tag{2.40}$$

The colimit is a universal cocone

$$(L, (u_i)) = \mathrm{Colim}((X_i), (u_{ij})), \tag{2.41}$$

which means that for every cocone $(Y, (f_i))$ there exists a unique $f \colon L \to Y$ in C such that $f_i = f u_i \colon X_i \to Y$, for all $i \in I$.

For sure we do not want to include here the case $I = \emptyset$, as it will be evident in 2.6.2(c).

2.6.2 Exercises and complements, I

Generally speaking, inductive limits in a concrete category C are easy to construct, and an important tool. The reader should find it interesting to complete the details of the following exercises. The first two already show that inductive limits can be much simpler than general colimits.

I is always a filtered set, and $U \colon \mathsf{C} \to \mathsf{Set}$ the forgetful functor of C.

(a) We are given a functor $X \colon I \to \mathsf{Set}$. One begins by forming the sum $S = \Sigma X_i$, whose elements are written as pairs (x, i), with $i \in I$ and $x \in X_i$ (as in (2.6)). The reader will prove that the relation $(x, i) \sim (y, j)$ in S defined by

$$\text{there exists } k \geqslant i, j \text{ in } I \text{ such that } u_{ik}(x) = u_{jk}(y), \tag{2.42}$$

is an equivalence relation, and that the colimit of X can be obtained as the quotient:

$$L = S/\sim, \qquad u_i \colon X_i \to L, \quad u_i(x) = [x, i]. \tag{2.43}$$

Loosely speaking, the colimit L is constructed by putting together all the elements of the system (X_i), and identifying them when they 'eventually coincide', along I.

(b) For a functor $X \colon I \to \mathsf{Ab}$, one takes the colimit L of the 'underlying functor' $UX \colon I \to \mathsf{Set}$. The reader will check that L has a structure of abelian group, defined as follows

$$[x, i] + [y, j] = [u_{ik}(x) + u_{jk}(y), k], \qquad \text{for } i, j \leqslant k \text{ in } I. \tag{2.44}$$

Moreover this is the unique structure of abelian group of L that makes all $u_i \colon X_i \to L$ into homomorphisms. Finally $(L, (u_i))$ is the colimit of X in Ab, *created* by $U \colon$ Ab $\to$ Set (as defined in 2.2.8).

(c) The reader will note that the previous construction fails for $I = \emptyset$: the colimit of $X \colon \emptyset \to$ Ab is the trivial group $\{0\}$, and is not created by $U \colon$ Ab $\to$ Set.

(d) The construction in (b) works, quite similarly, in many categories of 'algebras', like Mon, Gp, Rng, R Mod, etc. In fact, it works in every variety of algebras, as we shall see in Theorem 4.4.1.

(e) For a functor $X \colon I \to$ Top, one takes the colimit L of the 'underlying functor' $UX \colon I \to$ Set, and endows L with the finest topology making all mappings $u_i \colon X_i \to L$ continuous. But this is nothing new, because one can do the same for any colimit, as we have observed in 2.2.5(b).

(f) One would have a similar situation in the category pOrd of preordered sets (see 2.2.5(c)). For *ordered sets* we come back to something of interest: the forgetful functor $U \colon$ Ord $\to$ Set does not preserve general colimits (see 3.2.3(b)), but does preserve all inductive limits (without creating them).

In fact, for a functor $X \colon I \to$ Ord, one takes the colimit L of the 'underlying functor' $UX \colon I \to$ Set, and endows L with the finest (or 'smallest') ordering making all mappings $u_i \colon X_i \to L$ monotone

$$\xi \leqslant \eta \quad \text{if } \xi = [x, i], \ \eta = [y, i] \ \text{ for some } i \in I \text{ and } x \leqslant y \text{ in } X_i. \quad (2.45)$$

2.6.3 Exercises and complements, II

(a) In Set every object X is the inductive limit of the filtered set $\mathcal{P}_f(X)$ of its finite subsets, ordered by inclusion. More precisely

$$(X, (u_K \colon K \subset X)) = \mathrm{Colim}((K), (u_{KK'} \colon K \subset K')) \quad (2.46)$$

where the family of inclusions $u_{KK'} \colon K \subset K'$ is indexed by the pairs $K \subset K'$ in $\mathcal{P}_f(X)$.

(b) In Ab every object A is the inductive limit of the filtered set $\mathrm{Sub}_{\mathrm{fg}}(A)$ of its *finitely generated* subgroups, ordered by inclusion.

(c) The reader can easily extend this fact to various categories of 'algebras', like Mon, Gp, Rng, R Mod, etc. Or to any variety of algebras (after reading Chapter 4, if needed).

*(d) A compactly generated space is the inductive limit of the filtered set of its compact subspaces. (We recall that a topological space X is said to be *compactly generated* if every subset C that intersects every compact subspace K of X in a set closed in K, is itself closed in X.)

This point is an extension of point (a), viewing a set as a discrete space and a finite subset as a compact subspace.

2.6.4 Filtered colimits

More generally, a *filtered category* S is a (small) *non-empty* category such that (writing $I = \mathrm{Ob}\mathsf{S}$):

(i) for each pair $i, j \in I$ there exist a pair of arrows $a\colon i \to k$, $b\colon j \to k$ in S,

(ii) for each pair of arrows $a, b\colon i \to j$ there exists an arrow $c\colon j \to k$ such that $ca = cb$.

If S is an ordered set, this coincides with the definition of a filtered set, since condition (ii) is trivially satisfied.

A *filtered colimit* is the colimit of a functor $X\colon \mathsf{S} \to \mathsf{C}$ defined on a filtered category. Writing again the functor X in index notation

$$X = ((X_i), (u_a)), \quad u_a\colon X_i \to X_j, \qquad \text{for } i \text{ and } a\colon i \to j \text{ in S}, \quad (2.47)$$

a cocone $(Y, (f_i))$ of X is an object Y equipped of a family of morphisms in C such that

$$f_i\colon X_i \to Y, \quad f_j u_a = f_i\colon X_i \to Y, \qquad \text{for } i \text{ and } a\colon i \to j \text{ in S}. \quad (2.48)$$

The colimit is a universal cocone $(L, (u_i))$: for every cocone $(Y, (f_i))$ there exists a unique $f\colon L \to Y$ in C such that $f_i = fu_i\colon X_i \to Y$, for all $i \in I$.

2.6.5 Exercises and complements

Filtered colimits in a concrete category C are constructed by extending what we have seen in 2.6.2.

(a) Let $X\colon \mathsf{S} \to \mathsf{Set}$ be a functor defined on a filtered category. One forms again the sum $S = \Sigma X_i$, with elements (x, i), for $i \in I$ and $x \in X_i$. The reader will prove that the relation $(x, i) \sim (y, j)$ in S defined by

$$\text{there exist } a\colon i \to k, \; b\colon j \to k \text{ in S such that } u_a(x) = u_b(y), \quad (2.49)$$

is an equivalence relation, and that the colimit of X can be obtained as:

$$L = S/\sim, \qquad u_i\colon X_i \to L, \quad u_i(x) = [x, i]. \quad (2.50)$$

(b) The other constructions of 2.6.2 are extended in the same way.

2.6.6 Theorem

Filtered colimits in Set *commute with finite products, in a sense made precise in the proof.*

Note. This statement is a particular instance of a much more general result: see 2.6.7(b).

Proof We begin by considering binary products. Let $X = ((X_i), (u_a))$ and $Y = ((Y_i), (v_a))$ be two functors $\mathsf{S} \to \mathsf{Set}$ defined on the same filtered category. Denoting by $X \times Y \colon \mathsf{S} \to \mathsf{Set}$ the functor

$$(X \times Y)_i = X_i \times Y_i, \quad w_a = u_a \times v_a \colon X_i \times Y_i \to X_j \times Y_j, \qquad (2.51)$$

we want to prove that $\mathrm{Colim}(X \times Y) \cong \mathrm{Colim} X \times \mathrm{Colim} Y$, by a canonical bijection.

In fact, there is a canonical mapping $(f, g) \colon \mathrm{Colim}(X \times Y) \to \mathrm{Colim} X \times \mathrm{Colim} Y$, with components f, g produced by the cocones

$$f_i = u_i p_i \colon X_i \times Y_i \to X_i \to \mathrm{Colim} X,$$
$$g_i = v_i q_i \colon X_i \times Y_i \to Y_i \to \mathrm{Colim} Y. \qquad (2.52)$$

(The cocone conditions are plainly satisfied: for an arrow $a \colon i \to j$ in S we have: $f_j w_a = u_j p_j w_a = u_j u_a p_i = u_i p_i = f_i$ and, similarly, $g_j w_a = g_i$.)

The mapping (f, g) acts as follows (for $i \in I$ and $(x, y) \in X_i \times Y_i$):

$$(f, g) \colon \mathrm{Colim}(X \times Y) \to \mathrm{Colim} X \times \mathrm{Colim} Y,$$
$$[x, y, i] \mapsto ([x, i], [y, i]). \qquad (2.53)$$

It is surjective, since an arbitrary pair $([x, i], [y, j])$ can be rewritten as $([u_a x, k], [v_b y, k])$, for any pair of arrows $a \colon i \to k$, $b \colon j \to k$ of S. To prove that it is injective, let us suppose that

$$[x, y, i], \ [x', y', j] \in \mathrm{Colim}(X \times Y),$$

$$[x, i] = [x', j] \text{ in } \mathrm{Colim} X, \qquad [y, i] = [y', j] \text{ in } \mathrm{Colim} Y.$$

Then there exists two pairs of arrows $a \colon i \to k$, $b \colon j \to k$ and $c \colon i \to k'$, $d \colon j \to k'$ such that

$$u_a x = u_b x' \text{ in } X_k, \qquad v_c y = v_d y' \text{ in } Y_{k'}.$$

By Lemma 2.6.8 (below) there exist two arrows $r \colon k \to k''$, $s \colon k' \to k''$ such that

$$ra = sc \colon i \to k'', \qquad rb = sd \colon j \to k''. \qquad (2.54)$$

Then

$$u_{ra}x = u_{rb}x' \text{ in } X_{k''}, \qquad v_{sc}y = v_{sd}y' \text{ in } Y_{k''}, \qquad (2.55)$$

which means that $(x,y) \in X_i \times Y_i$ and $(x',y') \in X_j \times Y_j$ coincide in $X_{k''} \times Y_{k''}$, under the arrows $ra = sc \colon i \to k''$ and $rb = sd \colon j \to k''$. Therefore $[x,y,i] = [x',y',j]$.

Finally, the 0-ary power of X is the constant functor $X^0 \colon \mathsf{S} \to \mathsf{Set}$, that sends everything to the terminal object $\top = \{*\}$. Its colimit is $\top = (\mathrm{Colim}X)^0$. □

2.6.7 Exercise and complements

(a) Extend this result to Mon, Gp, Rng, R Mod.

*(b) More generally, filtered colimits commute with finite limits in all varieties of algebras: see [M4, Bo1].

*(c) A category S is filtered if (and only if) every functor $\mathsf{X} \to \mathsf{S}$ defined on a finite category has a cocone. The reader can prove this fact, or find a proof in [Bo1], Lemma 2.13.2. Our Lemma below is a particular case, sufficient for the proof of Theorem 2.6.6.

2.6.8 Lemma

In the filtered category S, *four arrows* a,b,c,d *are given, as in the diagram below. Then there exist two arrows* $r \colon k \to k''$, $s \colon k' \to k''$ *making the diagram commutative*

$$ra = sc,$$

$$(2.56)$$

$$rb = sd.$$

Proof First we have two arrows $m \colon k \to k_1$, $n \colon k' \to k_1$

$$(2.57)$$

We add two arrows u,v such that $uma = unc$ and $vmb = vnd$. Then we add two arrows x, y making the right square commutative (which can be done in two steps, using 2.6.4(i), (ii)).

Now the maps $r = xum: k \to k''$ and $s = yvn: k' \to k''$ satisfy our conditions:

$$ra = xuma = xunc = yvnc = sc, \qquad rb = xumb = yvmb = yvnd = sd.$$

$\square$

2.7 Universal arrows and free objects

Universal arrows yield a general, fundamental way of formalising universal properties in a category X, with respect to a functor $U: \mathsf{A} \to \mathsf{X}$ and an object X of X. Obviously, there are two forms, dual to each other.

In particular, limits, colimits and free objects can be defined as universal arrows.

2.7.1 Universal arrows

A *universal arrow from the object X to the functor U* is a pair

$$(A, \eta: X \to UA),$$

consisting of an object A of A and an arrow η of X which is universal.

This means that every similar pair $(B, f: X \to UB)$ factorises uniquely through (A, η): there exists a unique map $g: A \to B$ in A such that the following triangle commutes in X

$$
\begin{array}{ccc}
X & \xrightarrow{\;\eta\;} & UA \\
 & {\scriptstyle f}\searrow & \downarrow{\scriptstyle Ug} \\
 & & UB
\end{array}
\qquad Ug.\eta = f.
\qquad (2.58)
$$

In other words we are saying that, for every B in A, the following mapping is bijective

$$\mathsf{A}(A, B) \to \mathsf{X}(X, UB), \qquad g \mapsto Ug.\eta. \qquad (2.59)$$

The pair (A, η) is then determined up to isomorphism: if also (B, f) is a solution, the unique A-morphism $g: A \to B$ such that $Ug.\eta = f$ is invertible.

Dually, a *universal arrow from the functor U to the object X* is a pair $(A, \varepsilon: UA \to X)$ consisting of an object A of A and an arrow ε of X such that every similar pair $(B, f: UB \to X)$ factorises uniquely through

(A, ε): there exists a unique $g : B \to A$ in A such that the following triangle commutes in X

$$UA \xrightarrow{\;\varepsilon\;} X$$
$$Ug \uparrow \quad \nearrow_{f}$$
$$UB$$

$$\varepsilon.Ug = f. \qquad (2.60)$$

2.7.2 Limits and colimits as universal arrows

Consider the category C^{S} of functors $\mathsf{S} \to \mathsf{C}$ and their natural transformations, for a small category S (see 1.6.1). The *diagonal functor*

$$D : \mathsf{C} \to \mathsf{C}^{\mathsf{S}}, \qquad (DA)_i = A, \quad (DA)_a = \mathrm{id}A, \qquad (2.61)$$

sends an object A to the constant functor at A, and a morphism $f : A \to B$ to the natural transformation $Df : DA \to DB : \mathsf{S} \to \mathsf{C}$ whose components are constant at f.

The limit of a functor $X : \mathsf{S} \to \mathsf{C}$ in C is the same as a universal arrow $(L, \varepsilon : DL \to X)$ from the functor D to the object X of C^{S} (see 2.7.1). Dually, the colimit of X in C is the same as a universal arrow $(L, \eta : X \to DL)$ from the object X of C^{S} to the functor D.

2.7.3 Free objects

If $U : \mathsf{A} \to \mathsf{Set}$ is a concrete category (see 1.4.8), the *free* A-*object over* the set X is defined as a universal arrow $(A, \eta : X \to UA)$ from the set X to the forgetful functor U. (It may exist or not, of course.)

The mapping $\eta : X \to UA$ can be called *the insertion of the basis*: in fact, the definition means that every mapping $f : X \to U(B)$ with values in the 'underlying set' of any object B in A can be uniquely 'extended' to a morphism $g : A \to B$ in A, such that $U(g).\eta = f$.

The examples below can help to make this clear. In particular, the injectivity of η is examined in 2.7.5.

One can extend this terminology to any functor $U : \mathsf{A} \to \mathsf{X}$ which is *viewed*, in a given context, as a 'forgetful functor', even if it is not faithful.

2.7.4 Exercises and complements, I

(a) A reader which is not familiar with these notions might begin by constructing the universal arrow from a set X to the forgetful functor $U : R\,\mathsf{Mod} \to \mathsf{Set}$, using the free left module generated by X

$$A = \bigoplus_{x \in X} R. \qquad (2.62)$$

This is a direct sum in $R\,\mathsf{Mod}$ of copies of the left module R (see 2.1.6(c)), so that an element of $F(X)$ is a *quasi-null* family $(\lambda_x)_{x \in X}$ of scalars $\lambda_x \in R$. The embedding

$$\eta \colon X \to UA, \qquad (\eta x)_x = 1, \quad (\eta x)_y = 0 \qquad (y \neq x), \qquad (2.63)$$

sends each element of X to the corresponding Kronecker family. Then every mapping $f \colon X \to U(B)$ with values in a left R-module can be uniquely extended to a homomorphism $g \colon A \to B$ in $R\,\mathsf{Mod}$, by linear combinations in B (with quasi-null scalar coefficients)

$$g((\lambda_x)_{x \in X}) = \Sigma_{x \in X}\, \lambda_x . f(x). \qquad (2.64)$$

Usually one identifies an element $x \in X$ with the family $\eta(x) \in A$, so that the quasi-null family $(\lambda_x) \in A$ can be uniquely rewritten as a *formal linear combination* $\Sigma\, \lambda_x . x$ of elements of the basis, with quasi-null coefficients in R.

(b) Similarly we can form the universal arrow from a set X to the forgetful functor $U \colon \mathsf{Mon} \to \mathsf{Set}$ using the free monoid $A = \Sigma_{n \in \mathbb{N}}\, X^n$ of *finite words* on the alphabet X. The reader can easily work out the details (or find most of them in 3.6.1).

Many other exercises of this kind will be listed and commented in Section 4.1.

(c) In a group G, the *subgroup of commutators* is the subgroup generated by all commutators

$$[x, y] = xyx^{-1}y^{-1},$$

for $x, y \in G$. Plainly, G is commutative if and only if this subgroup is trivial.

The reader will prove that this subgroup is necessarily normal in G, and that the canonical projection on the *abelianised* quotient group G^{ab}

$$G^{\mathrm{ab}} = G/[G, G], \qquad \eta \colon G \to G^{\mathrm{ab}}, \qquad (2.65)$$

gives the universal arrow from the group G to the embedding $U \colon \mathsf{Ab} \to \mathsf{Gp}$.

(d) A functor $U \colon \mathsf{A} \to \mathsf{Set}$ is representable (see 1.6.4) if and only if the singleton $\{*\}$ has a free object A_0 in A, so that

$$U(A) \cong \mathsf{Set}(\{*\}, U(A)) \cong \mathsf{A}(A_0, A). \qquad (2.66)$$

This property often holds in a concrete category, and ensures – by 2.3.5(a) – that U *preserves monomorphisms* (and reflects them, if it is faithful).

2.7.5 Exercises and complements, II

Let us come back to a general universal arrow $(A, \eta: X \to UA)$, for a functor $U: \mathsf{A} \to \mathsf{X}$.

(a) Prove that the arrow η is a monomorphism of X if and only if there exists a monomorphism $f: X \to UB$, for some object B in A.

(b) This condition obviously fails for the embedding $U: \mathsf{Ab} \to \mathsf{Gp}$, whenever X is a non-commutative group (see 2.7.4(c)).

(c) The forgetful functor $U: \mathsf{A} \to \mathsf{Set}$ of a concrete category 'usually' satisfies this condition, so that the insertion of the basis $\eta: X \to UA$ (if extant) is indeed an *embedding*.

For varieties of algebras it is always the case, with the exception of two 'trivial varieties', as we shall see in Corollary 4.4.3.

2.7.6 Exercises and complements, III

(a) Universal arrows compose: for a composed functor $UV: \mathsf{B} \to \mathsf{A} \to \mathsf{X}$, given

- a universal arrow $(A, \eta: X \to UA)$ from an object X to U,
- a universal arrow $(B, \zeta: A \to VB)$ from the previous object A to V,

we have a universal arrow from X to UV constructed as follows

$$(B, U\zeta.\eta: X \to UA \to UV(B)). \qquad (2.67)$$

For instance, the free ring on a set can be easily constructed in two steps, along the composed forgetful functor $\mathsf{Rng} \to \mathsf{Mon} \to \mathsf{Set}$ (as will be examined in 4.1.6(b)).

(b) Universal arrows for 'two-dimensional categories' will be considered in 7.1.5.

2.8 *Other universals

This miscellaneous section is about three kinds of universal solutions: the splitting of idempotents, Kan extensions and coends. These issues, even though important in category theory, will have a limited use in this book and can be skipped until needed.

2.8.1 Absolute limits

The limit $(L, (u_i: L \to X_i))$ of a functor $X: \mathsf{S} \to \mathsf{C}$ is said to be an *absolute limit* if it is preserved (as a limit) by every functor $\mathsf{C} \to \mathsf{D}$, with values in an arbitrary category.

Let us begin by noting that there are some trivial limits which always exist and are preserved by any functor, like unary products (with $S = 1$).

More generally, every constant functor $X \colon S \to C$ (with $S \neq 0$) has an absolute limit.

As another extension, if S has an initial object h any functor $X \colon S \to C$ has absolute limit X_h, with structural map $u_i \colon X_h \to X_i$ given by the unique arrow $h \to i$ of S. (This also holds for a large category S.)

But there is an interesting, non trivial case.

2.8.2 Exercises and complements (The splitting of idempotents)

As we have seen in 1.4.9(d), an idempotent endomorphism $e \colon X \to X$ can be equivalently described as a functor $F \colon E \to C$ defined on the idempotent monoid $E = \{1, e\}$.

(a) The reader should find interesting (and easy) to prove that the following conditions are equivalent:

(i) the idempotent endomorphism e *splits* in C, i.e. it has a factorisation $e = mp \colon X \to A \to X$, with $pm = 1_A$ (so that m is a split mono and p is a split epi),

(ii) the idempotent endomorphism e has an epi-mono factorisation $e = mp$,

(iii) the functor F has a limit $m \colon A \to X$,

(iv) the pair $1, e \colon X \to X$ has an equaliser $m \colon A \to X$,

(iii*) the functor F has a colimit $p \colon X \to A$,

(iv*) the pair $1, e \colon X \to X$ has a coequaliser $p \colon X \to A$.

(b) Now, every functor defined on C plainly preserves the existing splittings of idempotents, which are thus absolute limits and colimits.

(c) After biproducts (in 2.1.9, 2.2.6) this is another case where *the limit and colimit objects coincide*; here the coincidence is even necessary, as soon as our (co)limit exists. Again (and of course) the limit-cone $m \colon A \to X$ and the colimit-cocone $p \colon X \to A$ are distinct.

(d) All idempotents split in every category with equalisers (or coequalisers). But we shall see in 6.6.2(g) that this also holds in the category Rel Ab, which lacks general equalisers.

2.8.3 The idempotent completion

Every category C has a diagonal embedding (see 2.7.2)

$$D \colon C \to C^E, \qquad C^E = \mathsf{Cat}(E, C) = \mathsf{Idm}(C), \tag{2.68}$$

with values in the functor-category C^E, which is called the *idempotent completion* of C, as we motivate now.

As we know, an object $F\colon E \to C$ amounts to an idempotent endomorphism e of C, and will be written thus. A morphism of C^E

$$(e, f, e')\colon e \to e'$$

(where $e\colon X \to X$ and $e'\colon Y \to Y$) is determined by a morphism $f\colon X \to Y$ of C such that $fe = f = e'f$. The composition is induced by that of C

$$(e', g, e'').(e, f, e') = (e, gf, e'')\colon e \to e'', \qquad \mathrm{id}(e) = (e, e, e). \qquad (2.69)$$

The embedding $D\colon C \to \mathrm{Idm}(C)$ sends X to 1_X and $f\colon X \to Y$ to $(1_X, f, 1_Y)\colon 1_X \to 1_Y$.

In $\mathrm{Idm}(C)$ *all idempotents split*: in fact an idempotent $(e, f, e)\colon e \to e$ comes from a morphism

$$f\colon X \to X, \qquad fe = f = ef, \quad ff = f, \qquad (2.70)$$

and splits in $\mathrm{Idm}(C)$ as the composite

$$(e, f, e) = (f, f, e).(e, f, f)\colon e \to f \to e, \qquad (e, f, f).(f, f, e) = \mathrm{id}(f).$$

* The *weak universal property* of $\mathrm{Idm}(C)$ (a biuniversal property in Cat, see 7.1.5) is slightly more complicated to prove: every functor $F\colon C \to D$ with values in a category where all idempotents split can be extended to a functor $G\colon \mathrm{Idm}(C) \to D$ (so that $GD = F$), and the extension is determined up to functorial isomorphism.*

2.8.4 Kan extensions

Given two functors $F\colon S \to C$ and $T\colon S \to T$ with the same domain, a *right Kan extension* of F along T is a pair (R, ε) formed of a functor $R\colon T \to C$ and a natural transformation $\varepsilon\colon RT \to F$

$$ (2.71) $$

which is universal: for every functor $R'\colon T \to C$ and natural transformation $\varepsilon'\colon R'T \to F$ there is a unique natural transformation $\rho\colon R' \to R$ such that $\varepsilon' = \varepsilon.\rho T$.

The solution, written as $R = \mathrm{Ran}_T F$, is determined up to a (unique) natural isomorphism. In fact, leaving apart problems of smallness, the property above means that $(R, \varepsilon: RT \to F)$ is a universal arrow from the functor $C^T = -.T$ to the object F of the category C^S

$$C^T : C^T \to C^S, \qquad R \mapsto RT, \quad \rho \mapsto \rho T. \tag{2.72}$$

Dually (by reversing natural transformations), the *left Kan extension* $L = \operatorname{Lan}_T F \colon T \to C$ is a functor equipped with a natural transformation $\eta \colon F \to LT$, such that every similar pair $(L', \eta' \colon F \to L'T)$

$$\tag{2.73}$$

factorises as $\eta' = \lambda T.\eta$, by a unique natural transformation $\lambda \colon L \to L'$.

2.8.5 Limits and colimits as Kan extensions

In particular, the limit of a functor $X \colon S \to C$ amounts to the right Kan extension $R = \operatorname{Ran}_T X$ along the functor $T \colon S \to \mathbf{1}$ (with values in the terminal category)

$$\tag{2.74}$$

while the colimit of $X \colon S \to C$ amounts to the left Kan extension $L = \operatorname{Lan}_T \check{X} \colon \mathbf{1} \to C$.

2.8.6 Coends

Let us be given a functor $F \colon X^{\mathrm{op}} \times X \to C$. Its *coend* is an object C of C equipped with a family of morphisms $u_X \colon F(X, X) \to C$ indexed by the objects of X. This family must be:

(i) *coherent*, in the sense that for every morphism $f \colon X \to Y$ in X we have $u_X.F(f, 1) = u_Y.F(1, f) \colon F(Y, X) \to C$,

(ii) *universal*, in the sense that every coherent family $(v_X \colon F(X, X) \to C')_X$ factorises uniquely through the former: there exists precisely one morphism $c \colon C \to C'$ such that $u_X.c = v_X$, for all X.

The coend of F is written as $\int^X F(X,X)$.

If X is small and C is cocomplete, the coend of F exists and can be obtained as the coequalizer of the two morphisms

$$a,b\colon \textstyle\sum_{f\colon X\to Y} F(Y,X) \rightrightarrows \textstyle\sum_X F(X,X),$$

whose co-component on the morphism $f\colon X \to Y$ comes, respectively, from

$$a_f = F(f,1)\colon F(Y,X) \to F(X,X),$$
$$b_f = F(1,f)\colon F(Y,X) \to F(Y,Y).$$

(Dually, *ends* can be constructed by means of limits.)

A classical example from Algebraic Topology, the geometric realisation of simplicial sets and cubical sets, will be described in 5.3.5 and 5.4.6.

3

Adjunctions and monads

Adjunctions, a crucial step in category theory, were introduced by Kan in 1958 [K2]. They extend Galois connections (defined in Section 1.7) from ordered sets to general categories. The closely related notion of 'monad' is also studied here.

The Adjoint Functor Theorem of Freyd, dealt with in Section 3.5, is an effective tool in order to prove the existence of an adjoint, which might be difficult to construct.

Many concrete examples can be found in this chapter, then in Chapter 4 (for algebraic structures) and Chapter 5 (for topological structures).

3.1 Adjoint functors

We begin by two well known situations, which help us to introduce the adjunction of functors. Then we give a definition, in four equivalent forms that extend those of Galois connections, in 1.7.1.

3.1.1 Examples

(a) The forgetful functor $U\colon \mathsf{Top} \to \mathsf{Set}$ has two well known 'best approximations' to an inverse which does not exist, the functors $D, C\colon \mathsf{Set} \to \mathsf{Top}$, where DX (resp. CX) is the set X with the discrete (resp. codiscrete) topology. In fact they provide the set X with the finest (resp. the coarsest) topology.

For every set X and every space Y we can 'identify' the following hom-sets

$$\mathsf{Top}(D(X), Y) = \mathsf{Set}(X, U(Y)),$$
$$\mathsf{Set}(U(Y), X) = \mathsf{Top}(Y, C(X)),$$

$$(3.1)$$

since every mapping $X \to Y$ becomes continuous if we put on X the discrete topology, and every mapping $Y \to X$ becomes continuous if we put on X the codiscrete one.

These two facts will tell us that $D \dashv U$ and $U \dashv C$, respectively: D is left adjoint to U, and C is right adjoint to the latter. This presentation of the two adjunctions corresponds to form (i) of a Galois connection $f \dashv g$, in 1.7.1, which says that the hom-sets $Y(f(x), y)$ and $X(x, g(y))$ have the same cardinal: either 0 or 1.

(b) In other cases, it may be convenient to follow an approach corresponding to the form (ii) of 1.7.1, and based on universal arrows (defined in Section 2.7).

To wit, let us start from the forgetful functor $U \colon R\,\mathsf{Mod} \to \mathsf{Set}$. We have already seen, in 2.7.4(a), that, for every set X, there is a universal arrow from X to the functor U

$$F(X) = \bigoplus_{x \in X} R, \qquad \eta \colon X \to UF(X), \tag{3.2}$$

consisting of the free R-module $F(X)$ on the set X, with the insertion of the basis η.

The fact that this universal arrow exists *for every X in* Set allows us to construct a 'backward' functor $F \colon \mathsf{Set} \to R\,\mathsf{Mod}$, *left adjoint* to U.

F is already defined on the objects. For a mapping $f \colon X \to Y$ in Set

$$
\begin{array}{ccc}
X \xrightarrow{\;\eta X\;} UF(X) & \qquad & F(X) \\
\left\downarrow{\scriptstyle f}\right. \qquad \left\downarrow{\scriptstyle U(Ff)}\right. & & \left\downarrow{\scriptstyle Ff}\right. \\
Y \xrightarrow[\;\eta Y\;]{} UF(Y) & & F(Y)
\end{array}
\tag{3.3}
$$

the universal property of ηX implies that there is precisely one homomorphism $F(f) \colon F(X) \to F(Y)$ such that $U(F(f))$ makes the square above commutative.

(Concretely, $F(f)$ is just the R-linear extension of f; but the general pattern is better perceived if we proceed in a formal way.)

Further applications of the universal property would show that F preserves composition and identities (but we shall see this in a general context). Let us note that we have extended the given object-function $F(X)$ in the only way that makes the family (η_X) into a natural transformation $\eta \colon 1 \to UF \colon \mathsf{Set} \to \mathsf{Set}$, called the *unit* of the adjunction.

(c) The relationship between these functors U and F (assuming now that we have defined both) can be equivalently expressed in form (i), by a family of bijective mappings (for X in Set and A in $R\,\mathsf{Mod}$)

$$\varphi_{XA} \colon R\,\mathsf{Mod}(FX, A) \to \mathsf{Set}(X, UA), \tag{3.4}$$

which is natural in X and A, as will be made precise in 3.1.2.

Concretely, $\varphi_{XA}(g\colon FX \to A)$ is the restriction of the homomorphism g to the basis X of $F(X)$; its bijectivity means that every mapping $f\colon X \to UA$ has a unique extension to a homomorphism $g\colon FX \to A$.

Formally, given the family (ηX), we define

$$\varphi_{XA}(g\colon FX \to A) = Ug.\eta X\colon X \to UA.$$

The other way round, given the family (φ_{XA}), we define

$$\eta X = \varphi_{X,FX}(\mathrm{id}FX)\colon X \to UFX.$$

3.1.2 Definition

An *adjunction* $F \dashv G$, with a functor $F\colon \mathsf{C} \to \mathsf{D}$ *left adjoint* to a functor $G\colon \mathsf{D} \to \mathsf{C}$, can be equivalently presented in four main forms.

(i) We assign two functors $F\colon \mathsf{C} \to \mathsf{D}$ and $G\colon \mathsf{D} \to \mathsf{C}$ together with a family of bijections

$$\varphi_{XY}\colon \mathsf{D}(FX,Y) \to \mathsf{C}(X,GY) \qquad (X \text{ in } \mathsf{C}, Y \text{ in } \mathsf{D}), \qquad (3.5)$$

which is natural in X, Y. More formally, the family (φ_{XY}) is a functorial isomorphism

$$\varphi\colon \mathsf{D}(F(-),=) \to \mathsf{C}(-,G(=))\colon \mathsf{C}^{\mathrm{op}} \times \mathsf{D} \to \mathsf{Set}. \qquad (3.6)$$

(ii) We assign a functor $G\colon \mathsf{D} \to \mathsf{C}$ and, for every object X in C, a *universal* arrow (see 2.7.1)

$$(F_0X, \eta X\colon X \to GF_0X) \qquad \text{from the object } X \text{ to the functor } G.$$

(ii*) We assign a functor $F\colon \mathsf{C} \to \mathsf{D}$ and, for every object Y in D, a *universal* arrow

$$(G_0Y, \varepsilon Y\colon FG_0Y \to Y) \qquad \text{from the functor } F \text{ to the object } Y.$$

(iii) We assign two functors $F\colon \mathsf{C} \to \mathsf{D}$ and $G\colon \mathsf{D} \to \mathsf{C}$, together with two natural transformations

$$\eta\colon \mathrm{id}\mathsf{C} \to GF \quad (\text{the } unit), \qquad \varepsilon\colon FG \to \mathrm{id}\mathsf{D} \quad (\text{the } counit),$$

which satisfy the *triangular identities*:

$$\varepsilon F.F\eta = \mathrm{id}F, \qquad\qquad G\varepsilon.\eta G = \mathrm{id}G \qquad (3.7)$$

$$F \xrightarrow{F\eta} FGF \xrightarrow{\varepsilon F} F \qquad\qquad G \xrightarrow{\eta G} GFG \xrightarrow{G\varepsilon} G$$
$$\underset{\mathrm{id}F}{\longrightarrow} \qquad\qquad\qquad\qquad \underset{\mathrm{id}G}{\longrightarrow}$$

The proof of the equivalence will be given in Theorem 3.1.5. Essentially:

- given (i) one defines

$$\eta X = \varphi_{X,FX}(\mathrm{id}FX)\colon X \to GFX,$$

$$\varepsilon Y = (\varphi_{GY,Y})^{-1}(\mathrm{id}GY)\colon FGY \to Y,$$

- given (ii) one defines $F(X) = F_0 X$, the morphism $F(f\colon X \to X')$ by the universal property of ηX and the morphism $\varphi_{XY}(g\colon FX \to Y)$ as $Gg.\eta X\colon X \to GY$,

- given (iii) one defines the mapping $\varphi_{XY}\colon \mathsf{D}(F(X),Y) \to \mathsf{C}(X,G(Y))$ as above, and a backward mapping $\psi_{XY}(f\colon X \to GY) = \varepsilon Y.Ff$ (which are inverse to each other by the triangular identities).

Finally, the naturality of φ in (3.6) means that, for every pair of morphisms $u\colon X' \to X$ and $v\colon Y \to Y'$ (in C and D, respectively) we have a commutative square in Set

$$Gv.\varphi_{XY}(g).u = \varphi_{X'Y'}(v.g.Fu) \tag{3.8}$$

$$
\begin{array}{ccc}
\mathsf{D}(F(X),Y) & \xrightarrow{\ \varphi_{XY}\ } & \mathsf{C}(X,G(Y)) \\
{\scriptstyle v.-.Fu}\downarrow & & \downarrow{\scriptstyle Gv.-.u} \\
\mathsf{D}(F(X'),Y') & \xrightarrow[\varphi_{X'Y'}]{} & \mathsf{C}(X',G(Y'))
\end{array}
$$

3.1.3 Comments and complements

(a) The previous forms have different features.

Form (i) is the classical definition of an adjunction, and is at the origin of the name, by analogy with adjoint maps of Hilbert spaces.

Form (ii) is used when one starts from a 'known' functor G and wants to construct its left adjoint, possibly less easy to define. Form (ii*) is used in a dual way.

The 'algebraic' form (iii) is free of quantifiers and adequate to the formal theory of adjunctions (because it makes sense in an abstract 2-category, see 7.1.2).

*(b) Adjunctions are related to Kan extensions in a complex way. The reader can see [M4], Section X.7.

3.1.4 Exercises and complements

(a) An adjoint equivalence (F,G,η,ε), defined in 1.5.5, amounts to an adjunction where the unit and counit are invertible; then $F \dashv G \dashv F$.

(b) Various examples of Galois connections have ben discussed in Section 1.7.

(c) For a 'forgetful functor' $U: A \to X$ the existence of the left adjoint $F \dashv U$ is equivalent to say that every object X of X has a free object $(FX, \eta: X \to UFX)$ in A (as defined in 2.7.3).

We shall see, in Theorem 4.4.2, that the forgetful functor $U: A \to Set$ of any variety of algebras has always a left adjoint $F: Set \to A$, the *free-algebra functor* for A. In many cases there are simple constructions for F; we have seen a few cases in 2.7.4 and 3.1.1; many others will be analysed in Section 4.1.

(d) We have seen that the forgetful functor $U: Top \to Set$ has both adjoints $D \dashv U \dashv C$, where DX (resp. CX) is the set X with the discrete (resp. codiscrete) topology. Let us note that U has no left adjoint, while C has no right adjoint, as will be proved in 3.2.3(a). (In Section 3.8 we shall see that these adjunctions are idempotent.)

Many other adjoints of interest in Topology will be examined in Chapter 5.

(e) The forgetful functor $U: pOrd \to Set$ of the category of preordered sets has similar adjoints $D \dashv U \dashv C$, where DX is the set X with the discrete order $x = x'$, while CX is the same set with the codiscrete or chaotic preorder, namely $x, x' \in X$.

Here D has also a left adjoint $\pi_0: pOrd \to Set$, where $\pi_0(X)$ is the *set of connected components* of the preordered set X, namely the quotient of the underlying set $|X|$ modulo the equivalence relation generated by the preorder $x \prec x'$ (i.e. the least equivalence relation containing the former).

(f) Prove that the embedding $J: pOrd \to Cat$ of preorderd sets into small categories (see 1.2.3) has a left adjoint $po: Cat \to pOrd$, using the preordered set $po(C)$ defined in 1.4.5. Prove that J does not preserve co-equalisers, while po does not preserve equalisers.

(g) Starting from the (non-faithful) functor $Ob: Cat \to Set$, there is a chain of adjunctions

$$\pi_0 \dashv D \dashv Ob \dashv C,$$

which extends that of point (e).

In fact, by composing the adjunctions in (e) and (f), we get $\pi_0(C)$ ap-plying the functor $\pi_0: pOrd \to Set$ to the preordered set $po(C)$.

It may be more interesting to give a direct construction. First, we say that a category C is *connected* if it is not empty, and every pair of objects is linked by a 'zig-zag path of arrows'

$$\bullet \longrightarrow \bullet \longleftarrow \bullet \longrightarrow \bullet \ldots \bullet \longrightarrow \bullet$$

Now $\pi_0(\mathsf{C})$ can be defined as the set of the connected components of C, that is its maximal connected subcategories. The empty category has none, and a category is connected if and only if $\pi_0(\mathsf{C})$ is the singleton. Every category is the categorical sum of its connected components.

3.1.5   Theorem (Characterisation of adjoints)

The presentations (i)–(iii) of an adjunction in Definition 3.1.2 are equivalent.

Proof (i) $\Rightarrow$ (ii). Given (i), one defines $\eta X = \varphi_{X,FX}(\mathrm{id}FX)\colon X \to GFX$. The naturality of φ_{XY} on $\mathrm{id}X$ and $g\colon FX \to Y$ gives

$$\varphi_{XY}(g) = G(g).\varphi_{X,FX}(\mathrm{id}FX) = G(g).\eta X.$$

The bijectivity of φ_{XY} gives the universal property of ηX: for every $f\colon X \to GY$ in C there exists precisely one $g\colon FX \to Y$ in D such that $G(g).\eta X = f$.

(ii) $\Rightarrow$ (i). Conversely, given (ii), one defines $F(X) = F_0 X$. Then the morphism $F(f\colon X \to X')$ is defined by the universal property of ηX, as the unique morphism of D such that

$$
\begin{array}{ccc}
X & \xrightarrow{\ \eta X\ } & GF(X) \\
{\scriptstyle f}\downarrow & & \downarrow{\scriptstyle G(Ff)} \\
X' & \xrightarrow[\ \eta X'\ ]{} & GF(X')
\end{array}
\qquad G(Ff).\eta X = \eta X'.f. \qquad (3.9)
$$

The reader will verify that G is a functor, using the following diagrams

$$
\begin{array}{ccc}
X & \xrightarrow{\ \eta X\ } & GF(X) \\
{\scriptstyle f}\downarrow & & \downarrow{\scriptstyle G(Ff)} \\
X' & \xrightarrow{\ \eta X'\ } & GF(X') \\
{\scriptstyle f'}\downarrow & & \downarrow{\scriptstyle G(Ff')} \\
X'' & \xrightarrow[\ \eta X''\ ]{} & GF(X'')
\end{array}
\qquad
\begin{array}{ccc}
X & \xrightarrow{\ \eta X\ } & GF(X) \\
{\scriptstyle 1}\downarrow & & \downarrow{\scriptstyle G(F1_X)} \\
X & \xrightarrow[\ \eta X\ ]{} & GF(X)
\end{array}
\qquad (3.10)
$$

Finally we define $\varphi_{XY}(g\colon FX \to Y) = Gg.\eta X\colon X \to GY$. It is a bijective mapping, because of the universal property of ηX. The commutativity of the square (3.8) follows easily:

$$Gv.\varphi_{XY}(g).u = Gv.Gg.\eta X.u = Gv.Gg.GFu.\eta X'$$

$$= G(v.g.Fu).\eta X' = \varphi_{X'Y'}(v.g.Fu).$$

(i) $\Rightarrow$ (iii). Given (i), one defines

$$\eta X = \varphi_{X,FX}(\mathrm{id}FX)\colon X \to GFX,$$
$$\varepsilon Y = (\varphi_{GY,Y})^{-1}(\mathrm{id}GY)\colon FGY \to Y.$$

The naturality of both families is easily verified. As to the first triangular identity $\varepsilon F.F\eta = \mathrm{id}F$, we have:

$$\varphi_{X,FX}(\varepsilon FX.F\eta X) = \varphi_{X,FX}(\varepsilon FX).\eta X = \mathrm{id}(GFX).\eta X$$
$$= \eta X = \varphi_{X,FX}(\mathrm{id}FX),$$

whence $\varepsilon FX.F\eta X = \mathrm{id}FX$, for every X in C.

(iii) $\Rightarrow$ (i). Given (iii), one defines the mappings

$$\varphi_{XY}\colon \mathsf{D}(F(X),Y) \to \mathsf{C}(X,G(Y)), \qquad \varphi_{XY}(g) = Gg.\eta,$$
$$\psi_{XY}\colon \mathsf{C}(X,G(Y)) \to \mathsf{D}(F(X),Y), \qquad \psi_{XY}(f) = \varepsilon Y.Ff,$$

and proves that they are inverse to each other by the triangular identities.

The naturality of φ has already been proved in the point (ii) $\Rightarrow$ (i) (using the commutativity of square (3.9), which here comes out of the naturality of η, while in the previous point was a consequence of the definition of $F(f)$). $\qquad\square$

3.1.6 Theorem (Uniqueness)

Given a functor $G\colon \mathsf{D} \to \mathsf{C}$, its left adjoint is uniquely determined up to isomorphism.

More precisely, we suppose that there are two left adjoints $F\colon \mathsf{C} \to \mathsf{D}$ and $F'\colon \mathsf{C} \to \mathsf{D}$, with units $\eta\colon 1 \to GF$ and $\eta'\colon 1 \to GF'$. Then there is precisely one natural transformation $\varphi\colon F \to F'$ such that $G\varphi.\eta = \eta'$, and it is a functorial isomorphism.

Proof Let X be an object of C. By the uniqueness property of universal arrows, in 2.7.1, there is precisely one morphism $\varphi X\colon F(X) \to F'(X)$ in D such that the following square commutes

$$\begin{array}{ccc} X & \xrightarrow{\;\eta X\;} & GF(X) \\ {\scriptstyle 1}\downarrow & & \downarrow{\scriptstyle G\varphi X} \\ X & \xrightarrow{\;\eta'X\;} & GF'(X) \end{array} \qquad (3.11)$$

and φX is invertible. To prove that the family of all φX is natural, let us take a morphism $f \colon X \to X'$ in C. In the left diagram below, the left square and the outer rectangle commute, by naturality of η and η', respectively

$$
\begin{array}{ccccc}
X & \xrightarrow{\ \eta X\ } & GF(X) & \xrightarrow{\ G(\varphi X)\ } & GF'(X) \\
f\downarrow & = & \downarrow{\scriptstyle GFf} & ? & \downarrow{\scriptstyle GF'f} \\
X' & \xrightarrow[\ \eta X'\ ]{} & GF(X') & \xrightarrow[\ G(\varphi X')\ ]{} & GF'(X')
\end{array}
\tag{3.12}
$$

Therefore

$$
G(F'f.\varphi X).\eta X = G(\varphi X').\eta X'.f = G(\varphi X'.Ff).\eta X,
$$

and the naturality of φ follows from the universal property of ηX:

$$
F'f.\varphi X = \varphi X'.Ff.
$$

$\square$

3.2 Properties of adjunctions

This section takes on the general study of adjoint functors.

3.2.1 A category of adjunctions

As a consequence of the composition of universal arrows (in 2.7.6), two consecutive adjunctions

$$
\begin{aligned}
F \colon \mathsf{C} \rightleftarrows \mathsf{D} \colon G, && \eta \colon 1 \to GF, && \varepsilon \colon FG \to 1, \\
H \colon \mathsf{D} \rightleftarrows \mathsf{E} \colon K, && \rho \colon 1 \to KH, && \sigma \colon HK \to 1,
\end{aligned}
\tag{3.13}
$$

give a composed adjunction from the first to the third category

$$
HF \colon \mathsf{C} \rightleftarrows \mathsf{E} \colon GK,
$$

$$
G\rho F.\eta \colon 1 \to GF \to GK.HF,
\tag{3.14}
$$

$$
\sigma.H\varepsilon K \colon HF.GK \to HK \to 1.
$$

There is thus a category AdjCat of small categories and adjunctions, with morphisms written in one of the following forms, and *conventionally directed as the left adjoint*

$$
(F, G, \eta, \varepsilon) \colon \mathsf{C} \to \mathsf{D}, \qquad F \dashv G \colon \mathsf{C} \to \mathsf{D}.
$$

We also write $(\eta, \varepsilon) \colon F \dashv G$, but one should not view this as an arrow from F to G in some category.

Duality of categories interchanges left and right adjoint, unit and counit. AdjCat is thus a category with involution (as defined in 1.8.1)

$$\mathsf{C} \mapsto \mathsf{C}^{\mathrm{op}},$$

$$((F, G, \eta, \varepsilon) \colon \mathsf{C} \to \mathsf{D}) \mapsto ((G^{\mathrm{op}}, F^{\mathrm{op}}, \varepsilon^{\mathrm{op}}, \eta^{\mathrm{op}}) \colon \mathsf{D}^{\mathrm{op}} \to \mathsf{C}^{\mathrm{op}}).$$ (3.15)

3.2.2 The preservation of (co)limits

As another consequence of the composition of universal arrows, a left adjoint preserves (the existing) colimits, while a right adjoint preserves (the existing) limits. It follows that a left adjoint preserves epimorphisms and a right adjoint preserves monomorphisms (by 2.3.5(a)).

For a covariant Galois connection $f \dashv g$, this amounts to saying that f preserves the existing joins and g the existing meets, as already shown in 1.7.3.

As we know, an equivalence of categories $F \colon \mathsf{C} \to \mathsf{D}$ can always be completed to an adjoint equivalence, where $F \dashv G \dashv F$; therefore F preserves limits and colimits.

The standard way of proving that a functor cannot have a left (resp. a right) adjoint is to find a limit (resp. a colimit) that it does not preserve.

3.2.3 Exercises and complements

(a) We have seen that the forgetful functor $U \colon \mathsf{Top} \to \mathsf{Set}$ has both adjoints $D \dashv U \dashv C$. Accordingly U preserves all limits and colimits.

On the other hand D does not preserve infinite products and cannot have a left adjoint, while C does not preserve (even binary) sums and cannot have a right adjoint.

(b) *Colimits of ordered sets.* The embedding $U \colon \mathsf{Ord} \to \mathsf{pOrd}$ has a left adjoint F, that takes every preordered set X to the quotient set $|X|/\sim$, equipped with the induced order: $[x] \leqslant [y]$ if $x \prec y$ in X (as in 1.2.2).

This is of help in constructing a colimit in Ord, as a quotient of the corresponding colimit in pOrd.

Namely, given a functor $X \colon \mathsf{S} \to \mathsf{Ord}$, the colimit of UX in pOrd is easily obtained as the colimit $(L, (u_i \colon |X_i| \to L))$ in Set, equipped with the finest preordering that makes all mappings u_i monotone (as we have already seen in 2.2.5(c)). Then the colimit in Ord is the associated ordered set $L/\sim$, or more precisely the cocone $(F(L), (Fu_i \colon X_i \to FL))$ produced by the left adjoint $F \colon \mathsf{pOrd} \to \mathsf{Ord}$.

Colimits of Hausdorff spaces can be computed in a similar way, as quotients of colimits in Top (see 5.1.4(b)).

3.2.4 *Limit functors as right adjoints*

The description of the limit of a functor $X: \mathsf{S} \to \mathsf{C}$ as a universal arrow $(L(X), \varepsilon X: DL(X) \to X)$ from the diagonal functor $D: \mathsf{C} \to \mathsf{C}^{\mathsf{S}}$ to the object X of C^{S} (in 2.7.2) shows that the existence of all S-based limits in C amounts to the existence of a right adjoint $L: \mathsf{C}^{\mathsf{S}} \to \mathsf{C}$ (the *limit functor*) for D, with counit ε. The limit cone has components $(\varepsilon X)_i: L(X) \to X_i$, for $i \in \mathrm{Ob}\mathsf{S}$.

Dually, a *colimit functor* $L': \mathsf{C}^{\mathsf{S}} \to \mathsf{C}$ is left adjoint to D, with a unit $\eta X: X \to DL'(X)$ and cocone-components $(\eta X)_i: X_i \to L'(X)$.

3.2.5 *Theorem* (Faithful and full adjoints)

Suppose we have an adjunction $F \dashv G$, with counit $\varepsilon: FG \to 1$.

(i) G is faithful if and only if all the components εY of the counit are epi,

(ii) G is full if and only if all the components εY of the counit are split mono,

(iii) G is full and faithful if and only if the counit ε is invertible.

Proof (i) This fact is easily proved, using for both implications the following diagram of two commutative squares, produced by two maps $g, g': Y \to Y'$ in D

$$
\begin{array}{ccc}
FG(Y) & \xrightarrow{\;\varepsilon Y\;} & Y \\
{\scriptstyle FGg}\big\downarrow\big\downarrow{\scriptstyle FGg'} & & {\scriptstyle g}\big\downarrow\big\downarrow{\scriptstyle g'} \\
FG(Y') & \xrightarrow[\;\varepsilon Y'\;]{} & Y'
\end{array}
\qquad (3.16)
$$

(ii) If εY is a split mono, there is some $p: Y \to FGY$ in D such that $p.\varepsilon Y = 1_{FGY}$. For a morphism $f: GY \to GY'$ in C, we let

$$ g = \varepsilon Y'.Ff.p: Y \to Y'. $$

Then the two squares of the diagram below commute, and $G(g) = f$ by the universal property of $\varepsilon Y'$

$$
\begin{array}{ccc}
FG(Y) & \xrightarrow{\;\varepsilon Y\;} & Y \\
{\scriptstyle FGg}\big\downarrow\big\downarrow{\scriptstyle Ff} & & \big\downarrow{\scriptstyle g} \\
FG(Y') & \xrightarrow[\;\varepsilon Y'\;]{} & Y'
\end{array}
\qquad (3.17)
$$

Conversely, we suppose that G is full. Then there exists some $p: Y \to FGY$ in D such that $G(p) = \eta GY: GY \to GFGY$. To prove that $v =$

$p.\varepsilon Y$ coincides with 1_{FGY} we apply the universal property of ηGY in the following diagram

$$
\begin{array}{ccc}
G(Y) & \xrightarrow{\ \eta GY\ } & GFG(Y) \\
{\scriptstyle 1}\downarrow & & \scriptstyle{G(1)}\downarrow\downarrow\scriptstyle{G(v)} \\
G(Y) & \xrightarrow[\ \eta GY\ ]{} & GFG(Y)
\end{array}
\qquad (3.18)
$$

$$ G(v).\eta GY = \eta GY.G\varepsilon Y.\eta GY = \eta GY = G(1_{FGY}).\eta GY. $$

$\square$

3.2.6 *Reflective and coreflective subcategories*

A subcategory $\mathsf{D} \subset \mathsf{C}$ is said to be *reflective* (notice: not 'reflexive') if the inclusion functor $U\colon \mathsf{D} \to \mathsf{C}$ has a left adjoint, and *coreflective* if U has a right adjoint. For a full reflective subcategory the counit ε is invertible (by Theorem 3.2.5).

For instance Ab is reflective in Gp, with reflector $(-)^{\mathrm{ab}}\colon \mathsf{Gp} \to \mathsf{Ab}$ sending a group G to the abelianised group $G^{\mathrm{ab}} = G/[G,G]$ (see 2.7.4(c)). The unit-component $\eta G\colon G \to G^{\mathrm{ab}}$ is the canonical projection, and the counit $\varepsilon A\colon A \to A/[A,A]$ is invertible.

In Ab the full subcategory tAb formed by all torsion abelian groups is coreflective; the counit $\varepsilon A\colon tA \to A$ is the embedding of the torsion subgroup of an abelian group. On the other hand the full subcategory $\mathsf{tfAb} \subset \mathsf{Ab}$ formed by all torsion-free abelian groups is reflective in Ab, with unit the canonical projection $\eta A\colon A \to A/tA$.

The ordered set $\mathbb{Z}$ is reflective and coreflective in the real line $\mathbb{R}$ (see 1.7.2), while $\mathbb{Q}$ is neither.

3.2.7 *Lemma* (Lifting left adjoints)

Let us suppose that we have an adjunction $(\eta, \varepsilon)\colon F \dashv U$

$$
\begin{array}{ccc}
\mathsf{Y} & \underset{V}{\overset{G}{\rightleftarrows}} & \mathsf{A} \\
{\scriptstyle W}\downarrow & & \| \\
\mathsf{X} & \underset{U}{\overset{F}{\rightleftarrows}} & \mathsf{A}
\end{array}
\qquad (3.19)
$$

where the right adjoint factorises as $U = WV$, *with* W *full and faithful.*

Then $G = FW\colon \mathsf{Y} \to \mathsf{C}$ *is left adjoint to* V, *with unit the unique natural transformation* $\vartheta\colon 1 \to VG$ *such that* $W\vartheta = \eta W$.

Proof Since W is full and faithful, for every Y in Y there is precisely one morphism $\vartheta Y \colon Y \to VG(Y)$ in Y such that $W(\vartheta Y) = \eta(WY) \colon WY \to UFWY = W(VGY)$.

The pair $(G(Y), \vartheta Y \colon Y \to VG(Y))$ is a universal arrow from the object Y to the functor $V \colon \mathsf{A} \to \mathsf{Y}$. Indeed, for every similar pair $(A, f \colon Y \to VA)$, a morphism $g \colon GY \to A$ satisfies $Vg.\vartheta Y = f$ (as in the left diagram below) if and only if $W(Vg.\vartheta Y) = W(f)$

$$
\begin{array}{ccc}
Y \xrightarrow{\ \vartheta Y\ } VGY & \qquad & WY \xrightarrow{\ \eta WY\ } UFWY = W(VGY) \\
\ {}_{f}\searrow \ \ \downarrow {\scriptstyle Vg} & & \ {}_{Wf}\searrow \ \ \downarrow {\scriptstyle Ug} \\
\quad VA & & \quad UA = W(VA)
\end{array}
\tag{3.20}
$$

and there is precisely one solution, because of the universal property of $\eta(WY)$, as in the right diagram above.

Finally we only have to check that the action of the functor G on the morphisms of Y, as determined by the universal property of ϑ, coincides with that of the functor $FW \colon \mathsf{Y} \to \mathsf{C}$.

This amounts to saying that the family $\vartheta Y \colon Y \to VG(Y) = VFW(Y)$ is a natural transformation $1 \to VFW$; but this follows from the naturality of the family $W(\vartheta Y) = (\eta W)Y$, and the faithfulness of W. $\square$

3.2.8 *Exercises and complements* (Chains of adjunctions, II)

(a) Chains of Galois connections between ordered sets X, Y have been considered in 1.7.7. Each of them can be transformed into a chain of adjunctions between the categories C^X and C^Y, for any category C.

We now investigate some non-trivial chains of adjunctions of period 2, or even 1, between categories – a situation which does not exist for ordered sets, where we know that a relation $f \dashv g \dashv f$ forces f and g to be inverse to each other (see 1.7.2(b)).

(b) Let $D \colon \mathsf{C} \to \mathsf{C}^2$ be the diagonal functor of a category C into $\mathsf{C} \times \mathsf{C}$. By 3.2.4, D has a left adjoint S and a right adjoint P if (and only if) C has binary sums and products

$$
\begin{array}{ccc}
S, P \colon \mathsf{C}^2 \to \mathsf{C}, & \qquad & S \dashv D \dashv P, \\
S(X, Y) = X + Y, & & P(X, Y) = X \times Y.
\end{array}
\tag{3.21}
$$

When C is Abm, Ab or $R\,\mathsf{Mod}$, these two functors coincide (as we have seen in 2.1.9): the direct sum (or *biproduct*) $B(X, Y) = X \oplus Y$ is at the same time

- left adjoint to D, with a unit $\eta\colon (X,Y) \to (X \oplus Y, X \oplus Y)$ giving the two injections of the sum,

- right adjoint to D, with a counit $\varepsilon\colon (X \oplus Y, X \oplus Y) \to (X,Y)$ giving the two projections of the product.

We have thus a periodic chain $\dots B \dashv D \dashv B \dashv D \dots$

More generally, the same holds in every semiadditive category, see Section 6.4.

(c) In 2.8.2–2.8.3 we have considered the two-element idempotent monoid $E = \{1, e\}$ and the diagonal functor $D\colon \mathsf{C} \to \mathsf{C}^E$. We have seen that in C all idempotents split if and only C has all E-based limits, if and only if it has all E-based colimits.

Moreover, if this is the case, the limit-object and colimit-object coincide, and (by 3.2.4) we have one functor $L\colon \mathsf{C}^E \to \mathsf{C}$ which is:

- left adjoint to D, with a unit $p_e\colon e \to 1_{L(e)}$ giving the epimorphism of the splitting $e = m_e p_e$,

- right adjoint to D, with a counit $m_e\colon 1_{L(e)} \to e$ giving the monomorphism of the splitting $e = m_e p_e$.

Here $L(e)$ is (a choice of) the image of e in C. This gives another periodic chain of adjunctions

$$\dots L \dashv D \dashv L \dashv D \dots$$

(d) For a fixed set A there is an obvious endofunctor

$$F\colon \mathsf{Rel\,Set} \to \mathsf{Rel\,Set}, \qquad F(X) = X \times A, \tag{3.22}$$

given by the cartesian product (let us recall that this is *not* a categorical product in $\mathsf{Rel\,Set}$, see 2.2.6(b)). The reader will prove that $F \dashv F$, which gives a chain of adjunctions of period 1.

(One may prefer to write $- \times A \dashv A \times -$, so that the adjunction in 'form (i)' is even more evident.)

3.3 Comma categories

Comma categories are an important construction, with strong relationships with all the main tools of category theory, like universal arrows, limits and adjoints.

3.3.1 Comma categories and slice categories

For functors $F\colon \mathsf{X} \to \mathsf{Z}$ and $G\colon \mathsf{Y} \to \mathsf{Z}$ with the same codomain Z, one constructs a *comma category* $(F \downarrow G)$ equipped with functors P, Q and a natural transformation π

$$(F \downarrow G) \quad)\,\pi \qquad \overset{P}{\nearrow}\,\overset{X}{\searrow}\,\overset{F}{\searrow} \quad Z \qquad (3.23)$$

(The original notation was (F, G), whence the name.)

The objects of $(F \downarrow G)$ are the triples $(X, Y, z\colon FX \to GY)$ formed of an object of X, an object of Y and a morphism of Z. A morphism of $(F \downarrow G)$

$$(f, g)\colon (X, Y, z\colon FX \to GY) \to (X', Y', z'\colon FX' \to GY'),$$

comes from a pair of maps $f\colon X \to X'$, $g\colon Y \to Y'$ that form a commutative square in Z, namely $z'.Ff = Gg.z$.

Composition and identities come from those of $X \times Y$

$$(f', g').(f, g) = (f'f, g'g), \qquad \mathrm{id}(X, Y, z) = (\mathrm{id}X, \mathrm{id}Y). \qquad (3.24)$$

There is an obvious universal property, that makes the triple (P, Q, π) a sort of *directed 2-dimensional pullback*: for every category C equipped with similar data

$$P'\colon C \to X, \quad Q'\colon C \to Y, \quad \pi'\colon FP' \to GQ',$$

there is precisely one functor $W\colon C \to (F \downarrow G)$ which commutes with the structural data

$$
\begin{aligned}
W(C) &= (P'C, Q'C, \pi'C), \qquad W(c) = (P'c, Q'c), \\
PW &= P', \qquad QW = Q', \qquad \pi W = \pi'.
\end{aligned}
\qquad (3.25)
$$

As a matter of notation, one writes $(F \downarrow \mathrm{id}Z)$ as $(F \downarrow Z)$ and $(\mathrm{id}Z \downarrow G)$ as $(Z \downarrow G)$. Moreover an object Z_0 of the category Z can be viewed as a functor $Z_0\colon 1 \to Z$; therefore the comma category $(F \downarrow Z_0)$ has objects $(X, z\colon FX \to Z_0)$, while $(Z_0 \downarrow G)$ has objects $(Y, z\colon Z_0 \to GY)$.

In particular we have the *slice categories*

$$(Z_0 \downarrow Z) = Z \backslash Z_0, \qquad (Z \downarrow Z_0) = Z/Z_0, \qquad (3.26)$$

of *objects* $(Z, z\colon Z_0 \to Z)$ *below* Z_0 and *objects* $(Z, z\colon Z \to Z_0)$ *above* Z_0, respectively.

3.3.2 Exercises and complements

(a) The categories Set. and Top. can be identified with the slice categories Set$\backslash\{*\}$ and Top$\backslash\{*\}$, respectively.

(b) Less trivially the category gRng of 'general rings' (associative, possibly non-unitary) is equivalent to the category Rng/$\mathbb{Z}$ of *copointed unitary rings*, or rings over the initial object $\mathbb{Z}$, by the following weakly inverse functors

$$H\colon \text{gRng} \rightleftarrows \text{Rng}/\mathbb{Z} \colon K,$$
$$H(R) = (R^+, p\colon R^+ \to \mathbb{Z}), \qquad K(R, p\colon R \to \mathbb{Z}) = \text{Ker}\, p. \tag{3.27}$$

Starting from a 'general ring' R, the ring R^+ is obtained by adding a unit (without considering whether R has a unit or not): it has additive part $\mathbb{Z} \oplus R$, multiplication

$$(h, r)(k, s) = (hk, hs + kr + rs),$$

and unit $(1, 0)$; p is the first projection and $\text{Ker}\, p \cong R$.

(The reader may know that in Algebra it is often preferred to define R^+ by 'adding a unit when it does not already exist'; this 'ad hoc' procedure is idempotent, *but is not a functor* gRng $\to$ Rng.)

The other way round, starting from a copointed ring $(R, p\colon R \to \mathbb{Z})$, we get an isomorphic copointed ring $(\text{Ker}\, p)^+$, with a natural isomorphism

$$R \to (\text{Ker}\, p)^+, \qquad r \mapsto (p(r), r - p(r).1_R).$$

(In fact there is a short exact sequence $\text{Ker}\, p \rightarrowtail R \twoheadrightarrow \mathbb{Z}$ of general rings, which splits in Ab because $\mathbb{Z}$ is a free abelian group.)

3.3.3 Theorem (Limits in comma categories)

(a) If the categories X *and* Y *have all limits, preserved by* $F\colon$ X $\to$ Z *and* $G\colon$ Y $\to$ Z*, then the comma category* W $= (F \downarrow G)$ *has all limits, preserved by the functor* $(P, Q)\colon$ W $\to$ X$\times$Y*. The same is true of colimits.*

(b) If the category X *has all limits, preserved by* $F\colon$ X $\to$ Z*, and* $Z \in$ ObZ*, then the slice categories* $(F \downarrow Z)$ *and* $(Z \downarrow F)$ *have all limits, preserved by the forgetful functor with values in* X*. The same is true of colimits.*

Note. If in (a) we also assume that F, G create all limits, we can deduce that $(P, Q)\colon$ W $\to$ X$\times$Y also creates them. The proof is essentially the same as below.

Proof The argument is standard, and the reader may prefer to work it out. It suffices to prove (a), since (b) is a consequence. Below, it is understood that the index i varies in ObS.

A functor $W\colon$ S $\to (F \downarrow G)$ has vertices

$$W_i = (PW_i, QW_i, \pi W_i) = (X_i, Y_i, z_i\colon FX_i \to GY_i).$$

The composite $(X, Y) = (P, Q).W \colon \mathsf{S} \to \mathsf{X} \times \mathsf{Y}$ has a limit (L, M), with cone

$$w_i = (u_i, v_i) \colon (L, M) \to (X_i, Y_i).$$

Since F and G preserve limits, FL and GM are limits of FX and GY in Z, with cones

$$Fu_i \colon FL \to FX_i, \qquad Gv_i \colon GM \to GY_i.$$

The natural transformation $\pi W \colon FX \to GY \colon \mathsf{S} \to \mathsf{Z}$ (of components $\pi W_i = z_i$) determines a unique morphism $z \colon FL \to GM$ in Z consistent with these cones:

$$Gv_i.z = z_i.Fu_i \colon FL \to GY_i \qquad \text{(for all } i \in \mathrm{ObS}).$$

Finally, the object $(L, M, h \colon FL \to GM)$ of $(F \downarrow G)$ is the limit of W, with cone $(u_i, v_i) \colon (L, M, z) \to (X_i, Y_i, z_i)$. $\qquad\qquad\square$

3.3.4 Comma categories and universal arrows

A universal arrow from the object X to the functor $U \colon \mathsf{A} \to \mathsf{X}$ (see 2.7.1) is the same as an initial object $(A, \eta \colon X \to UA)$ of the comma category $(X \downarrow U)$.

Dually, a universal arrow from the functor U to the object X is the same as a terminal object $(A, \varepsilon \colon UA \to X)$ of the comma category $(U \downarrow X)$.

3.3.5 Comma categories and limits

We have seen that limits and colimits in the category C, based on a small category S, can be viewed as universal arrows of the diagonal functor $D \colon \mathsf{C} \to \mathsf{C}^\mathsf{S}$ (see 2.7.2). Therefore they also amount to terminal or initial objects in comma categories of D.

In fact a cone $(A, (f_i \colon A \to X_i))$ of the functor $X \colon \mathsf{S} \to \mathsf{C}$ is the same as an object of the comma category $(D \downarrow X)$ and the limit of X is the same as the terminal object of $(D \downarrow X)$.

Dually a cocone $(A, (f_i \colon X_i \to A))$ of X is an object of the comma category $(X \downarrow D)$, and the colimit of X is the initial object of $(D \downarrow X)$.

3.3.6 Comma categories and adjunctions

As an extension of the case of Galois connections in 1.7.6, the *graph* of the adjunction $(F, G, \eta, \varepsilon) \colon \mathsf{X} \to \mathsf{Y}$ will be the comma category

$$\mathsf{G}(F, G) = (F \downarrow \mathsf{Y}), \tag{3.28}$$

with objects $(X, Y, c\colon FX \to Y)$ and morphisms

$$(f, g)\colon (X, Y, c\colon FX \to Y) \to (X', Y', c'\colon FX' \to Y'), \qquad c'.Ff = g.c.$$

The adjunction $F \dashv G$ has a *graph factorisation* in AdjCat

$$\mathsf{X} \underset{G'}{\overset{F'}{\rightleftarrows}} \mathsf{G}(F, G) \underset{G''}{\overset{F''}{\rightleftarrows}} \mathsf{Y} \qquad\qquad (3.29)$$

$$F'(X) = (X, FX, 1_{FX}), \quad G'(X, Y, d) = X, \quad \eta'X = 1_X,$$

$$F''(X, Y, c) = Y, \quad G''(Y) = (GY, Y, \varepsilon Y\colon FG(Y) \to Y), \quad \varepsilon''Y = 1_Y.$$

This forms, also here, a *natural weak factorisation system* in AdjCat, in the sense of [GT] (see 2.5.5), which is *mono-epi* (because $G'F'$ and $F''G''$ are identities).

One can replace $\mathsf{G}(F, G)$ with the comma category $\mathsf{G}'(F, G) = (\mathsf{X} \downarrow G)$, which is isomorphic to the former because of the adjunction (and coincides with it when X and Y are order categories). Or one can replace (3.29) with a factorisation in three adjunctions, where the central one is a pair of isomorphisms, inverse to each other

$$\mathsf{X} \rightleftarrows \mathsf{G}(F, G) \rightleftarrows \mathsf{G}'(F, G) \rightleftarrows \mathsf{Y} \qquad\qquad (3.30)$$

3.3.7 A second digression on categorical equivalence

Going on with the discusion of 1.5.9, let us note that we have given a strict construction for $(F \downarrow G)$, and a characterisation up to isomorphism by the universal property of 3.3.1, as a sort of 'directed 2-dimensional pullback'.

Weakening this item up to categorical equivalence seems to complicate things without any advantage, as discussed in [G8], Subsection 1.8.9, and briefly reviewed below. Some notions of homotopy theory are required to follow the argument.

*A reader familiar with homotopy pullbacks [Mat] will recognise a comma category as similar to a *standard homotopy pullback*, defined by its universal property up to homeomorphism; and will know that a *general homotopy pullback* is defined up to homotopy equivalence (the 'topological version' of categorical equivalence).

Both notions are important: the standard form gives a functor, and the weak definition is closed under pasting – while a pasting of standard homotopy pullbacks is only homotopy equivalent to the appropriate standard solution.

Now, the second fact fails in the case of comma categories (essentially because a natural transformation is a kind of 'directed homotopy', generally non reversible): a pasting of comma squares gives a category which – in general – is not categorically equivalent to the appropriate comma, but a full subcategory of the latter, reflective or coreflective according to the direction of comma squares (see [G8], loc. cit.).

(In fact, a more appropriate categorical version of a homotopy pullback would be an 'iso-comma', where we replace the natural transformation of the universal property by a functorial isomorphism; the solution is an obvious full subcategory of $(F \downarrow G)$. In this case the E-notion would indeed be closed under pasting. However, iso-commas are of no interest here.)*

3.4 Monoidal categories and exponentials

Loosely speaking, a monoidal category is a category equipped with a 'tensor product', like $A \otimes B$ in Ab or – more simply – the cartesian product $X \times Y$ in Set. These structures are related with 'internal homs' (or exponentials), via adjunctions.

The examples dealt with in 3.4.7 include the categories Set, Ord, Cat, Set, and $\mathcal{S}$, Ab and R Mod, K Prj, Rel Set, Ban and Ban$_1$.

Monoidal categories were introduced as 'categories with multiplication', in two 1963 articles by by Bénabou [Be1] and Mac Lane [M3], then renamed by Eilenberg. Other standard references for this topic are Mac Lane [M4], Eilenberg–Kelly [EiK] and Kelly [Kl2].

A reader interested in Computer Science will likely know that the notion of cartesian closed category (exposed in 3.4.5) is 'equivalent' to typed λ-calculus, as shown in Lambek and Scott [LaS], or Barr and Wells [BarW].

3.4.1 Monoidal categories

A *monoidal category* is a category C equipped with a system $(\otimes, E, \kappa, \lambda, \rho)$. It consists of a functor in two variables, often called a *tensor product*

$$\mathsf{C} \times \mathsf{C} \to \mathsf{C}, \qquad (A, B) \mapsto A \otimes B, \tag{3.31}$$

and an object E, called the *unit*. The operation $- \otimes -$ is assumed to be associative, up to a functorial isomorphism of components

$$\kappa(A, B, C): A \otimes (B \otimes C) \to (A \otimes B) \otimes C \qquad (associator),$$

and the object E is assumed to be an identity, up to functorial isomorphisms

$$\lambda(A): E \otimes A \to A, \qquad \rho(A): A \otimes E \to A \qquad (unitors).$$

All these isomorphisms must satisfy coherence conditions, that will be analysed in the more general framework of weak double categories (see 7.2.3); this allows us to 'forget' them and write

$$A \otimes (B \otimes C) = (A \otimes B) \otimes C, \qquad E \otimes A = A = A \otimes E.$$

The case of a *strict* monoidal category, where the comparison isomorphisms are identities, is less frequent but has important examples: see 3.4.2(g), 7.3.3(d) and 7.3.4.

A monoidal category with small hom-sets has a canonical forgetful functor, represented by its unit

$$U = \mathsf{C}(E, -) \colon \mathsf{C} \to \mathsf{Set}, \tag{3.32}$$

which need not be faithful (as some examples in 3.4.2 will show). For every pair of objects A, B in C there is a canonical mapping

$$\varphi_{AB} \colon U(A) \times U(B) \to U(A \otimes B), \qquad (a, b) \mapsto a \otimes b \tag{3.33}$$

where $a \colon E \to A$, $b \colon E \to B$ and $a \otimes b \colon E = E \otimes E \to A \otimes B$. It is easy to prove that the family (φ_{AB}) is a natural transformation between two functors $\mathsf{C} \times \mathsf{C} \to \mathsf{Set}$.

A *symmetric* monoidal category is further equipped with a symmetry isomorphism, coherent with the other ones

$$s(A, B) \colon A \otimes B \to B \otimes A. \tag{3.34}$$

The latter cannot be omitted: note that $s(A, A) \colon A \otimes A \to A \otimes A$ is not the identity, in general.

3.4.2 Exercises and complements

(a) A category C with finite products has a symmetric monoidal structure given by the categorical product $A \times B$, with unit the terminal object $\top$. The reader will define the comparison isomorphisms, using the universal property of products. This structure is called the *cartesian* (monoidal) structure of C.

It is the *important* monoidal structure of many categories like Set, Ord, Cat, Top,... The reader will determine the canonical forgetful functor $\mathsf{C}(\top, -)$ of each of them, and show that it is not faithful when C is Cat.

However, if a category C with finite products *is pointed*, its cartesian structure will likely be of little interest, since the associated forgetful functor $U = \mathsf{C}(0, -) \colon \mathsf{C} \to \mathsf{Set}$ is constant at the singleton. In fact $R\,\mathsf{Mod}$, $\mathsf{Set}_{\bullet}$, Ban, etc. also have an important symmetric monoidal structure – examined below – which is not cartesian. In many cases such structures arise out of the notion of a 'bi-morphism' $A \times B \to C$, that 'preserves the structure in each variable'.

(b) The *important* symmetric monoidal structure of the category Ab of abelian groups is the usual tensor product $A \otimes B$, with unit the additive group $\mathbb{Z}$.

More generally the important symmetric monoidal structure of the category $R\,\mathsf{Mod}$, of modules on a commutative unitary ring R, is the tensor product over R, which the reader will likely have already seen in some text dealing with multilinear algebra (e.g. [Bor1]).

$A \otimes_R B$ is characterised up to isomorphism by the fact that this module is equipped with an *R-bilinear* mapping $\varphi \colon A \times B \to A \otimes_R B$ (linear in each variable) such that every bilinear mapping $\psi \colon A \times B \to C$ factorises through φ by a unique homomorphism h

$$
\begin{array}{cc}
\begin{array}{ccc}
A \times B & \xrightarrow{\ \psi\ } & C \\
{\scriptstyle\varphi}\big\downarrow & \nearrow{\scriptstyle h} & \\
A \otimes_R B & &
\end{array}
&
\begin{array}{ccc}
U(A) \times U(B) & \xrightarrow{\ \psi\ } & U(C) \\
{\scriptstyle\varphi}\big\downarrow & \nearrow{\scriptstyle Uh} & \\
U(A \otimes_R B) & &
\end{array}
\end{array}
\qquad (3.35)
$$

(Note that this diagram lives in Set; to be more precise – or perhaps pedantic – one can rewrite it as at the right hand.)

The proof of the existence and the construction of the associator require some work (and can be found in any text on multilinear algebra), while the unitors and the symmetry $A \otimes_R B \to B \otimes_R A$ follow easily from the universal property. Many other properties of the functors $-\otimes_R A$, like the preservation of colimits, will follow easily from the adjunction of 3.4.6.

The reader will note that the forgetful functor of $(R\,\mathsf{Mod}, \otimes, R)$ is the usual one (up to isomorphism), while that of the cartesian structure is trivial (a constant functor), as we have already remarked.

(c) The important symmetric monoidal structure of Set$_\bullet$ is given by the *smash product* $(X, x_0) \wedge (Y, y_0)$.

It can be realised as the quotient of the product $(X, x_0) \times (Y, y_0) = (X \times Y, (x_0, y_0))$ that identifies all the points of the sum $(X, x_0) + (Y, y_0)$, computed as a subset of the product (see 2.1.6)

$$(X, x_0) \wedge (Y, y_0) = (X \times Y)/((X \times \{y_0\}) \cup (\{x_0\} \times Y)). \qquad (3.36)$$

The base point of the quotient is the equivalence class of the base point of the product

$$[(x_0, y_0)] = (X \times \{y_0\}) \cup (\{x_0\} \times Y).$$

The unit is the pointed set $(\{0, 1\}, 0)$. The canonical forgetful functor Set$_\bullet$ $\to$ Set is the usual one.

The smash product in Top. is constructed in the same way, as a topological quotient of the topological product. *It is an important construction in Topology; for instance the n-dimensional sphere $\mathbb{S}^n$ (pointed at some point) is homeomorphic to the n-th smash power $\mathbb{S}^1 \wedge \ldots \wedge \mathbb{S}^1$. Note also that the unit is the pointed 0-dimensional sphere $\mathbb{S}^0$.*

(d) The reader will note that the tensor product of Set. (or Top.) satisfies a universal property similar to that of the tensor product of modules, with respect to 'bipointed mappings' $X \times Y \to Z$, separately pointed in each variable.

(e) The reader will also note that a similar approach for Ord would simply give the cartesian product, since a mapping $X \times Y \to Z$ which preserves the order in each variable is monotone.

(f) Describe the symmetric monoidal structure of S produced by the equivalence with Set. (in 1.8.6). Determine the associated forgetful functor $S \to$ Set.

(g) The category End(C) of endofunctors of a category C and their natural transformations has a strict monoidal structure given by the composition of endofunctors; it is non-symmetric, generally.

3.4.3 The exponential law for sets

For a fixed set A, we show now that the functor $F = - \times A$

$$F: \text{Set} \to \text{Set}, \qquad F(X) = X \times A, \quad F(f) = f \times A = f \times \text{id} A, \qquad (3.37)$$

admits, as a right adjoint, the functor $G = \text{Set}(A, -)$

$$G: \text{Set} \to \text{Set}, \qquad G(Y) = \text{Set}(A, Y), \quad G(g) = \text{Set}(A, g) = g. - . \quad (3.38)$$

The adjunction $F \dashv G$ (depending on the *parameter* A), is expressed by a natural isomorphism

$$\varphi: \text{Set}(F(-), =) \to \text{Set}(-, G(=)): \text{Set}^{\text{op}} \times \text{Set} \to \text{Set}. \qquad (3.39)$$

In fact we have a family of bijections, for X, Y in Set

$$\varphi_{XY}: \text{Set}(X \times A, Y) \to \text{Set}(X, \text{Set}(A, Y)) \quad (\textit{exponential law}),$$
$$(u: X \times A \to Y) \mapsto (v: X \to \text{Set}(A, Y)), \qquad (3.40)$$

where the mappings u and v are related by the identity $u(x, a) = v(x)(a)$, for all $x \in X$, $a \in A$.

This family is natural: for two mappings $f: X' \to X$, $g: Y \to Y'$, the following diagram commutes

$$\begin{array}{ccc} \mathsf{Set}(X \times A, Y) & \xrightarrow{\varphi_{XY}} & \mathsf{Set}(X, \mathsf{Set}(A, Y)) \\ {\scriptstyle g.-.Ff}\downarrow & & \downarrow{\scriptstyle Gg.-.f} \\ \mathsf{Set}(X' \times A, Y') & \xrightarrow[\varphi_{X'Y'}]{} & \mathsf{Set}(X', \mathsf{Set}(A, Y')) \end{array} \qquad (3.41)$$

because, for any mapping $u \colon X \times A \to Y$, the mappings

$$g.u.(f \times 1_A) \colon X' \times A \to Y',$$

$$(\varphi_{X'Y'})^{-1}(Gg.\varphi_{XY}(u).f) \colon X' \times A \to Y',$$

take the same value at any $(x', a) \in X' \times A$, namely $g(u(fx', a))$.

3.4.4 Comments

The set $\mathsf{Set}(X, Y)$ is often written as Y^X, consistently with the calculus of cardinals (and with the exponential of small discrete categories, in 1.6.1).

In this way, the previous natural isomorphism can be rewritten as

$$\mathsf{Set}(X \times A, Y) \cong \mathsf{Set}(X, Y^A), \qquad Y^{X \times A} \cong (Y^A)^X. \qquad (3.42)$$

The second form motivates the name of *exponential law* for these isomorphisms (and their version in other categories).

The existence of the right adjoint $(-)^A$ of the functor $F = - \times A$ will be expressed saying that the object A is 'exponentiable' in Set (as a category equipped with its cartesian product). The fact that all of them are exponentiable will be expressed saying that Set is 'cartesian closed'.

3.4.5 Exponentiable objects and closed structures

More generally, in a *symmetric* monoidal category C an object A is said to be *exponentiable* (with respect to the tensor product that we are considering) if the functor $- \otimes A \colon \mathsf{C} \to \mathsf{C}$ has a right adjoint. The latter is generally written as $\mathrm{Hom}(A, -) \colon \mathsf{C} \to \mathsf{C}$ or $(-)^A$, and called an *internal hom*, or an *exponential*.

There is thus a family of bijections, natural in the variables X, Y (and depending on the parameter A)

$$\varphi_{XY}^A \colon \mathsf{C}(X \otimes A, Y) \to \mathsf{C}(X, \mathrm{Hom}(A, Y)) \qquad (X, Y \text{ in C}). \qquad (3.43)$$

Since adjunctions compose, all the tensor powers $A^{\otimes n} = A \otimes ... \otimes A$ are also exponentiable, with

$$\mathrm{Hom}(A^{\otimes n}, -) = (\mathrm{Hom}(A, -))^n. \qquad (3.44)$$

A symmetric monoidal category is said to be *closed* if all its objects are exponentiable.

Then one can form a hom-functor Hom: $C^{op} \times C \to C$ in two variables. Without entering in details, a morphism $f: A \to A'$ gives a natural transformation $- \otimes f: - \otimes A \to - \otimes A': C \to C$, which has a 'mate' (see 7.1.6) $\text{Hom}(f, -): \text{Hom}(A', -) \to \text{Hom}(A, -).$

A category C with finite products is said to be *cartesian closed* if all its objects are exponentiable for the cartesian monoidal structure.

In the non-symmetric case one should consider a left and a right internal hom-functor, as is the case with cubical sets (see [K1, BroH2, G7]).

Instead of starting from a monoidal category and introducing the internal hom as a right adjoint, one can define a 'closed category' [EiK], by axiomatising an internal hom-functor Hom: $C^{op} \times C \to C$, and derive the tensor product as a left adjoint. Concretely, this procedure is even easier in many cases considered below; formally, the axioms for Hom are less evident than those of a tensor product.

3.4.6 The exponential law for modules

Modules on a commutative unitary ring R have also, as sets, an exponential law (and family of adjunctions), based on well known functors of two variables: the tensor product (recalled in 3.4.2(b)) and the internal hom-functor Hom_R

$$- \otimes_R -: R\,\text{Mod} \times R\,\text{Mod} \to R\,\text{Mod},$$
$$\text{Hom}_R: R\,\text{Mod}^{op} \times R\,\text{Mod} \to R\,\text{Mod}. \tag{3.45}$$

Here $\text{Hom}_R(A, B)$ is the set of homomorphisms $R\,\text{Mod}(A, B)$ equipped with the pointwise sum and scalar product:

$$(u + v)(a) = u(a) + v(a), \qquad (\lambda u)(a) = \lambda.u(a).$$

On the morphisms we have the usual action: $\text{Hom}_R(f, g) = g. - .f.$

In fact, for a fixed R-module A, there is again a natural family of isomorphisms indexed by X, Y in $R\,\text{Mod}$

$$\varphi_{XY}: \text{Hom}_R(X \otimes_R A, Y) \to \text{Hom}_R(X, \text{Hom}_R(A, Y)),$$
$$\varphi_{XY}(u)(x) = u(x \otimes -): A \to Y, \tag{3.46}$$

for $u: X \otimes_R A \to Y$ and $x \in X$. The underlying bijections

$$\varphi_{XY}: R\,\text{Mod}(X \otimes_R A, Y) \to R\,\text{Mod}(X, \text{Hom}_R(A, Y)),$$

give an adjunction $F \dashv G$ (depending on the parameter A) between the endofunctors:

$$F: R\,\mathsf{Mod} \to R\,\mathsf{Mod},$$
$$F(X) = X \otimes_R A, \qquad F(f) = f \otimes \mathrm{id}A,$$
$$G: R\,\mathsf{Mod} \to R\,\mathsf{Mod},$$
$$G(Y) = \mathrm{Hom}_R(A, Y), \qquad G(g) = \mathrm{Hom}_R(A, g) = g. - . \tag{3.47}$$

With this structure $R\,\mathsf{Mod}$ is a symmetric monoidal closed category.

3.4.7 Exercises and complements

(a) We have seen that Set is cartesian closed, with $\mathrm{Hom}(A, Y) = \mathsf{Set}(A, Y)$. The category Ord is also cartesian closed, with internal hom

$$\mathrm{Hom}(A, Y) = (\mathsf{Ord}(A, Y), \leqslant),$$

equipped with the pointwise order (1.4).

The category Cat of small categories is cartesian closed, with the internal hom $\mathrm{Hom}(\mathsf{S}, \mathsf{C}) = \mathsf{C}^\mathsf{S}$ described in 1.6.1. Every category of presheaves of sets (see 1.6.3) is cartesian closed.

(b) Ab is not cartesian closed: for every non-trivial abelian group A the product $- \times A$ does not preserves sums and cannot have a right adjoint. But we have seen that Ab and $R\,\mathsf{Mod}$ are symmetric monoidal closed, with respect to the usual tensor product and Hom functor.

(c) Similarly $\mathsf{Set}_\bullet$ is symmetric monoidal closed, with respect to the smash product described in 3.4.2(c): $\mathrm{Hom}(A, Y)$ is the set of pointed mappings $\mathsf{Set}_\bullet(A, Y)$, pointed at the zero morphism $A \to Y$.

Describe the associated symmetric monoidal closed structure of $\mathcal{S}$, whose tensor product was the subject of Exercise 3.4.2(f).

(d) The internal hom-functor $\mathrm{Hom}_R: R\,\mathsf{Mod}^{\mathrm{op}} \times R\,\mathsf{Mod} \to R\,\mathsf{Mod}$ produces the *dual-module* functor

$$\mathrm{Hom}_R(-, R): R\,\mathsf{Mod}^{\mathrm{op}} \to R\,\mathsf{Mod}, \quad X \mapsto X^* = \mathrm{Hom}_R(X, R). \tag{3.48}$$

Many important 'dual-object' functors arise in this way, from an internal hom-functor, fixing the covariant variable and leaving free the contravariant one.

(e) We shall see, in 5.1.1, that Top is not cartesian closed, but *every locally compact Hausdorff space A is exponentiable*: Y^A is the set of maps $\mathsf{Top}(A, Y)$ endowed with the *compact-open* topology (for an arbitrary space Y).

(f) For a commutative field K, we shall see that the symmetric monoidal closed structure of the category $K\,\mathsf{Vct}$ ($= K\,\mathsf{Mod}$) of vector spaces induces a similar structure on the category $K\,\mathsf{Prj}$ of projective spaces on K (in 6.5.6).

(g) Use exercise 3.2.8(d) to prove that the category $\mathsf{Rel\,Set}$ is symmetric monoidal closed, with tensor product $X \times Y$ (the cartesian product of sets) and the same internal hom: $\mathrm{Hom}(X, Y) = X \times Y$.

*(h) The categories Ban and Ban_1 are both symmetric monoidal closed, with unit the scalar field K ($\mathbb{R}$ or $\mathbb{C}$). *In both cases*, $\mathrm{Hom}(A, Y)$ is the set $\mathsf{Ban}(A, Y)$ of all continuous linear mappings $A \to Y$, with the usual K-linear structure and norm

$$||f|| = \sup\{||f(x)|| \mid x \in A,\ ||x|| \leqslant 1\}.$$

The tensor product is obtained by a metric completion of the linear tensor product, see [Se].

The canonical forgetful functors have already been considered in (1.34):

$$
\begin{aligned}
U &= \mathsf{Ban}(K, -)\colon \mathsf{Ban} \to \mathsf{Set}, & U(X) &= |X|, \\
B_1 &= \mathsf{Ban}_1(K, -)\colon \mathsf{Ban}_1 \to \mathsf{Set}, & B_1(X) &= \{x \in X \mid ||x|| \leqslant 1\}.
\end{aligned}
\tag{3.49}
$$

Note that, applying each of them to the Banach space $\mathrm{Hom}(A, Y)$, we get the correct set of morphisms

$$\mathsf{Ban}(A, Y) = |\mathrm{Hom}(A, Y)|, \qquad \mathsf{Ban}_1(A, Y) = B_1(\mathrm{Hom}(A, Y)). \tag{3.50}$$

3.4.8 Internal monoids

Let $\mathsf{C} = (\mathsf{C}, \otimes, E)$ be a monoidal category; the comparison isomorphisms are left understood.

An *internal monoid* in C is a triple (M, e, m) consisting of an object M and two morphisms $e\colon E \to M$, $m\colon M \otimes M \to M$ of C, called the *unit* and *multiplication*, which make the following diagrams commute

$$
\begin{array}{ccc}
M \xrightarrow{\ e \otimes M\ } M^{\otimes 2} \xleftarrow{\ M \otimes e\ } M & \qquad & M^{\otimes 3} \xrightarrow{\ M \otimes m\ } M^{\otimes 2} \\
\searrow \quad \downarrow{m} \quad \swarrow & & {\scriptstyle m \otimes M}\downarrow \qquad \downarrow{m} \\
M & & M^{\otimes 2} \xrightarrow{\ m\ } M
\end{array}
\tag{3.51}
$$

These axioms are called *unitarity* and *associativity*.

A monoid in Set (resp. Top or Ord), with respect to the cartesian product, is an ordinary monoid (resp. a topological or ordered monoid). A monoid in Ab (resp. R Mod), with respect to the usual tensor product, is a ring (resp. an R-algebra).

One can define a topological group as an 'internal group' in Top, by adding to an internal monoid M an *inversion* morphism $i\colon M \to M$ satisfying suitable axioms. Lie groups can also be defined in a similar way.

But the reader is warned that *this approach would not work in* Ord, because the inversion mapping necessarily reverses the order: we should consider Ord as a category with reversor, as in 1.8.1, and introduce the inversion morphism as $i\colon M^{\mathrm{op}} \to M$. This also works in the previous cases, equipping Top (or the category of differentiable manifolds) with the trivial reversor, the identity.

3.4.9 Internal comonoids

Dually, an *internal comonoid* in $(\mathsf{C}, \otimes, E)$ is a triple (X, e, d) where the morphisms $e\colon X \to E$ (*counit*) and $d\colon X \to X^{\otimes 2}$ (*comultiplication*) satisfy dual axioms

$$E \xleftarrow{\; e \;} X \xrightarrow{\; d \;} X^{\otimes 2} \qquad\qquad (3.52)$$

$$(e \otimes X)d = 1_X = (X \otimes e)d, \qquad (d \otimes X)d = (X \otimes d)d.$$

For a cartesian structure $(\mathsf{C}, \times, \top)$, every object X has a structure of internal comonoid, where $e\colon X \to \top$ is determined and $d = (1,1)\colon X \to X^2$ is the diagonal.

In Cat, letting X be the arrow category $\mathbf{2}$, we get a structure of interest (see 7.6.2).

3.5 The Adjoint Functor Theorem

Freyd's Adjoint Functor Theorem is one of the main tool to prove the existence of an adjoint, in a case where a (more or less) explicit construction would be difficult, or too complicated to be useful, or unknown (unless by repeating the argument of the theorem itself).

Another existence theorem for adjoints, the Special Adjoint Functor Theorem, by P. Freyd again, can be found in [M4], Section V.8.

3.5.1 The Initial Object Theorem (Freyd)

Let C *be a complete category with small hom-sets.* C *has an initial object if and only if it satisfies the following condition:*

(Solution Set Condition) *there exists a solution set, i.e. a family W_i of objects of* C *indexed by a* small *set I, such that:*

- for every object X there exists some $i \in I$ and some arrow $W_i \to X$.

Note. One can drop the local-smallness condition on C, and only require that the product W of the family W_i has a small set of endomorphisms $C(W, W)$. The proof is the same as below.

Proof The necessity of the previous condition is obvious: an initial object W forms, by itself, a solution set (with $C(W, W)$ a singleton).

Conversely, let us suppose that the above condition is satisfied and let $W = \Pi W_i$ with projections $p_i \colon W \to W_i$. The object W is *weakly initial*: in fact, for every object X there exists *some* arrow $W \to W_i \to X$.

We want to prove that the initial object Z can be obtained from the *global equaliser* $h \colon Z \to W$ of the whole set of endomorphisms of W, i.e. the limit of the (small) diagram formed by them. This means that:

(a) for every $f \colon W \to W$, $fh = h$,

(b) every arrow $h' \colon Z' \to W$ which equalises all the endomaps of W factorises as $h' = kh$, by a unique $k \colon Z' \to Z$.

The morphism h is mono and the object Z is weakly initial, because of the arrow $h \colon Z \to W$. To prove that it is the initial object of the category, we suppose to have two maps $f, g \colon Z \to X$, and prove that they coincide.

Let $e \colon Z_0 \to Z$ be their equaliser. Since W is weakly initial, we have a map $k \colon W \to Z_0$ and a diagram

$$W \xrightarrow{\;k\;} Z_0$$

$$h \searrow \quad \downarrow e$$

$$Z$$

(3.53)

Now, the maps $\mathrm{id}W$ and $hek \colon W \to W$ must be equalised by $h \colon Z \to W$, whence $h = (hek)h = h(ekh)$. Cancelling the monomorphism h we get $ekh = \mathrm{id}Z$. Therefore e is (a split) epi and $f = g$. $\qquad\square$

3.5.2 *The Adjoint Functor Theorem* (Freyd [Fr1])

Let $G \colon D \to C$ be a functor defined on a category D *with small hom-sets and all small limits. Then G has a left adjoint if and only if it satisfies the following conditions:*

(i) G preserves all small limits,

(ii) (Solution Set Condition) *for every object X in* C *there exists a family*

of objects Y_i in D *and arrows $w_i \colon X \to G(Y_i)$ in* C, *indexed by a small set I_X, such that every map $f \colon X \to G(Y)$ (with Y in* D*) factorises as*

$$G(g).w_i \colon X \to G(Y_i) \to G(Y),$$

for some $i \in I_X$ and $g \in$ D(Y_i, Y).

If C *has also small hom-sets, one can simplify the* Solution Set Condition:

(iii) for every object X in C *there exits a small set $S(X) \subset$* ObD, *such that every map $f \colon X \to G(Y)$ (with Y in* D*) factorises as*

$$G(g).w \colon X \to G(Y_0) \to G(Y),$$

for some $Y_0 \in S(X)$, $w \colon X \to G(Y_0)$ in C *and $g \in$* D(Y_0, Y).

Proof The necessity of (i), (ii) is obvious: a right adjoint preserves all the existing limits, and a universal arrow $(Y_0, \eta \colon X \to G(Y_0))$ from X to G gives a solution set formed by a single object of D and a single arrow of C.

Conversely, assuming (i) and (ii), it is sufficient to prove that, for every object X in C, the comma category $(X \downarrow G)$ has an initial object (as we know from 3.3.4); this follows from the previous Initial Object Theorem.

In fact $(X \downarrow G)$ has small hom-sets, because D has. It is small complete (by Theorem 3.3.3), because C is and G preserves small limits. Finally condition (ii) precisely gives a solution set $(Y_i, w_i \colon X \to G(Y_i))$ for $(X \downarrow G)$, in the sense of Theorem 3.5.1.

The simplified version is obvious. $\square$

3.5.3 Some points on cardinality

We recall some basic points of the theory of cardinal sets, that are useful in order to construct solution sets and prove the existence of an adjoint. Definitions and proofs can be found in [Je].

(a) We write $\sharp X$ the *cardinal* of a set X, and $\aleph_0 = \sharp \mathbb{N}$. The cardinal sets are totally ordered by the relation $\alpha \leqslant \beta$, meaning that there exists an injective mapping $\alpha \to \beta$.

(b) For cardinals α, β, the symbols $\alpha + \beta$ and $\alpha.\beta$ usually denote the cardinal of their sum or product, respectively. If one at least of α, β is infinite, then $\alpha + \beta = \alpha.\beta = \max(\alpha, \beta)$. One always has $\alpha < 2^\alpha = \sharp(\mathcal{P}\alpha)$.

(c) For a small cardinal α, the cardinal sets $\leqslant \alpha$ are the elements of a *small set*.

3.5.4 Exercises and comments

(a) The reader will find useful to apply the Adjoint Functor Theorem to prove the existence of the free group on an arbitrary set X, or equivalently of the left adjoint to the forgetful functor $U \colon \mathsf{Gp} \to \mathsf{Set}$. The argument sets a pattern that can be followed in many other cases.

One can fill-in the following outline. First we know that Gp has small hom-sets and small limits, preserved by U. Let us fix a (small) set X.

Every mapping $f \colon X \to U(G)$ with values in a group factorises through the subgroup $G_0 = \langle f(X) \rangle$ generated by $f(X)$ in G. But $\sharp f(X) \leqslant \sharp X$, whence the cardinal $\alpha = \sharp G_0$ satisfies:

$$\alpha \leqslant \beta_X = \max\{\sharp X, \aleph_0\}.$$

An arbitrary bijection $|G_0| \to \alpha$ allows us to make the cardinal α into a group G_1 isomorphic to G_0, so that f factorises as

$$U(g).f_1 \colon X \to U(G_1) \to U(G),$$

for a monomorphism $g \colon G_1 \to G_0 \subset G$.

The reader will now consider the small set $C(X)$ of all cardinals $\leqslant \beta_X$, and prove that the groups whose underlying set *belongs* to $C(X)$, like the previous G_1, form a *small set* $S(X)$. This gives a solution set for X, in the simplified form 3.5.2(iii) – legitimate because also Set has small hom-sets.

The usual construction of the free group on the set X will be sketched in 4.1.3(c).

(b) A similar procedure can be followed for any variety of algebras, as we shall do in Theorem 4.4.2.

(c) A similar, even easier, procedure can also be followed to prove that the full subcategory $\mathsf{Hsd} \subset \mathsf{Top}$ of Hausdorff spaces is reflective. A 'construction' of the reflector can be found in Proposition 5.1.3.

*(d) Use the Adjoint Functor Theorem to prove that Gp has small sums – the 'free products'.

3.6 Monads and algebras

Monads and their algebras yield a far-reaching formalisation of the 'algebraic character' of a category over another – typically used over the category of sets but also of interest in many other cases.

Given a functor $G \colon \mathsf{A} \to \mathsf{X}$, this theory answers the question: is A 'algebraic' over X, via G? More precisely, we say that the functor G is 'monadic' if it has a left adjoint $F \colon \mathsf{X} \to \mathsf{A}$ and the monad $T = FG \colon \mathsf{X} \to \mathsf{X}$ (i.e. the trace left by the adjunction on the category X) allows us to reconstruct the category A as a category of T-algebras.

We follow the classical terminology and notation of Mac Lane [M4], Chapter VI. Monads are also called 'triples' or 'dual standard constructions' in other texts, as [BarB, Bc, Ds].

The reader is warned that the 'algebraic character' considered here can be linked with infinitary operations (see 3.6.6). *Finitary algebraic theories*, or Lawvere theories, are studied in Lawvere's thesis (1963), recently reprinted with comments and a supplement in [Lw2]; see also [Bo2], Chapter 3.

3.6.1 An example

As an informal presentation of this subject, we start from the forgetful functor of monoids, here written as G: Mon $\to$ Set. The free monoid on a set X can be constructed as a sum in Set of the cartesian powers X^n

$$F(X) = \Sigma_{n\in\mathbb{N}}\, X^n,$$
$$(x_1, ..., x_p).(y_1, ..., y_q) = (x_1, ..., x_p, y_1, ..., y_q), \tag{3.54}$$

so that an element of $F(X)$ is a *word* $(x_1, ..., x_p)$ in the alphabet X, and two words are multiplied by concatenation. The empty word $e = (-) \in X^0$ acts as the identity element. We have thus the left adjoint $F \dashv G$, with the following unit and counit (for a set X and a monoid M)

$$\eta_X : X \to GF(X), \qquad \eta_X(x) = (x),$$
$$\varepsilon_M : FG(M) \to M, \qquad \varepsilon_M(m_1, ..., m_n) = m_1 * ... * m_n. \tag{3.55}$$

Here η_X sends each element of X to the corresponding unary word in the underlying set of $F(X)$, while ε_M sends a word of the free monoid on the set $G(M)$ to the product of its elements in M (writing this multiplication as $*$).

The adjunction leaves, on the domain Set of the left adjoint, a 'trace' formed of an endofunctor T and two natural transformations, called the *unit* and the *multiplication*

$$T = GF : \mathsf{Set} \to \mathsf{Set},$$
$$\eta : \mathrm{id} \to T, \qquad \mu = G\varepsilon F : T^2 \to T. \tag{3.56}$$

We shall see that these data (T, η, μ) form a *monad* on Set, namely the monad associated to the adjunction, often simply written as T. The important fact is that *the monad allows us to reconstruct the category of monoids* (and the whole adjunction).

Loosely speaking, the monad T has a category of T-*algebras* Set^T, where an object is a pair $(X, a: TX \to X)$ consisting of a set X and a mapping $a: TX \to X$ satisfying some axioms. We have thus a binary multiplication $x * y = a(x, y)$ on X, and the axioms of T-algebras force this multiplication to be a monoid-law, with $a(x_1, ..., x_n) = x_1 * ... * x_n$.

Finally, we conclude that Mon is, up to isomorphism of categories, a category of T-algebras over Set, for the monad asociated to the functor $G:$ Mon $\to$ Set, and we express this fact saying that Mon is monadic over Set, via G.

Let us note that the free semigroup on the empty set is empty: the category of 'non-empty semigroups', discussed in 1.1.6, lacks such a free object and is not monadic over Set. The same argument applies in other 'artificial cases', like the category of 'non-abelian groups'.

We expose now the theory of monads and their links with adjunctions. The exercises are deferred to the next section.

3.6.2 Monads

A *monad* on the category X is a triple (T, η, μ) where $T: \mathsf{X} \to \mathsf{X}$ is an endofunctor, while $\eta: 1 \to T$ and $\mu: T^2 \to T$ are natural transformations (called the *unit* and *multiplication* of the monad) which make the following diagrams commute

$$
\begin{array}{ccc}
T \xrightarrow{\eta T} T^2 \xleftarrow{T\eta} T & \qquad & T^3 \xrightarrow{T\mu} T^2 \\
\searrow \quad \downarrow \mu \quad \swarrow & & \mu T \downarrow \qquad \downarrow \mu \\
T & & T^2 \xrightarrow{\mu} T
\end{array} \qquad (3.57)
$$

These axioms are called *unitarity* and *associativity*. In fact, a monad on the category X is an internal monoid (see 3.4.8) in the category $\mathsf{C} = \mathrm{End}(\mathsf{X})$ of endofunctors of X and their natural transformations, equipped with the strict (non-symmetric) monoidal structure given by the composition of endofunctors (already mentioned in 3.4.2(g)).

For an ordered set X, viewed as a category, a monad $T: X \to X$ is just a monotone function which is *inflationary* $(1_X \leqslant T)$ and *idempotent* $(T^2 = T)$; this is also called a *closure* operator on X. Obvious examples come from the topological closure $\mathcal{P}S \to \mathcal{P}S$ of a topological space S (which, moreover, has to preserve finite unions). Idempotent monads on arbitrary categories will be studied in Section 3.8.

3.6.3 From adjunctions to monads

It is easy to verify that an adjunction

$$F: \mathsf{X} \rightleftarrows \mathsf{A} : G, \qquad \eta: 1 \to GF, \quad \varepsilon: FG \to 1, \qquad (3.58)$$

has an *associated monad* (T, η, μ) on X (the domain of the left adjoint), where $T = GF: \mathsf{X} \to \mathsf{X}$, η is the unit of the adjunction and

$$\mu = G\varepsilon F: GF.GF \to GF.$$

Typically, a variety of algebras A has an associated monad $T = UF$: $\mathsf{Set} \to \mathsf{Set}$ produced by the forgetful functor $U: \mathsf{A} \to \mathsf{Set}$ and its left adjoint $F: \mathsf{Set} \to \mathsf{A}$, the *free-algebra functor* (as we shall see in Theorem 4.4.2).

3.6.4 Eilenberg–Moore algebras

The other way round, from monads to adjunctions, there are two main constructions; we begin from the more important one.

Given a monad (T, η, μ) on X one defines the category X^T of *T-algebras*, or *Eilenberg–Moore algebras* for T: these are pairs $(X, a: TX \to X)$ consisting of an object X of X and a map a (called the *algebraic structure*) satisfying two coherence axioms:

$$a.\eta X = 1_X, \qquad a.Ta = a.\mu X, \qquad (3.59)$$

A morphism of T-algebras $f: (X, a) \to (Y, b)$ is a morphism $f: X \to Y$ of X which preserves the algebraic structures, in the sense that $f.a = b.Tf$. They compose as in X.

The category X^T is thus a full subcategory of the comma category $(T \downarrow \mathsf{X})$.

One can find this category written as $\mathsf{Alg}(T)$, but we prefer to follow Mac Lane's notation. Moreover, we shall reserve the symbol $\mathsf{Alg}(T)$ for the category of algebraic objects of an *idempotent* monad, which is isomorphic to X^T (in Section 3.8).

3.6.5 From monads to adjunctions and back

Given a monad (T, η, μ) on X there is an adjunction with the category X^T of T-algebras

$$F^T : \mathsf{X} \rightleftarrows \mathsf{X}^T : G^T,$$

$$\eta^T = \eta : 1 \to G^T F^T, \qquad \varepsilon^T : F^T G^T \to 1, \tag{3.60}$$

whose associated monad coincides with the given one. We give an outline of the procedure, and the reader is invited to complete the (easy) missing verifications.

First we have an obvious (faithful) forgetful functor

$$G^T : \mathsf{X}^T \to \mathsf{X},$$

$$G^T(X, a) = X, \qquad G^T(f : (X, a) \to (Y, b)) = (f : X \to Y). \tag{3.61}$$

Backwards we have the functor

$$F^T : \mathsf{X} \to \mathsf{X}^T,$$

$$F^T(X) = (TX, \mu X : T^2 X \to TX), \tag{3.62}$$

$$F^T(f : X \to Y) = Tf : (TX, \mu X) \to (TY, \mu Y).$$

Now $G^T F^T = T : \mathsf{X} \to \mathsf{X}$. The unit of the adjunction is $\eta^T = \eta : 1 \to T$, while the counit is defined as follows

$$\varepsilon^T : F^T G^T \to 1, \qquad \varepsilon^T(X, a) = a : (TX, \mu X) \to (X, a). \tag{3.63}$$

The triangular identities of (3.7) are satisfied, so that $F^T \dashv G^T$, and the functor F^T gives indeed the *free T-algebra* on an object X of X.

It is now easy to see that the monad associated to this adjunction is the original one, because the new multiplication also coincides with the old one:

$$G^T \varepsilon^T F^T(X) = G^T \varepsilon^T(TX, \mu X) = \mu X.$$

3.6.6 Monadicity

On the other hand, if we start from an adjunction $(F, G, \eta, \varepsilon) : \mathsf{X} \to \mathsf{A}$ as in 3.6.3, form the associated monad (T, η, μ) and then the associated adjunction $(F^T, G^T, \eta, \varepsilon^T) : \mathsf{X} \to \mathsf{X}^T$, we get a comparison between the two adjunctions which may be an isomorphism or not; if it is the case, we think of A as a 'category of algebras' over X.

In fact there is a *comparison functor*:

$$K : \mathsf{A} \to \mathsf{X}^T, \qquad K(A) = (GA, G\varepsilon A : GFGA \to GA),$$

$$K(f : A \to B) = Gf : (GA, G\varepsilon A) \to (GB, G\varepsilon B). \tag{3.64}$$

K links the two adjunctions (which share the unit), in the sense that – as one can easily verify

$$
\begin{array}{ccc}
\mathsf{X} \underset{G}{\overset{F}{\rightleftarrows}} \mathsf{A} & KF = F^T, & G^T K = G, \\
\parallel \qquad \downarrow K & & \\
\mathsf{X} \underset{G^T}{\overset{F^T}{\rightleftarrows}} \mathsf{X}^T & \varepsilon^T K = K\varepsilon & (\eta^T = \eta).
\end{array}
\tag{3.65}
$$

Now a functor $G\colon \mathsf{A} \to \mathsf{X}$ is said to be *monadic* (or *algebraic*), or to make A *monadic over* X, if it has a left adjoint $F\colon \mathsf{X} \to \mathsf{A}$ and moreover the comparison functor $K\colon \mathsf{A} \to \mathsf{X}^T$ defined above is an *isomorphism* of categories. (This point is commented in the Remark below.)

We shall see, in 4.4.6, that the forgetful functor $U\colon \mathsf{A} \to \mathsf{Set}$ of any variety of algebras is monadic. But the present *formalisation of the algebraic character of a category* is much broader: for instance the category of compact Hausdorff spaces is monadic over Set (see [M4], Section VI.9), and this monadicity depends on the closure operator of such spaces, a sort of 'infinitary operation'.

There are various 'monadicity theorems' (also called 'tripleability theorems') that give sufficient (or necessary and sufficient) conditions for a functor to be monadic: the interested reader can see [M4], Chapter VI, [Bo2], Section 4.4, and [Bc, Ds].

Remark. We are following Mac Lane ([M4], Section VI.3) in defining monadicity in a strong sense, *up to isomorphism of categories*, because concrete examples, based on structured sets, generally fall in this case.

On the other hand [Bo2], in Section 4.4, and various other text only ask that the comparison K be an equivalence of categories – another aspect of the opposition that we have considered in 1.5.9 and 3.3.7.

The difference can be better appreciated at the light of this example. The category $\mathsf{Set}_\bullet$ of pointed sets is a variety of algebras, produced by a single zeroary operation under no axioms, and is thus monadic over Set, via its forgetful functor U. The equivalent category $\mathcal{S}$ of sets and partial mappings (see 1.8.6) is equipped with the composed functor

$$
US\colon \mathcal{S} \to \mathsf{Set}_\bullet \to \mathsf{Set},
$$

and the associated comparison functor is now an equivalence of categories.

We prefer to view $\mathcal{S}$ as 'weakly algebraic' over Set, rather than as 'algebraic'.

3.6.7 *Theorem* (Limits of T-algebras)

Let S *be a small category. Given a monad* (T, η, μ) *on the category* X, *we suppose that* X *has the limits of all functors* S $\to$ X.

Then X^T *also has such limits, created by the forgetful functor* $G^T : X^T \to$ X *(in the sense of 2.2.8).*

Proof The argument is standard; writing it down makes a useful exercise.

For a functor $X : $ S $\to$ X we write $(LX, (u_i : LX \to X_i))$ the limit cone (for $i \in S = $ ObS). For a functor $Y : $ S $\to X^T$, we let

$$X = G^T Y : S \to X, \qquad Y_i = (X_i, a_i : TX_i \to X_i).$$

There is an obvious structure of algebra on LX, defined by the universal property of the limit

$$a : T(LX) \to LX, \qquad u_i a = a_i.T(u_i) : T(LX) \to X_i, \qquad (3.66)$$

because the maps $a_i.T(u_i)$ form a cone of X. In fact, if $s : i \to j$ is in S, we have a commutative diagram

$$
\begin{array}{ccccc}
T(LX) & \xrightarrow{Tu_i} & TX_i & \xrightarrow{a_i} & X_i \\
\| & & \downarrow{TX(s)} & & \downarrow{X(s)} \\
T(LX) & \xrightarrow{Tu_j} & TX_j & \xrightarrow{a_j} & X_j
\end{array}
\qquad (3.67)
$$

The axioms for the structure a are 'detected' by the jointly monic morphisms u_i of the limit cone

$$u_i(a.\eta_{LX}) = a_i.T(u_i).\eta_{LX} = a_i.\eta_{X_i}.u_i = u_i,$$

$$u_i(a.\mu_{LX}) = a_i.T(u_i).\mu_{LX} = a_i.\mu_X i.T^2 u_i = a_i.Ta_i.T^2 u_i = u_i(a_i.Ta_i).$$

The cone-morphisms give a family of morphisms of T-algebras

$$u_i : (LX, a) \to (X_i, a_i),$$

because of (3.66), and a is the only structure on LX that makes all u_i morphisms of X^T. This family is a cone of Y, because the faithful functor G^T reflects commutative diagrams.

Finally the universal property of this cone follows easily: if

$$f_i : (X', a') \to (X_i, a_i)$$

is a cone of Y, there is precisely one map $f : X' \to LX$ such that $u_i f = f_i$ (for all i), and this f lifts to the unique $f : (X', a') \to (LX, a)$ that solves the same condition in X^T. $\qquad \square$

3.6.8 Kleisli algebras and their adjunction

The second construction of an adjunction from a monad (T, η, μ) on X is based on *Kleisli algebras* for T and their category X_T.

An object of X_T is an object of X, *viewed as the basis of a free T-algebra*. A morphism $f^\sharp \colon X \to Y$ of X_T is 'represented' by an arbitrary morphism $f \colon X \to TY$ of X (which tells us as $f^\sharp$ acts on the basis).

By definition, the composite $g^\sharp f^\sharp$ of $f^\sharp$ with $g^\sharp \colon Y \to Z$ is represented by the map

$$\mu Z.Tg.f \colon X \to TY \to T^2 Z \to TZ, \tag{3.68}$$

while the identity $\mathrm{id} X \colon X \to X$ is represented by $\eta X \colon X \to TX$. The axioms of unitarity and associativity for X_T are easily proved.

The forgetful functor G_T of Kleisli algebras and its left adjoint F_T are now:

$$\begin{aligned}
G_T \colon \mathsf{X}_T \to \mathsf{X}, \qquad G_T(X) &= TX, \\
G_T(f^\sharp \colon X \to Y) &= \mu Y.Tf \colon TX \to TY, \\
F_T \colon \mathsf{X} \to \mathsf{X}_T, \qquad F_T(X) &= X, \\
F_T(f \colon X \to Y) &= (\eta Y.f)^\sharp \colon X \to Y.
\end{aligned} \tag{3.69}$$

Again $G_T F_T = T$. The unit and counit of the adjunction $F_T \dashv G_T$ are

$$\begin{aligned}
\eta_T = \eta \colon 1 &\to T = G_T F_T, \\
\varepsilon_T(X) = (1_{TX})^\sharp \colon F_T G_T(X) = TX &\to X.
\end{aligned} \tag{3.70}$$

Also here the monad associated to this adjunction is the original one, since the new multiplication coincides with the previous one:

$$(G_T.\varepsilon_T.F_T)(X) = G_T(1_{TX}) = \mu X.$$

If we start from an adjunction $(F, G, \eta, \varepsilon) \colon \mathsf{X} \to \mathsf{A}$, as in 3.6.3, form the associated monad (T, η, μ) and then the Kleisli adjunction

$$(F_T, G_T, \eta, \varepsilon_T) \colon \mathsf{X} \to \mathsf{X}_T,$$

there is again a comparison functor which links the two adjunctions

$$\begin{aligned}
H \colon \mathsf{X}_T \to \mathsf{A}, \qquad H(X) &= FX, \\
H(f^\sharp \colon X \to Y) &= \varepsilon FY.Ff \colon FX \to FY, \\
HF_T = F, \quad GH = G_T, \quad &H\varepsilon_T = \varepsilon H.
\end{aligned} \tag{3.71}$$

3.6.9 Comonads and coalgebras

A *comonad* on the category A is a triple (S, ε, δ) formed of an endofunctor $S\colon A \to A$ with natural transformations $\varepsilon\colon S \to 1$ (*counit*) and $\delta\colon S \to S^2$ (*comultiplication*) which make the following diagrams commute

$$(3.72)$$

A comonad on the category A is an internal comonoid (see 3.4.9) in the monoidal category End(A).

The category $^S A$ of *S-coalgebras* $(A, c\colon A \to SA)$ is also defined by duality.

An adjunction $(F, G, \eta, \varepsilon)\colon X \to A$, as in 3.6.3, gives a comonad (S, ε, δ) on A (the domain of the right adjoint G), where $S = FG\colon A \to A$, the natural transformation ε is the counit of the adjunction and

$$\delta = F\eta G\colon FG \to FG.FG.$$

3.7 Exercises and complements on monads

This sections contains exercises on monads, their (Eilenberg–Moore) algebras and Kleisli algebras.

3.7.1 Exercises and complements

(a) The forgetful functor $U\colon \mathsf{Mon} \to \mathsf{Set}$ is monadic: the reader can prove directly this fact, following the outline of 3.6.1, or take it as granted by a general result on varieties of algebras (Theorem 4.4.6).

(b) Similarly, the forgetful functor $U\colon \mathsf{Ab} \to \mathsf{Set}$ is monadic. But we are also interested in the Kleisli category Set_T of the associated monad: an object is a set and a morphism $f^\sharp\colon X \to Y$ can be described as a $\mathbb{Z}$-*weighted mapping between sets*, that associates to each $x \in X$ some formal linear combination $f^\sharp(x) = \sum \lambda_y.y$ of elements of Y, with quasi-null integral coefficients.

The reader can now easily compute the Kleisli composition with another weighted mapping $g^\sharp\colon Y \to Z$.

(c) Loosely speaking, this Kleisli category Set_T can be viewed as the category of free abelian groups with an assigned basis, and arbitrary homomorphisms between them. More precisely, the reader can prove that Set_T is equivalent to the full subcategory of Ab formed by the free abelian groups.

(d) One can extend all this to *R-weighted mappings*, for a unitary ring R. Or to other variety of algebras.

(e) These facts are better understood proving the following general fact about the comparison functor $H\colon \mathsf{X}_T \to \mathsf{A}$, constructed in (3.71) (starting from an adjunction $(F, G, \eta, \varepsilon)\colon \mathsf{X} \to \mathsf{A}$, with associated monad $T = GF$).

If we restrict the codomain of H to the full subcategory B of A formed by the objects isomorphic to some 'free object' $F(X)$ (for X in X), we get an equivalence of categories $H'\colon \mathsf{X}_T \to \mathsf{B}$.

3.7.2 Exercises (The monad of upper-complete lattices)

(a) Define a suitable monad $T\colon \mathsf{Set} \to \mathsf{Set}$, where $T(X) = |\mathcal{P}(X)|$ is the *set* of subsets of X, so that the category Set^T of T-algebras is isomorphic to the *category* VSlt *of upper-complete semilattices*, i.e. complete lattices with mappings that preserve arbitrary joins.

(b) The reader may find easier to start from the forgetful functor $U\colon \mathsf{VSlt} \to \mathsf{Set}$, define its left adjoint and deduce the structure of the associated monad T.

(c) Prove that the Kleisli category Set_T is isomorphic to $\mathsf{Rel\,Set}$.

(d) Study a similar monad $T'\colon \mathsf{Set} \to \mathsf{Set}$, where $T'(X) = |\mathcal{P}_f(X)|$ is the set of finite subsets of X, determining its categories of algebras and Kleisli algebras.

3.7.3 Exercises (Categories as algebras)

(a) We write as Gph the *category of small graphs* and their morphisms (in the sense of 1.1.2). Prove that the forgetful functor $U\colon \mathsf{Cat} \to \mathsf{Gph}$ has a left adjoint F, where the free category $F(X)$ on the graph X has the same objects, and a morphism $\xi = (\xi_1, ..., \xi_n)\colon x \to x'$ is a finite paths of consecutive arrows of X

$$x = x_0 \to x_1 \to ... \to x_n = x'. \tag{3.73}$$

This includes the empty path $\mathrm{id}(x)$ at any vertex x of X.

(b) Prove that Cat is monadic over Gph.

(c) In the category Gph_T of Kleisli algebras (of the associated monad), a morphism $f\colon X \to Y$ between graphs sends an arrow $\xi\colon x \to x'$ of X to a formal composite $f(\xi) = (\eta_1, ..., \eta_n)\colon f(x) \to f(x')$ of arrows of Y.

3.8 *Idempotent monads and idempotent adjunctions

Loosely speaking, being an algebra for an idempotent monad $T\colon \mathsf{X} \to \mathsf{X}$ reduces to a *property* of the objects of X. For instance, an algebra for the abelianisation monad $\mathsf{Gp} \to \mathsf{Gp}$ (examined in 3.8.4(c)) is 'the same' as a commutative group. Thus, the category X^T of T-algebras can be replaced with an isomorphic category $\mathsf{Alg}(T)$, which is the full subcategory of X formed by its 'algebraic objects'.

The theory of idempotent monads and idempotent adjunctions is well known in the domain of category theory, but it may be difficult to find it in a single text. (Some of the main facts are exposed in [Bo2] Section 4.2, [LaS] Lemma 4.3 and [AT] Section 6.)

We give here a rather detailed treatment of this topic, even though this part will be marginally used in the rest of the book; a reader might be satisfied with browsing definitions and statements, and omitting the proofs.

An adjunction is always 'equipped' with the associated monad (defined in 3.6.3).

3.8.1 Idempotent monads

A monad $T = (T, \eta, \mu)$ on the category X is said to be *idempotent* if $\mu\colon T^2 \to T$ is invertible. Then the inverse of μ is $T\eta = \eta T\colon T \to T^2$.

Equivalently one can define an idempotent monad on X as a pair (T, η) formed of a functor $T\colon \mathsf{X} \to \mathsf{X}$ and a natural transformation $\eta\colon 1 \to T$ such that $T\eta = \eta T\colon T \to T^2$ is invertible.

In this approach one defines $\mu = (T\eta)^{-1}\colon T^2 \to T$. The first axiom $\mu.\eta T = 1 = \mu.T\eta$ obviously holds; for the second it suffices to cancel $T^2\eta = T\eta T$ from the identity

$$(\mu.T\mu).T^2\eta = \mu = (\mu.\mu T).T\eta T.$$

Dually we define *idempotent comonads*.

3.8.2 Theorem and Definition (Idempotent adjunctions)

Let $F \dashv G$ be a pair of adjoint functors (or more generally an adjunction in a 2-category, see 7.1.2)

$$F\colon \mathsf{X} \rightleftarrows \mathsf{A} \colon G, \qquad \eta\colon 1_{\mathsf{X}} \to GF, \quad \varepsilon\colon FG \to 1_{\mathsf{A}},$$
$$\varepsilon F.F\eta = 1_F, \qquad G\varepsilon.\eta G = 1_G. \tag{3.74}$$

The following conditions are equivalent:

(i) one of the four natural transformations which appear in the triangular identities (namely $F\eta, \varepsilon F, \eta G$ and $G\varepsilon$) is invertible,

(ii) all of them are invertible, i.e. $F\eta.\varepsilon F = 1_{FGF}$ and $\eta G.G\varepsilon = 1_{GFG}$,

(iii) the associated monad $T = GF \colon \mathsf{X} \to \mathsf{X}$ is idempotent, i.e. its multiplication $\mu = G\varepsilon F \colon T^2 \to T$ is invertible (and then $T\eta = \eta T$ is its inverse),

(iv) $T\eta = GF\eta$ is invertible,

(v) $\eta T = \eta GF$ is invertible,

(iii*) the associated comonad $S = FG \colon \mathsf{A} \to \mathsf{A}$ is idempotent, i.e. its comultiplication $\delta = F\eta G \colon S \to S^2$ is invertible (and then $S\varepsilon = \varepsilon S$ is its inverse),

(iv*) $\varepsilon S = \varepsilon FG$ is invertible,

(v*) $S\varepsilon = FG\varepsilon$ is invertible.

When these conditions hold we say that the adjunction is idempotent. Then the same is true of the opposite adjunction $G^{\mathrm{op}} \dashv F^{\mathrm{op}}$.

Proof The triangle identities plainly show that $F\eta$ is invertible if and only if εF is; the same holds for ηG and $G\varepsilon$.

The main point is proving that (iv) implies that $F\eta$ and εF are invertible, i.e. $F\eta.\varepsilon F = 1_{FGF}$. The natural transformation $F\eta.\varepsilon F$ forms the upper row of the following commutative diagram

$$
\begin{array}{ccccc}
FGF & \xrightarrow{\ \varepsilon F\ } & F & \xrightarrow{\ F\eta\ } & FGF \\
{\scriptstyle FGF\eta}\downarrow & & \downarrow{\scriptstyle F\eta} & & \downarrow{\scriptstyle FGF\eta} \\
FGFGF & \xrightarrow[\ \varepsilon FGF\ ]{} & FGF & \xrightarrow[\ F\eta GF\ ]{} & FGFGF
\end{array}
\qquad (3.75)
$$

The lower row $F\eta GF.\varepsilon FGF = (F\eta.\varepsilon F)GF$ is the identity, because $F\eta GF$ is invertible by hypothesis and the 'reversed composite' $(\varepsilon F.F\eta)GF$ is the identity, by (3.74). Therefore $FGF\eta.(F\eta.\varepsilon F) = FGF\eta$ and we get our conclusion by cancelling the transformation $FGF\eta$, which is invertible by hypothesis.

Adding obvious implications we have:

- (ηG and $G\varepsilon$ are invertible) $\Rightarrow$ (iv) $\Rightarrow$ ($F\eta$ and εF are invertible),

- (iii) $\Leftrightarrow$ (iv) $\Leftrightarrow$ (v).

By duality all the properties of the statement are equivalent. $\qquad\square$

3.8.3 Proposition and Definition (Algebraic objects)

Let (T, η) be an idempotent monad on X as defined in 3.8.1: $T\eta = \eta T \colon T \to T^2$ is invertible and $\mu = (T\eta)^{-1} \colon T^2 \to T$.

(a) The following conditions on an object X of X are equivalent, and we say that X is an algebraic object (with respect to T) when they hold:

(i) ηX is invertible,

(ii) there exists a morphism $h \colon TX \to X$ such that $h.\eta X = 1_X$,

(iii) there exists an (Eilenberg–Moore) T-algebra (X, h) over X,

(iv) ηX is invertible and $\hat{X} = (X, (\eta X)^{-1})$ is the only T-algebra on X,

(v) X is isomorphic to TX in X.

In this case the algebra $\hat{X} = (X, (\eta X)^{-1})$ is isomorphic to the free T-algebra $(TX, \mu X)$ on X. (See the last point below.)

(b) An arrow of T-algebras $f \colon \hat{X} \to \hat{Y}$ is an ordinary X-arrow $f \colon X \to Y$ where X and Y satisfy the equivalent conditions above. The forgetful functor $G^T \colon \mathsf{X}^T \to \mathsf{X}$ of the category of T-algebras is thus a full embedding

$$G^T(\hat{X}) = X, \qquad G^T(f \colon \hat{X} \to \hat{Y}) = (f \colon X \to Y), \qquad (3.76)$$

and makes X^T isomorphic to the full reflective subcategory $\mathsf{Alg}(T) \subset \mathsf{X}$ of algebraic objects defined above. The counit ε^T is invertible.

(c) Using this isomorphism the adjunction $F^T \dashv G^T$ associated to T can be rewritten in the following equivalent form, where G' is a full embedding with reflector F'

$$F' \colon \mathsf{X} \rightleftarrows \mathsf{Alg}(T) \colon G', \qquad \eta' = \eta \colon 1 \to G'F' = T, \qquad \varepsilon' \colon F'G' \to 1,$$

$$F'(X) = TX, \qquad F'(f \colon X \to Y) = Tf \colon TX \to TY,$$

$$\varepsilon'(X) = (\eta X)^{-1} \colon F'G'(X) = TX \to X.$$

$$(3.77)$$

If T is the monad of an adjunction $F \dashv G \colon \mathsf{X} \to \mathsf{A}$, the comparison $K \colon \mathsf{A} \to \mathsf{X}^T$ can be rewritten as $K' \colon \mathsf{A} \to \mathsf{Alg}(T)$, a codomain-restriction of $G \colon \mathsf{A} \to \mathsf{X}$, so that $G = G'K'$. If G is full, then K' is also (and K as well).

(d) The composed functor from X^T to the category X_T of Kleisli algebras

$$F_T G^T \colon \mathsf{X}^T \to \mathsf{X} \to \mathsf{X}_T, \qquad F_T G^T(X, (\eta X)^{-1}) = X, \qquad (3.78)$$

is an equivalence of categories (not an isomorphism, generally).

Proof (a) (i) $\Rightarrow$ (ii) Obvious.

(ii) $\Rightarrow$ (iii) If $h.\eta X = 1_X$ then (X, h) is a T-algebra because the second axiom is here a consequence:

$$h.Th.T\eta X = h = h.\mu X.T\eta X.$$

(iii) $\Rightarrow$ (iv) The naturality of η on the morphism h and the property $\acute{T}\eta = \eta T$ give

$$\eta X.h = Th.\eta TX = Th.T\eta X = T(h.\eta X) = 1_{TX},$$

so that ηX and h are inverses.

(iv) $\Rightarrow$ (v) Obvious.

(v) $\Rightarrow$ (i) The existence of an isomorphism $i\colon X \to TX$ gives $Ti.\eta X = \eta TX.i$, whence ηX is invertible.

The last assertion follows from the isomorphism

$$\eta X\colon (X, (\eta X)^{-1}) \to (TX, \mu X).$$

(b) and (c). It is an obvious consequence of (a), applying the naturality of η.

(d) As in 3.6.8 we write as $f^\sharp\colon X \to Y$ the X_T-arrow represented by the X-morphism $f\colon X \to TY$; recall that $\mathrm{id}(X) = (\eta X)^\sharp$ and the composite of $f^\sharp$ with $g^\sharp\colon Y \to Z$ is represented by $\mu Z.Tg.f\colon X \to TZ$.

The functor $F_T G^T\colon \mathsf{X}^T \to \mathsf{X}_T$ sends the arrow $f\colon \hat{X} \to \hat{Y}$ to

$$(\eta Y.f)^\sharp\colon X \to Y.$$

Since ηY is invertible, this mapping $\mathsf{X}^T(\hat{X}, \hat{Y}) \to \mathsf{X}_T(X, Y)$ is bijective. It is now sufficient to prove that $F_T G^T$ is essentially surjective on the objects.

Let us take an object X of X_T, i.e. an arbitrary object of X, and consider the T-algebra $(TX, \mu X)$; for an arbitrary monad this is the free T-algebra on X, with $F_T G^T(TX, \mu X) = TX$. Here X and TX are *isomorphic objects in* X_T (not in X, generally), by the inverse morphisms

$$(T\eta X.\eta X)^\sharp\colon X \to TX, \qquad (1_X)^\sharp\colon TX \to X,$$

since

$$(\mu X.T1_X.T\eta X.\eta X)^\sharp = (\eta X)^\sharp = \mathrm{id}(X),$$
$$(\mu TX.T^2\eta X.T\eta X.1_X)^\sharp = (\mu TX.T\eta TX.T\eta X)^\sharp = (\eta TX)^\sharp = \mathrm{id}(TX).$$

$\square$

3.8.4 Exercises and complements

(a) We know, from 3.6.2, that a monad T on an ordered set X is always idempotent and amounts to a closure operator; an algebraic object for T is the same as a *closed* element $x = Tx$. Therefore every adjunction

$$(F, G, \eta, \varepsilon) \colon X \to \mathsf{A}$$

from the ordered set X to an arbitrary category is idempotent; by duality this also holds for every adjunction $\mathsf{A} \to X$.

(b) Let A be a *full reflective subcategory* of X which is *replete* in X: any isomorphism $i \colon A \to X$ of X that involves an object of A belongs to A.

Then the embedding $G \colon \mathsf{A} \subset \mathsf{X}$ has a left adjoint $F \dashv G$, and the counit $\varepsilon \colon FG \to 1$ is invertible (by 3.2.5), which implies that the adjunction is idempotent. Moreover A coincides with $\mathsf{Alg}(GF)$ and G is monadic.

(In fact, every $A \in \mathrm{Ob}\mathsf{A}$ is an algebraic object of X, because $\eta(GA)$ is invertible, and conversely an algebraic object $X \cong GF(X)$ belongs to A.)

(c) In particular we have seen (in 3.2.6) that Ab is full reflective in Gp, and obviously replete. The groups which are algebraic for the associated *abelianisation* monad

$$T \colon \mathsf{Gp} \to \mathsf{Gp}, \qquad \eta G \colon G \to T(G) = G/[G, G], \qquad (3.79)$$

are precisely the abelian ones. Similar facts hold for many embeddings of algebraic varieties, like $\mathsf{Gp} \subset \mathsf{Mon}$ or $\mathsf{CRng} \subset \mathsf{Rng}$. But also for $\mathsf{Ord} \subset \mathsf{pOrd}$ (see 3.2.3(b)) and $\mathsf{Hsd} \subset \mathsf{Top}$ (see 5.1.3).

(d) On the other hand the embedding $\mathsf{Rng} \subset \mathsf{gRng}$ (see 3.3.2(b)) is not full, since a homomorphism in gRng need not preserve units, when they exist; thus its left adjoint $(-)^+ \colon \mathsf{gRng} \to \mathsf{Rng}$ (that universally adds a unit, as described in 3.3.2) does not yield an idempotent monad: the ring R^{++} will have a new unit added with respect to R^+.

The embedding of monoids into semigroups behaves in a similar way.

(e) The interested reader can prove the following extension of (b). We have an adjunction $F \dashv G$ where the functor G is full and faithful (whence $\varepsilon \colon FG \to 1$ is invertible and the adjunction is idempotent). It follows that the comparison $K \colon \mathsf{A} \to \mathsf{X}^T$ is an equivalence of categories, with quasi inverse $FG^T \colon \mathsf{X}^T \to \mathsf{A}$, and G is *monadic in the weak sense* mentioned in the Remark of 3.6.6.

3.8.5 Idempotent comonads and coalgebras

An idempotent comonad (S, ε) on the category A has a dual characterisation, as an endofunctor $S \colon \mathsf{A} \to \mathsf{A}$ with a natural transformation $\varepsilon \colon S \to 1$

such that $S\varepsilon = \varepsilon S\colon S^2 \to S$ is invertible; the comultiplication $\delta\colon S \to S^2$ is the inverse of the latter.

Now an S-coalgebra amounts to a *coalgebraic object* A of A, characterised by the following equivalent conditions:

(i) εA is invertible,

(ii) there exists a morphism $k\colon A \to SA$ such that $\varepsilon A.k = 1_A$,

(iii) there exists an S-coalgebra $(A, k\colon A \to SA)$ over A,

(iv) εA is invertible and $\hat{A} = (A, (\varepsilon A)^{-1})$ is the unique S-coalgebra over A,

(v) A is isomorphic to SA in A.

In this case $\hat{A} = (A, (\varepsilon A)^{-1})$ is isomorphic to the cofree S-coalgebra $(SA, \delta A)$ on A. The category of S-coalgebras ^{S}A is linked to the full coreflective subcategory $\mathsf{Coalg}(S) \subset$ A of coalgebraic objects, by the isomorphism

$$U\colon {}^S\mathsf{A} \to \mathsf{Coalg}(S), \qquad U(A, k) = A,$$
$$U(f\colon (A, k) \to (B, k')) = f\colon A \to B. \tag{3.80}$$

The adjunction associated to S

$$^SF\colon {}^S\mathsf{A} \rightleftarrows \mathsf{A} : {}^SG,$$
$$^S\eta\colon 1 \to {}^S G^S F, \qquad ^S\varepsilon = \varepsilon\colon {}^S F^S G \to 1, \tag{3.81}$$

is computed as follows:

$$^SF(A, k) = A, \qquad ^SF(f\colon (A, k) \to (B, k')) = f\colon A \to B,$$
$$^SG(A) = (SA, \delta\colon SA \to S^2A),$$
$$^SG(f\colon A \to B) = Sf\colon {}^SG(A) \to {}^SG(B),$$
$$^S\eta(A, k) = k\colon (A, k) \to (SA, \delta\colon SA \to S^2A).$$

It can be rewritten in the following equivalent form, where F'' is a full embedding with coreflector G''

$$F''\colon \mathsf{Coalg}(S) \rightleftarrows \mathsf{A} : G'',$$
$$\eta''\colon 1 \to G''F'', \qquad \varepsilon'' = \varepsilon\colon F''G'' \to 1,$$
$$G''(A) = SA, \qquad G''(f\colon A \to B) = Sf\colon SA \to SB,$$
$$\eta''(A) = (\varepsilon A)^{-1}\colon A \to SA, \qquad \varepsilon''(A) = \varepsilon A\colon SA \to A. \tag{3.82}$$

3.8.6 Examples

Dualising 3.8.4(b), the inclusion functor $U\colon \mathsf{A} \to \mathsf{X}$ of a full coreflective subcategory has a left adjoint $G\colon \mathsf{X} \to \mathsf{A}$ with invertible unit η.

We have now an idempotent comonad $S = GU \colon \mathsf{A} \to \mathsf{A}$; the category of coalgebras $^S\mathsf{A}$ is isomorphic to the full reflective subcategory $\mathsf{Coalg}(S) \subset \mathsf{A}$ of coalgebraic objects, namely the objects A of A such that εA is invertible.

In particular the category tAb of torsion abelian groups is full coreflective in Ab, with counit $\varepsilon A \colon tA \to A$ given by embedding of the torsion subgroup of A (see 3.2.6); the abelian groups which are coalgebraic for the associated comonad $t \colon \mathsf{Ab} \to \mathsf{Ab}$ are precisely the torsion ones.

3.8.7 Theorem (Factorising idempotent adjunctions)

We start from an idempotent adjunction $(F, G, \eta, \varepsilon) \colon \mathsf{X} \to \mathsf{A}$, *with associated idempotent monad* (T, η) *and associated idempotent comonad* (S, ε), *where* $T = GF \colon \mathsf{X} \to \mathsf{X}$ *and* $S = FG \colon \mathsf{A} \to \mathsf{A}$.

Rewriting the category of T-algebras as the full reflective subcategory $\mathsf{Alg}(T) \subset \mathsf{X}$ *(as in 3.8.3) and the category of S-coalgebras as the full coreflective subcategory* $\mathsf{Coalg}(S) \subset \mathsf{A}$ *(as in 3.8.5), the given adjunction can be factorised as follows up to isomorphism*

$$\mathsf{X} \; \underset{G'}{\overset{F'}{\rightleftarrows}} \; \mathsf{Alg}(T) \; \underset{G^\sharp}{\overset{F^\sharp}{\rightleftarrows}} \; \mathsf{Coalg}(S) \; \underset{G''}{\overset{F''}{\rightleftarrows}} \; \mathsf{A} \tag{3.83}$$

Proof The reflective adjunction $(\eta', \varepsilon') \colon F' \dashv G'$ is described in (3.77). G' is the full embedding of algebraic objects, with reflector $F'(X) = TX$ and invertible counit $\varepsilon'(X) = (\eta X)^{-1}$. The comparison functor K is now computed as G

$$K \colon \mathsf{A} \to \mathsf{Alg}(T), \qquad K(A) = GA, \quad K(f \colon A \to B) = Gf,$$
$$KF = F', \quad G'K = G, \quad K\varepsilon = \varepsilon'K. \tag{3.84}$$

The coreflective adjunction $(\eta'', \varepsilon'') \colon F'' \dashv G''$ is described in (3.82). F'' is the full embedding of coalgebraic objects, with coreflector $G''(A) = SA$ and invertible unit $\eta''(A) = (\varepsilon A)^{-1}$. The comparison functor H is computed as F

$$H \colon \mathsf{X} \to \mathsf{Coalg}(S), \qquad H(X) = FX, \quad H(f \colon X \to Y) = Ff,$$
$$HG = G'', \quad F''H = F, \quad H\eta = \eta'H. \tag{3.85}$$

The given adjunction restricts to an adjunction between our full subcategories

$$(\eta^\sharp, \varepsilon^\sharp) \colon F^\sharp \dashv G^\sharp \colon \mathsf{Alg}(T) \to \mathsf{Coalg}(S),$$
$$F^\sharp X = FX, \quad G^\sharp A = GA, \tag{3.86}$$

because, for every X in X, FX is a coalgebraic object of A (with structure $F\eta X\colon FX \to SFX$), and dually. This is actually an *adjoint equivalence*, because $\eta^{\sharp}X = \eta X$ is invertible for every algebraic object X, and dually. It is even an isomorphism if T is the identity over all algebraic objects and S is the identity over all the coalgebraic ones.

Finally, when we compose the three adjunctions above we get $F''F^{\sharp}F' = FGF$, isomorphic to F (by $F\eta = (\varepsilon F)^{-1}$) and $G'G^{\sharp}G'' = GFG$, isomorphic to G (by $\eta G = (G\varepsilon)^{-1}$). $\qquad\square$

4

Applications in Algebra

We begin by studying, in Section 4.1, the free structures for various categories of algebraic kind. Regular and Barr-exact categories are introduced in Section 4.2; their categories of relations in Section 4.5.

Varieties of algebras are briefly studied in Sections 4.3 and 4.4, proving that these categories $\mathsf{Alg}(\Omega, \Xi)$ are complete and Barr-exact. The fact that such algebras can be viewed as Eilenberg–Moore algebras for the monad associated to the forgetful functor $\mathsf{Alg}(\Omega, \Xi) \to \mathsf{Set}$ is stated in Theorem 4.4.6; the proof is only referred to [M4].

Other notions of 'exact category' will be presented in 6.3.1.

4.1 Free algebraic structures

This section examines many explicit constructions of free algebraic structures, over weaker structures or sets, as left adjoints of forgetful functors. The reader should find useful to consider these exercises, completing the missing details. General results on the existence of such adjoints for varieties of algebras will be given in Section 4.4.

4.1.1 Algebraic structures and forgetful functors

Various categories of sets equipped with a kind of algebraic structure (and their homomorphisms) have been listed in 1.1.1, like the categories Set (of sets), Ab (of abelian groups), Gp (of groups), Mon (of monoids), Abm (of abelian monoids), Rng (of unitary rings), CRng (of commutative unitary rings), $R\,\mathsf{Mod}$ (of left modules on a fixed unitary ring), $R\mathsf{Alg}$ (of unitary algebras on a fixed commutative unitary ring), $\mathsf{Set_{\bullet}}$ (of pointed sets).

This list includes Set and $\mathsf{Set_{\bullet}}$, where the algebraic structure of the object X is void or reduced to a 0-ary operation $X^0 = \{*\} \to X$, giving the basepoint 0_X.

Each of these categories A is made concrete by an obvious (faithful) forgetful functor

$$U : A \to \mathsf{Set}. \tag{4.1}$$

But there are also 'intermediate' forgetful functors $U : A \to B$ that forget part of the structure or part of its properties, like

$$\begin{array}{ll} \mathsf{Ab} \to \mathsf{Gp} \to \mathsf{Mon}, & \mathsf{Ab} \to \mathsf{Abm} \to \mathsf{Mon}, \\ R\mathsf{Alg} \to R\,\mathsf{Mod} \to \mathsf{Ab}, & R\mathsf{Alg} \to \mathsf{Rng} \to \mathsf{Mon}. \end{array} \tag{4.2}$$

All these functors have left adjoints, giving free algebraic structures over sets or – more generally – over 'weaker' algebraic structures; many of these constructions are reviewed below.

4.1.2 Free objects

As we have already seen in 2.7.3, given a functor $U : A \to \mathsf{Set}$, the free A-object over a set X is an object $F(X)$ (or $F_A(X)$) equipped with a mapping $\eta : X \to U(F(X))$ satisfying the usual universal property. The existence of all free objects amounts to the existence of the left adjoint $F \dashv U$.

These definitions can be extended to any functor $U : A \to B$ between arbitrary categories.

The first case, of a left adjoint F of a functor $U : A \to \mathsf{Set}$, has a peculiar property deriving from the fact that every set X is isomorphic to a categorical sum $\sum_{x \in X} \{*\}$ (in Set) of copies of the singleton. The functor F, as a left adjoint, preserves sums, and $F(X)$ must be isomorphic to the categorical sum $\sum_{x \in X} F(\{*\})$ (in A) of copies of the free A-object on the singleton (independently of the existence of all sums in A). A stronger result in this sense will be given in 4.1.3(e).

In fact, many free 'algebras' are concretely built, below, as categorical sums of the free 'algebra' on one generator.

4.1.3 Exercises and complements, I

(a) Let us recall once more (after 2.7.4 and 3.1.1) that the free left R-module $F(X) = RX$ can be constructed as a direct sum in $R\,\mathsf{Mod}$ of copies of the left module R

$$F(X) = \bigoplus_{x \in X} R. \tag{4.3}$$

In particular, the free abelian group $F(X) = \bigoplus_{x \in X} \mathbb{Z}$ is constructed as a direct sum in Ab of copies of the additive group $\mathbb{Z}$.

(b) The free abelian monoid $F(X) = \mathbb{N}X$ has a similar construction, as a direct sum $\bigoplus_{x \in X} \mathbb{N}$ in Abm of copies of the additive monoid $\mathbb{N}$.

*This can be generalised to left modules over any unitary *semiring* R (where the existence of opposites is not assumed).*

(c) The reader may know that the free group $F(X)$ can be constructed as a 'free product', i.e. a categorical sum in Gp, of copies of the additive group $\mathbb{Z}$

$$F(X) = \Sigma_{x \in X}\, \mathbb{Z}. \tag{4.4}$$

The elements of $F(X)$ can be described as equivalence classes $[w] \in W/\sim$ of 'words'

$$w = x_1^{k_1} x_2^{k_2} ... x_p^{k_p},$$

in the alphabet X, with integral exponents. Each word $w \in W$ has a 'reduced form' $R(w)$, obtained by adding up the exponents when possible and cancelling all occurrences of type x^0. The result does not depend on the sequence of reductions; this seemingly obvious point requires some care, but the rest is easy.

We say that $w \sim w'$ when $R(w) = R(w')$. Words are multiplied by concatenation, and the quotient set $W/\sim$ has a well defined product: $[w].[w'] = [w.w']$; indeed the concatenation $w.w'$ is equivalent to $R(w).R(w')$, because $R(w.w')$ can be obtained as $R(R(w).R(w'))$. Then we prove that $F(X)$ is a group, with basis the family of the equivalence classes $\eta(x) = [(x)]$.

(d) The existence of the free group $F(X)$ was proved in 3.5.4, in a non-constructive way, by the Adjoint Functor Theorem. The reader may note that, once we know that the universal arrow $\eta X\colon X \to UF(X)$ exists, it is straightforward to show that each element of $F(X)$ can be written in the form $x_1^{k_1} x_2^{k_2} \ldots x_p^{k_p}$ (with $x_i \in X$ and $k_i \in \mathbb{Z}$); it is also obvious that such a form can be reduced as above, but less obvious that no further identification is needed.

(e) As an overview of the previous cases, let us be given a functor $U\colon \mathsf{A} \to$ Set. We assume that:

(i) the singleton $\{*\}$ has a free A-object A, with universal mapping $\eta_0\colon \{*\} \to UA$,

(ii) for every (small) set X, the sum $F(X) = \Sigma_{x \in X}\, A$ exists in A, with injections $u_x\colon A \to F(X)$.

Then one can prove that $F(X)$ is the free A-object over X, with universal map

$$\eta\colon X \to UF(X), \qquad \eta(x) = (Uu_x)(\eta_0(*)). \tag{4.5}$$

4.1.4 Exercises and complements, II

The procedure we have followed above is effective when we know how to construct sums in A; in other cases there may be simpler direct constructions.

(a) We have seen in 3.6.1 that the free monoid on the set X has a simple construction as a sum in Set, $F(X) = \sum_{n\in\mathbb{N}} X^n$, with multiplication by concatenation.

Now, the free monoid on the singleton is obviously the additive monoid $\mathbb{N}$, generated by 1. It follows that $F(X)$ 'is' the categorical sum $\sum_{x\in X}\mathbb{N}$ in Mon.

*In fact $F(X)$ can be constructed in a similar way as the free group in 4.1.3(c), by equivalence classes $[w]$ of words $w = x_1^{k_1} x_2^{k_2} ... x_p^{k_p}$, in the alphabet X with *natural* exponents. Rewriting the class of the word $x^2 xyz^2 x$ as $xxxyzzx$ we find again the simpler construction that we have used above.*

(b) The free commutative unitary ring $F(X)$ can be realised as the polynomial ring $\mathbb{Z}[x \mid x \in X]$ with integral coefficients and indeterminates belonging to the set X (possibly infinite). The basis X is embedded as a set of monomials, by $\eta(x) = x$.

$F(X)$ is thus the sum $\sum_{x\in X}\mathbb{Z}[x]$ in CRng, of polynomial rings in one indeterminate,

The reader may know that the polynomial ring $\mathbb{Z}[x_1, ..., x_n]$ is the n-th tensor power of the ring $\mathbb{Z}[x]$, and that finite sums in CRng are constructed as tensor products – as we have already recalled in 2.1.6(g). An infinite set X produces an inductive limit of finitely generated commutative rings, see 2.6.3(c).

(c) More generally, for a commutative unitary ring R, the free commutative unitary R-algebra can be realised as the polynomial ring $R[x \mid x \in X]$, with indeterminates in X.

Free rings are examined below, in 4.1.6(b).

4.1.5 Exercises and complements, III

Many interesting constructions in algebra arise as a left adjoint to a forgetful functor $U : A \to B$, between categories of algebraic structures of which the second is 'weaker' than the first – in the sense of having less structure, or less properties, or both.

We sketch some cases but the reader can compute many more. (A general result will be given in Theorem 4.4.5.)

(a) As we have already seen (in 3.2.6 and 3.8.4) the embedding Ab $\subset$ Gp has a left adjoint, given by the abelianised group $G^{\mathrm{ab}} = G/[G,G]$.

(b) The forgetful functor $U\colon \mathsf{Rng} \to \mathsf{Mon}$ has a left adjoint that takes a monoid M to a ring whose underlying abelian group is the free abelian group $\mathbb{Z}M$ generated by the *set* M. Its elements are the formal linear combinations $\sum_{x \in X} \lambda_x . x$ of elements of M, with quasi-null integral coefficients.

The multiplication of $\mathbb{Z}M$ is the bilinear extension of the multiplication of M

$$(\textstyle\sum_y \lambda_y.y).(\sum_z \mu_z.z) = \sum_{y,z} \lambda_y \mu_z(yz) = \sum_{x \in X} (\sum_{yz=x} \lambda_y \mu_z).x. \qquad (4.6)$$

This construction is more often considered for a group G, and known as the *group ring* $\mathbb{Z}G$, or $\mathbb{Z}[G]$.

(c) More generally, for a commutative unitary ring R, the forgetful functor $U\colon \mathsf{RAlg} \to \mathsf{Mon}$ has a left adjoint that takes a monoid M to the R-algebra whose linear part is the free R-module RM over the set M; the multiplication is defined as above, in (4.6), with scalars in R. Again this construction is more often considered for a group G, and known as the *group algebra* RG, or $R[G]$.

(d) The embedding $U\colon \mathsf{Ab} \to \mathsf{Abm}$ has a left adjoint, which can be built extending the well-known construction of $\mathbb{Z}$ as a quotient of $\mathbb{N} \times \mathbb{N}$; some caution is required, since an abelian monoid need not satisfy the cancellation law.

Given an abelian monoid M, the abelian group $F(M)$ can be obtained as the quotient of the monoid $M \times M$ modulo the congruence where $(x, y) \sim (x', y')$ when

$$x + y' + z = x' + y + z, \qquad (4.7)$$

for some $z \in M$. The unit-component $\eta M\colon M \to UF(M)$ sends x to the equivalence class $[x, 0]$; it is injective if and only if M satisfies the cancellation law.

4.1.6 Exercises and complements, IV

(a) Composing the forgetful functors $\mathsf{Ab} \to \mathsf{Gp} \to \mathsf{Set}$ we get the forgetful functor $\mathsf{Ab} \to \mathsf{Set}$. Since adjoints compose, contravariantly, we deduce that the abelianised group of the free group $\sum_{x \in X} \mathbb{Z}$ is the free abelian group $\bigoplus_{x \in X} \mathbb{Z}$, actually a rather obvious fact.

(b) It is more interesting to examine the composed forgetful functor

$$\mathsf{Rng} \to \mathsf{Mon} \to \mathsf{Set}.$$

The free ring on the set X can thus be constructed in two steps: first the free monoid M on X (by finite words on this alphabet, or 'non-commutative monomials', where $xyx \neq x^2y$, see 3.6.1); then the free ring $\mathbb{Z}M$ on this monoid (by $\mathbb{Z}$-linear combinations of these monomials, see 4.1.5(b)).

The result is the ring of 'non-commutative polynomials' with indeterminates in X and integral coefficients, where the 'polynomial' $2 + xyx - 3x^2y$ cannot be reduced to $2 - x^2y$.

A different (less effective) construction of the 'same' ring can be obtained from the factorisation Rng $\to$ Ab $\to$ Set.

4.2 Regular and Barr-exact categories

This is a brief introduction to the theory of regular and Barr-exact categories. Further information can be found in [Bar] and Borceux [Bo2], Chapter 2.

4.2.1 Definition

A *regular* category is a category with finite limits, where the kernel pair $R \rightrightarrows A$ of every map $f: A \to B$ (see 2.3.4) has a coequaliser; moreover, the pullback of any regular epi (see 2.4.5) along any arrow must be a regular epi.

(The texts [Bar, Bo2] give a more general definition, where the only limits assumed to exist are the kernel pairs and the pullback of any regular epi along any arrow; but this extension is presented in [Bo2] as "essentially a matter of personal taste".)

4.2.2 Theorem and Definition (Canonical factorisation)

In a regular category C *every map* f *has a canonical factorisation* $f = mp$, *unique up to isomorphism, formed by a regular epimorphism* p *(the coequaliser of its kernel pair) and a monomorphism* m.

Proof We form a diagram

$$
\begin{array}{ccccc}
R & \underset{v}{\overset{u}{\rightrightarrows}} & A & \overset{f}{\longrightarrow} & B \\
q\downarrow & & \downarrow p & \nearrow m & \\
S & \underset{s}{\overset{r}{\rightrightarrows}} & P & &
\end{array}
\tag{4.8}
$$

where (u, v) is the kernel pair of f and p is the coequaliser of (u, v), a regular epimorphism; then $fu = fv$, and there is a unique factorisation $f = mp$ through p. To prove that m is mono, we let (r, s) be its kernel pair, and it will be sufficient to prove that $r = s$ (by 2.3.5(a)).

From $m(pu) = m(pv)$ it follows that there is precisely one map $q\colon R \to S$ such that $rq = pu$ and $sq = pv$. Therefore $rq = sq$, and it will be sufficient to prove that q is epi.

We form a commutative diagram where the four horizontal squares are pullbacks and $v = v''v'$

$$(4.9)$$

Trivially, the back squares are also pullbacks. Then the two front squares are pullbacks as well, by Lemma 2.3.7. But a pullback of the regular epi p is a regular epi, whence also p', q' and q'' are, and $q''q'\colon R \to S$ is an epimorphism; it coincides with the morphism q of the first diagram because $r(q''q') = pu = rq$ and $s(q''q') = pv = sq$.

Finally the essential uniqueness of a factorisation regular epi - mono is obvious: if $f = mp$ is such a factorisation, then (u, v) is also the kernel pair of p, and the latter is easily seen to be the coequaliser of (u, v), whence our factorisation is determined up to isomorphism. □

4.2.3 Corollary

In a regular category, regular epimorphisms coincide with strong epimorphisms and are closed under composition.

As a consequence, regular epimorphisms and general monomorphisms form a factorisation system (rEpi, Mono) *(as defined in Section 2.5).*

Proof It is an easy consequence of the previous theorem and Section 2.4.

In fact, every regular epimorphism is strong. Conversely, if $f = mp$ is the canonical factorisation of a strong epimorphism, the monomorphism m is a strong epi, and therefore invertible, while f is a regular epi. □

4.2.4 Definition (**Equivalence relations**)

In a category C with finite limits, an (internal) *equivalence relation* in the object A will be a jointly monic pair $u, v\colon R \to A$ (or equivalently a monomorphism $(u, v)\colon R \rightarrowtail A \times A$) that satisfies the following conditions:

(i) (*reflexivity*) the diagonal $d_A = (1, 1)\colon A \rightarrowtail A \times A$ is a smaller monomorphism: namely, there exist a (unique mono) $h\colon A \to R$ such that $uh = 1_A = vh$,

(ii) (*symmetry*) the monomorphism $(u, v)\colon R \rightarrowtail A \times A$ is equivalent to the monomorphism (v, u): namely, there exist a morphism $h\colon R \to R$ such that $uh = v$ and $vh = u$ (and then h is an involutive isomorphism: $hh = 1_R$),

(iii) (*transitivity*) in the following diagram, where (P, r, s) is a pullback

$$A \xleftarrow{\;u\;} R \xrightarrow{\;v\;} A \xleftarrow{\;u\;} R \xrightarrow{\;v\;} A$$

$$\begin{array}{ccc} & R \searrow_{r} & P & \nearrow_{s} R \end{array} \tag{4.10}$$

there exist a morphism $h\colon P \to R$ such that $uh = ur$ and $vh = vs$.

4.2.5 Exercises

(a) Verify that the conditions (i)–(iii) above amount to the usual ones, in Set.

(b) Prove that, in a category with finite limits, a kernel pair (u, v) is always an equivalence relation.

4.2.6 Definition

A *Barr-exact* category [Bar] is a regular category C where every equivalence relation is *effective*, i.e. a kernel pair of a morphism.

Equivalently, one can require that every equivalence relation be the kernel pair of its coequaliser. This is plainly true in Set, but we shall see that every variety of algebras is Barr-exact, in Theorem 4.4.4.

4.3 Varieties of algebras

Universal algebra is studied in well known texts, like Grätzer [Gr1] and Cohn [Coh].

As we have already recalled in 1.1.6, and will be made precise below, a 'variety of algebras' includes all algebraic structures of a given signature, satisfying a given set of equational axioms (or universally quantified identities).

We also recall that, as in [Coh], we do not assume that the underlying set of an algebra should be non empty: a convention which would destroy all the main results below, from the existence of all limits and all free algebras, to the monadicity of the forgetful functor with values in Set.

However, adapting a result of [Gr1] to the present approach is quite obvious, generally speaking.

4.3.1 Signatures and algebras

An (algebraic) *signature* (Ω, ar) is a set Ω equipped with a mapping $\mathrm{ar} \colon \Omega \to \mathbb{N}$. Each $\omega \in \Omega$ with $\mathrm{ar}(\omega) = n$ is called an *operator* of *arity n*, or an *n-ary* operator; their set is denoted as $\Omega(n)$. The signature will be written as Ω, unless this leads to ambiguity.

An Ω-*algebra* is a set A equipped with a family of *operations* (ω_A) indexed by Ω

$$\omega_A \in \mathsf{Set}(A^n, A), \qquad \text{for } n \geqslant 0 \text{ and } \omega \in \Omega(n), \tag{4.11}$$

and $\omega_A \colon A^n \to A$ is called the *realisation* of the operator ω in A. The underlying set of A will be written as $|A|$ when useful.

A *homomorphism* $f \colon A \to B$ is a mapping $|A| \to |B|$ that preserves all the operations

$$f.\omega_A = \omega_B.f^n \colon A^n \to B, \qquad \text{for } \omega \in \Omega(n). \tag{4.12}$$

We have thus the category $\mathsf{Alg}(\Omega)$ of Ω-algebras and their homomorphisms, made concrete by the underlying-set functor, which is faithful

$$U \colon \mathsf{Alg}(\Omega) \to \mathsf{Set}, \qquad U(A) = |A|. \tag{4.13}$$

A singleton $\{x\}$ has a unique structure of Ω-algebra, and gives the terminal object $\top$ of $\mathsf{Alg}(\Omega)$.

A 0-ary operator $\omega \in \Omega(0)$ is also called a *constant*, and provides every Ω-algebra A with an element $\omega_A \in A$, so that $|A|$ cannot be empty. If Ω does not contain any 0-ary operator, the empty set $\emptyset$ has a unique structure of Ω-algebra and gives the initial object $\bot$ of $\mathsf{Alg}(\Omega)$.

As an example, the signature Ω of semigroups consists of a single binary operator μ, called *multiplication*. An Ω-algebra is *any* set A equipped with a binary operation $\mu_A(x, y) = xy$; it is a semigroup when the associativity axiom is satisfied.

This is expressed as a pair of 'derived operators', the ternary multiplications

$$\xi' = \mu(p_1^3, \mu(p_2^3, p_3^3)), \qquad \xi'' = \mu(\mu(p_1^3, p_2^3), p_3^3), \tag{4.14}$$

which must give the same 'derived operations' on A

$$\xi_A' \colon A^3 \to A, \qquad \xi_A'(x,y,z) = x(yz),$$
$$\xi_A'' \colon A^3 \to A, \qquad \xi_A''(x,y,z) = (xy)z. \tag{4.15}$$

All this will be formalised defining the clone $\overline{\Omega}$ of derived operators of a signature Ω.

It is important to note that the axioms which one can express in this form are 'universally quantified equations' like, in the previous case:

$$\text{for all } x, y, z \text{ in } A \colon \qquad x(yz) = (xy)z. \tag{4.16}$$

The derived operators will be constructed making use of new formal n-ary operators p_i^n (for $n > 0$ and $i = 0, 1, ..., n$) called *projectors*. They operate on every set A as cartesian projections, or the terminal map

$$p_i^n \colon A^n \to A, \qquad p_i^n(x_1, ..., x_n) = x_i, \quad \text{for } i = 1, ..., n,$$
$$p_0^n \colon A^n \to \{*\}, \qquad p_0^n(x_1, ..., x_n) = *, \tag{4.17}$$

In particular we write p_1^1 as id. The reader will note that the projector p_0^1 plays a role in expressing the axiom $xx^{-1} = 1$ of groups.

4.3.2 A variety of algebras

A signature Ω has a *clone of derived operators* $\overline{\Omega}$ inductively constructed as follows. We start from the signature $(\Omega_0, \mathrm{ar}_0)$, formed by adding to (Ω, ar) all the projectors p_i^n $(0 \leqslant i \leqslant n)$, with arity $n \geqslant 0$. Then $(\Omega_{k+1}, \mathrm{ar}_{k+1})$ is obtained by adding to $(\Omega_k, \mathrm{ar}_k)$, in a disjoint way, new operators of the following form

$$\xi = \omega(\xi_1, ..., \xi_p) \tag{4.18}$$
$$\text{for } \omega, \xi_1, ..., \xi_p \in \Omega_k, \quad \mathrm{ar}_k(\omega) = p, \quad \mathrm{ar}_k(\xi_i) = n,$$

and extending ar_k by letting $\mathrm{ar}_{k+1}(\xi) = n$.

Finally $\overline{\Omega}$ is the union of the increasing sequence of sets (Ω_k), with the extended arity ar. The elements of $\overline{\Omega}$ are called the *derived operators* of Ω.

Every Ω-algebra A can be viewed as an $\overline{\Omega}$-algebra, with the following *clone of derived operations*:

- the projector p_i^n is realised on the set A as specified above, in (4.17),

- having defined the realisation in A of all the operators of Ω_k, we realise the operator ξ of formula (4.18) as:

$$\xi_A \colon A^n \to A,$$
$$\xi_A(x_1, ..., x_n) = \omega_A(\xi_{1A}(x_1, ..., x_n), ..., \xi_{pA}(x_1, ..., x_n)). \tag{4.19}$$

An *axiom* for the signature Ω is a pair (ξ', ξ'') of derived operators with the same arity. An *equational signature* (Ω, Ξ) is a signature equipped with a set Ξ of axioms.

We write $\mathsf{Alg}(\Omega, \Xi)$ the full subcategory of $\mathsf{Alg}(\Omega)$ of all (Ω, Ξ)-*algebras*, i.e. all Ω-algebras A that satisfy the axioms of Ξ. This plainly means that $\xi'_A = \xi''_A$, for all the axioms $(\xi', \xi'') \in \Xi$.

The category $\mathsf{Alg}(\Omega, \Xi)$ is called a *variety of algebras*. It is made concrete by the restriction $\mathsf{Alg}(\Omega, \Xi) \to \mathsf{Set}$ of the forgetful functor of $\mathsf{Alg}(\Omega)$; the latter coincides with $\mathsf{Alg}(\Omega, \emptyset)$.

Let us note that each singleton $\{x\}$ satisfies any axiom, and gives the terminal object $\top$ of any variety $\mathsf{Alg}(\Omega, \Xi)$.

4.3.3 Exercises and complements

(a) The forgetful functor $\mathsf{Alg}(\Omega, \Xi) \to \mathsf{Set}$ reflects the isomorphisms; in other words a homomorphism $f : A \to B$ is invertible if and only if the underlying mapping $|f|$ is bijective. It also reflects the identities.

(b) Set is the category of algebras for $\Omega = \emptyset$.

(c) Describe $\mathsf{Set}_\bullet$ as a category of algebras (up to isomorphism of categories, of course).

(d) Describe Ab as a variety $\mathsf{Alg}(\Omega, \Xi)$, where the signature Ω consists of three operators: the *unity* ζ, the *opposite* ι and the *addition* σ (of arity 0, 1 and 2, respectively).

(e) Describe $R\,\mathsf{Mod}$ as a variety $\mathsf{Alg}(\Omega, \Xi)$, where the signature Ω consists of the previous operators, with the addition of a unary operator λ for every scalar $\lambda \in R$.

(f) Describe the category Lth of (bounded) lattices and homomorphisms as a variety $\mathsf{Alg}(\Omega, \Xi)$, where the signature Ω has two constants $0, 1$ and two binary operations $\vee$, $\wedge$.

(g) Describe the category of boolean algebras as a variety $\mathsf{Alg}(\Omega, \Xi)$.

(h) The category Fld of (commutative) fields has poor categorical properties. Let us begin by observing that a homomorphism $f : F \to F'$ is necessarily injective (its kernel must be the null ideal) and induces an isomorphism between the minimal subfields of F and F'; the latter have thus the same characteristic p, which is 0 or a prime number. Fld is the categorical sum of its full subcategories Fld_p, of the fields of characteristic p.

The reader will prove that Fld and each Fld_p lack a terminal object. Therefore, by a remark in 4.3.1, these categories cannot be isomorphic, or even equivalent, to any variety $\mathsf{Alg}(\Omega, \Xi)$.

4.3.4 Trivial varieties

There are two *trivial* varieties of algebras. The first is formed by all the sets of cardinality 0 or 1, and can be obtained as $\mathsf{Alg}(\emptyset, \Xi)$, where Ξ contains precisely one axiom, namely (p_1^2, p_2^2) (or $x = y$, informally). In fact, this axiom gives the full subcategory of Set of all the sets X whose projections $X^2 \to X$ coincide.

The second is formed by all the sets of cardinality 1. The reader can describe it as a variety $\mathsf{Alg}(\Omega, \Xi)$, where Ω contains one 0-ary operator.

We shall see in 4.4.3 that there are no other varieties having finitely many isomorphism-types of algebras, or equivalently a finite skeleton.

(Marginally, we also note that the category $\mathbf{1}$, consisting of the empty set and its identity, is not – strictly speaking – a variety of algebras, because we have alredy remarked that the singletons belong to any variety. But of course $\mathbf{1}$ is *equivalent* to the trivial variety of all singletons.)

4.3.5 Subalgebras

For an (Ω, Ξ)-algebra A, a *subalgebra* is a subset $C \subset |A|$ closed under the operations of A:

- for every n-ary operator $\omega \in \Omega$ and every n-tuple $(x_1, ..., x_n) \in C^n$, $\omega_A(x_1, ..., x_n) \in C$.

The set C is viewed as an (Ω, Ξ)-algebra, by means of the restricted operations ω_C

$$\omega_C(x_1, ..., x_n) = \omega_A(x_1, ..., x_n), \qquad \text{for } (x_1, ..., x_n) \in C^n, \qquad (4.20)$$

which automatically satisfy the axioms of Ξ. This is the unique structure of C that makes the inclusion $C \subset A$ into a homomorphism, and therefore a monomorphism of $\mathsf{Alg}(\Omega, \Xi)$.

The set-theoretical intersection of any family (A_i) of subalgebras of an algebra A is a subalgebra $\cap A_i$. Each subset $S \subset |A|$ *generates* a subalgebra $\langle S \rangle$, namely the intersection of all subalgebras containing it. $\langle S \rangle$ can be obtained as the union of all subsets $S_k \subset A$, where $S_0 = S$ and

$$S_{k+1} = \{\omega_A(x_1, ..., x_n) \mid n \geqslant 0, \ \omega \in \Omega(n), \ x_1, ..., x_n \in S_k\}.$$

This proves that

$$\sharp \langle S \rangle \ \leqslant \ \max\{\sharp S, \sharp \Omega, \aleph_0\}, \qquad (4.21)$$

where $\sharp$ denotes the cardinal of a set (see 3.5.3).

A homomorphism $f : A \to B$ of (Ω, Ξ)-algebras has a (set-theoretical) image $f(A)$ which is a subalgebra of B, since

$$\omega_B(fx_1, ..., fx_n) = f(\omega_A(x_1, ..., x_n)) \qquad \text{for } x_i \in A.$$

4.3.6 Congruences

A *congruence* R in an (Ω, Ξ)-algebra A is an equivalence relation in the set $|A|$ which is consistent with the operations of A:

- for every n-ary operator $\omega \in \Omega$ and every pair $(\underline{x}, \underline{y}) \in R^n$, we have $(\omega_A(\underline{x}), \omega_A(\underline{y})) \in R$,

where (by an abuse of notation) $R^n \subset (|A| \times |A|)^n = |A|^n \times |A|^n$ is the associated equivalence relation of the set $|A|^n$. (The correspondence of this notion with an 'internal' equivalence relation, in the sense of 4.2.4, will be established in Theorem 4.4.4.)

The quotient set $C = |A|/R$ becomes an (Ω, Ξ)-algebra, by means of the induced operations ω_C

$$\omega_C([x_1], ..., [x_n]) = [\omega_A(x_1, ..., x_n)], \qquad \text{for } (x_1, ..., x_n) \in A^n, \qquad (4.22)$$

which automatically satisfy the axioms Ξ. This is the unique structure of C that makes the projection $p \colon |A| \to |A|/R$ into a homomorphism $p \colon A \to A/R$, and therefore an epimorphism of $\mathsf{Alg}(\Omega, \Xi)$.

We prove below, in Theorem 4.4.4, that $p \colon A \to C$ is a regular epimorphism of $\mathsf{Alg}(\Omega, \Xi)$.

For any family (R_i) of congruences of an algebra A, their set-theoretical intersection $\cap R_i$ is again a congruence. Therefore each relation R in the set $|A|$ *generates* a congruence $\overline{R}$, namely the intersection of all congruences containing R.

A homomorphism $f \colon A \to B$ of (Ω, Ξ)-algebra has a (set-theoretical) equivalence relation R in $|A|$, which is a congruence of A: if $f^n(\underline{x}) = f^n(\underline{y})$ then

$$f\omega_A(\underline{x}) = \omega_B(f^n\underline{x}) = \omega_B(f^n\underline{y}) = f\omega_A(\underline{y}).$$

4.3.7 Canonical factorisation

Putting together the previous results, every homomorphism $f \colon A \to B$ of (Ω, Ξ)-algebras has a canonical factorisation (derived from the canonical factorisation of the underlying mapping $|f|$), where

$$f = mgp \colon A \to A/R \to f(A) \to B, \qquad (4.23)$$

- $p \colon A \to A/R$ is the projection on the quotient of A modulo the congruence associated to f,

- $m\colon f(A) \to B$ is the inclusion of the subalgebra $f(A) \subset B$,

- g is the homomorphism produced by the bijection $[x] \to f(x)$, which preserves all operations:

$$g(\omega_C([x_1], ..., [x_n]) = g[\omega_A(x_1, ..., x_n)] = f(\omega_A(x_1, ..., x_n))$$
$$= \omega_B(fx_1, ..., fx_n) = \omega_{f(A)}(g[x_1], ..., g[x_n]),$$

and is an isomorphism of (Ω, Ξ)-algebras, by 4.3.3(a).

4.3.8 *Complements

A famous Birkhoff Theorem proves that a full subcategory C of $\mathsf{Alg}(\Omega)$ (containing the singleton) is a variety of (Ω, Ξ)-algebras (for some set Ξ of axioms) if and only if C is closed in $\mathsf{Alg}(\Omega)$ under *subalgebras, homomorphic images and cartesian products.*

The interested reader can see the proof in [Gr1], Section 2.6, Theorem 3 or [Coh], Chapter 4, Theorem 3.5.

4.4 Limits and free algebras

This section takes on the study of the categories $\mathsf{Alg}(\Omega, \Xi)$, showing that they are complete and Barr-exact. Moreover all of them are monadic over Set, and cocomplete.

4.4.1 Theorem (Limits in a variety)

(a) The category $\mathsf{Alg}(\Omega, \Xi)$ *has all limits and coequalisers of kernel pairs, created by the forgetful functor* $U\colon \mathsf{Alg}(\Omega, \Xi) \to \mathsf{Set}$ *(in the sense of 2.2.8).*

(b) The forgetful functor U *also creates all filtered colimits.*

Proof (a) Let S be a small category and $X\colon \mathsf{S} \to \mathsf{Alg}(\Omega, \Xi)$ a functor, written in index notation, for $i \in S = \mathrm{Ob}\mathsf{S}$ and $a\colon i \to j$ in S:

$$X\colon \mathsf{S} \to \mathsf{C}, \qquad i \mapsto X_i, \quad a \mapsto (u_a\colon X_i \to X_j). \qquad (4.24)$$

Let $(L, (u_i))$ be the limit-cone of UX in Set, with structural mappings $u_i\colon L \to |X_i|$ $(i \in S)$. For every $n \in \mathbb{N}$, L^n is the limit of the functor $(UX)^n$, with the mappings $(u_i)^n\colon L^n \to |X_i|^n$. For every n-ary operator $\omega \in \Omega$ we have a natural transformation

$$\omega\colon (UX)^n \to UX\colon \mathsf{S} \to \mathsf{Set}, \qquad \omega_i = \omega X_i\colon |X_i|^n \to |X_i|, \qquad (4.25)$$

(because each homomorphism $u_a \colon X_i \to X_j$ must be consistent with ω), and therefore a unique mapping ω_L (in Set) consistent with the cones

$$\omega_L \colon L^n \to L, \quad u_i.\omega_L = \omega X_i.(u_i)^n \colon L^n \to |X_i| \qquad \text{for } i \in S. \qquad (4.26)$$

There is thus a unique structure $(\omega_L)_{\omega \in \Omega}$ of Ω-algebra on the set L which makes all the mappings $u_i \colon L \to X_i$ into homomorphisms of Ω-algebras. It is easy to see that the limit is actually in $\mathsf{Alg}(\Omega, \Xi)$, and therefore a limit in the latter.

Finally, every homomorphism $f \colon A \to B$ has a kernel pair $R \rightrightarrows A$, where R is the set-theoretical congruence of $|f|$, with the structure of subalgebra of the product $A \times A$. We have seen in 4.3.6 that the coequaliser $|A|/R$ in Set has one algebraic structure that makes the projection $p \colon A \to A/R$ a homomorphism; it is easy to deduce that p is the algebraic coequaliser, which is thus created by U.

(b) If S is a filtered category, one can repeat the previous argument, taking into account that filtered colimits in Set commute with finite products (by Theorem 2.6.6). $\qquad \square$

4.4.2 Theorem (Free algebras)

The forgetful functor $U \colon \mathsf{Alg}(\Omega, \Xi) \to \mathsf{Set}$ has a left adjoint: for every set X there exists a free (Ω, Ξ)-algebra $F(X)$, equipped with a universal arrow $\eta X \colon X \to UF(X)$.

Proof We use the Adjoint Functor Theorem, extending the argument of Exercise 3.5.4(a) for the free group.

We know that $\mathsf{Alg}(\Omega, \Xi)$ has small hom-sets and all small limits, preserved by U. Moreover, every set X has a solution set $S(X)$ (in the simplified form 3.5.2(iii)), consisting of the small set of all (Ω, Ξ)-algebras whose underlying set is (precisely) a cardinal $\alpha \leqslant \beta_X = \max\{\sharp X, \sharp \Omega, \aleph_0\}$.

The details are left to the reader. $\qquad \square$

4.4.3 Corollary (The embedding of the basis)

The following conditions on a variety $\mathsf{Alg}(\Omega, \Xi)$ are equivalent:

(i) there exists an algebra with more than one element,

(ii) for every set X the insertion of the basis $\eta X \colon X \to UF(X)$ is injective,

(iii) $\mathsf{Alg}(\Omega, \Xi)$ has an infinite skeleton, with infinitely many types of isomorphism,

(iv) $\mathsf{Alg}(\Omega, \Xi)$ *is none of the two 'trivial varieties' of 4.3.4.*

Proof Let us assume (i). The existence of small products of algebras preserved by U proves that there are algebras of cardinality higher than any given small cardinal. Therefore their cardinals (and their types of isomorphism) cannot form a finite set. Moreover any set X can be embedded in the underlying set of some algebra, and ηX is injective (by 2.7.5(a)).

Conversely, if no algebra has more than one element, then $\mathsf{Alg}(\Omega, \Xi)$ can only contain the empty set and the singletons, with their unique (Ω, Ξ)-structure; since all the singletons must be there, we fall in one of the trivial varieties of 4.3.4. Then conditions (ii) and (iii) fail as well.   □

4.4.4 *Theorem* (Varieties and Barr-exactness)

A variety $\mathsf{Alg}(\Omega, \Xi)$ *is a complete Barr-exact category.*

More particularly, we have the following points, for every algebra A.

(i) The forgetful functor $U\colon \mathsf{Alg}(\Omega, \Xi) \to \mathsf{Set}$ *gives a bijective correspondence between equivalence relations* $R \rightarrowtail A^2$ *(as defined in 4.2.4) and congruences of A (as defined in 4.3.6). Every quotient* $p\colon A \to A/R$ *modulo a congruence is a regular epimorphism, whence a strong one. Every strong (or regular) epimorphism defined on A is of this kind, up to equivalence of epimorphisms.*

(ii) A monomorphism $i\colon C \to A$ *is the same as an injective homomorphism, and is equivalent to the embedding of a subalgebra. The subobjects of A can (and will) be identified to subalgebras, and form a complete lattice* $\mathsf{Sub}(A)$.

(iii) The factorisation $f = m.(ip)$ *of 4.3.7 is the canonical factorisation regular epi - mono of a regular category. It is strictly determined, if we require that m be the inclusion of a subalgebra.*

Proof We already know, by Theorem 4.4.1, that $\mathsf{Alg}(\Omega, \Xi)$ has all limits and coequalisers of kernel pairs, created by the forgetful functor with values in Set. It follows that regular epimorphisms are closed under pullback, as in Set.

Let $f = (u, v)\colon R \rightarrowtail A^2$ be an equivalence relation, as in 4.2.4. Then $|f|\colon |R| \rightarrowtail |A|^2$ is an equivalence relation of $|A|$, and we can think of $|f|$ as an inclusion. Moreover, $|R|$ is a congruence of algebras, because for every n-ary operator ω and $(\underline{x}, \underline{y}) \in |R|^n \subset |A|^n \times |A|^n$

$$(\omega_A(\underline{x}), \omega_A(\underline{y})) = \omega_{A^2}(f^n(\underline{x}, \underline{y})) = f(\omega_R(\underline{x}, \underline{y})) \in |R|.$$

The quotient $p\colon A \to A/R$ (described in 4.3.6) is turned by U into the coequaliser of the kernel pair $|f|\colon |R| \rightarrowtail |A|^2$ of Up; whence $(u,v)\colon R \rightarrowtail A^2$ is the kernel pair of p, which is the coequaliser of (u,v).

We have proved that $\mathsf{Alg}(\Omega, \Xi)$ is complete and Barr-exact. We have also proved point (i), taking into account that in a regular category strong and regular epimorphisms coincide.

Point (ii) follows from Theorem 4.4.2: the free (Ω, Ξ)-algebra on the singleton exists, therefore the (faithful) forgetful functor $U\colon \mathsf{Alg}(\Omega, \Xi) \to \mathsf{Set}$ is representable, and preserves and reflects monomorphisms (by 2.3.5(a)). Moreover, we have already seen in 4.3.5 that the subalgebras of A are closed under arbitrary intersections.

Point (iii) is a consequence. $\square$

4.4.5 Theorem (Comparing algebras)

Let us be given two equational signatures (Ω, Ξ), (Ω', Ξ'), with $\Omega' \subset \Omega$ and $\Xi' \subset \Xi$. Then the obvious forgetful functor

$$W\colon \mathsf{Alg}(\Omega, \Xi) \to \mathsf{Alg}(\Omega', \Xi')$$

has a left adjoint. W creates limits and coequalisers of kernel pairs.

Proof This is an extension of Theorems 4.4.2 and 4.4.1, since $\mathsf{Set} = \mathsf{Alg}(\emptyset, \emptyset)$.

The proof of the existence of the left adjoint of W is similar to that of Theorem 4.4.2. For the second part of the statement, we consider the forgetful functors

$$\mathsf{Alg}(\Omega, \Xi) \xrightarrow{\ W\ } \mathsf{Alg}(\Omega', \Xi') \xrightarrow{\ V\ } \mathsf{Set} \qquad U = VW. \qquad (4.27)$$

We know that U and V create all limits. Let S be a small category with $S = \mathsf{ObS}$; all families below have indices $i \in S$.

For a functor $A\colon \mathsf{S} \to \mathsf{Alg}(\Omega, \Xi)$, consider the composites WA and VWA. The latter has a limit $(L'', (r_i\colon X \to VWA_i))$. The limit of A in $\mathsf{Alg}(\Omega, \Xi)$ is the unique cone $(L, (p_i\colon X \to WA_i))$ taken by $U = VW$ to the limit cone $(L'', (r_i))$. Similarly, the limit of WA in $\mathsf{Alg}(\Omega', \Xi')$ is the unique cone $(L', (q_i\colon X \to WA_i))$ taken by V to the cone $(L'', (r_i))$.

The functor W takes $(L, (p_i))$ to a cone of WA, and there is a unique arrow $g\colon WL \to L'$ such that $q_i g = W(p_i)$, for all $i \in S$. But $Vp_i = r_i = V(Wp_i)$, for all i, whence the underlying mapping $V(g)$ is the identity and the homomorphism $g\colon WL \to L'$ is also.

The same argument holds for coequalisers of kernel pairs, which are also created by U and V. $\square$

*4.4.6 *Theorem* (Varieties and monadicity)

The forgetful functor $U \colon \mathsf{Alg}(\Omega, \Xi) \to \mathsf{Set}$ *is monadic.*

Proof This important result proves that each variety of algebras is a category of Eilenberg–Moore algebras, for the monad associated to the forgetful functor.

The proof can be found in Mac Lane's text [M4], Section VI.8. The argument is based on Beck's Monadicity Theorem (or Tripleability Theorem), in [M4], Section VI.7. □

*4.4.7 *Exercises and complements* (Colimits)

The interested reader can prove that every variety $\mathsf{Alg}(\Omega, \Xi)$ is also cocomplete.

(a) The existence of arbitrary coequalisers follows easily from 4.3.6.

(b) The existence of small sums can be proved with the Adjoint Functor Theorem, using the completeness of the lattices of subalgebras and the upper bound (4.21) for the cardinal of a subalgebra generated by a subset.

(c) In particular $\mathsf{Alg}(\Omega, \Xi)$ always has an initial object, whose underlying set can take different forms: for instance in Set, Mon, Rng, RAlg, etc.

4.5 Relations for regular categories

We begin by re-examining the ordered category Rel Set, viewed as a *locally ordered 2-category*. This construction is extended to a regular category C, forming the ordered category of relations Rel (C), where a relation $X \to Y$ is a subobject of the product $X \times Y$ (as in [Grt, Bo1]).

Marginally, we briefly sketch a second approach to relations (followed for instance in [Me]), where a relation $X \to Y$ is *any* monomorphism with values in $X \times Y$. In this way one gets a laxer structure, called a *locally preordered bicategory*, which is written here as Rel′(C).

These procedures can also be useful as an introduction to general 2-categories and bicategories, that will be dealt with in Section 7.1.

4.5.1 Cartesian relations and locally ordered 2-categories

Let us recall that R = Rel Set has been constructed in 1.8.3, as an involutive ordered category.

A relation $r \colon X \to Y$ is a subset $r \subset X \times Y$, and will be called a *cartesian relation* when we want to distinguish it from the similar item of 4.5.5 (a monomorphism $r\colon \bullet \rightarrowtail X \times Y$). Their composition has been recalled in (1.64). Given two parallel relations $r, s\colon X \rightarrow Y$, the ordering $r \leqslant s$ means that $r \subset s$ as subsets of $X \times Y$; it agrees with composition.

In this way, each hom-set $\mathsf{R}(X, Y)$ is the ordered set $\mathcal{P}(X \times Y)$, and the relation $r\colon X \rightarrow Y$ can be represented as in the diagram below, where the vertical map is the inclusion of a subset

$$X \xleftarrow{\ p'\ } X \times Y \xrightarrow{\ p''\ } Y$$

$$\uparrow r \tag{4.28}$$

An ordered category can be viewed as an elementary form of a 2-category, and an introduction to this higher dimensional notion that will be presented in Section 7.1 – in the same way as an ordered set is an elementary form of a category. In this perspective, we think of the relationship $r \leqslant s$ in $\mathsf{R}(X, Y)$ as a 2-*cell* $r \rightarrow s$ (determined by r, s), i.e. a higher arrow between parallel morphisms.

These higher arrows have operations of vertical composition and whisker composition as in 1.5.3 (for natural transformations), but much simpler – since here the result is determined by the 'boundaries'

$$X \;\substack{\xrightarrow{r} \\ \downarrow \\ \xrightarrow{} \\ \downarrow \\ \xrightarrow{t}}\; Y \qquad . \qquad X' \xrightarrow{u} X \;\substack{\xrightarrow{r} \\ \downarrow \\ \xrightarrow{s}}\; Y \xrightarrow{v} Y' \tag{4.29}$$

Namely, the vertical composition of $r \rightarrow s$ with $s \rightarrow t$ is the (unique) 2-cell $r \rightarrow t$, and the whisker composition above, at the right, is the (unique) 2-cell $vru \rightarrow vsu$.

This 'elementary form' is called a *locally ordered 2-category*, where 'locally' refers to the hom-sets. (The definition of a general 2-category will be seen in 7.1.2; here most of the axioms are automatically satisfied, because of the uniqueness of a 2-cell between two given arrows.)

4.5.2 The basic choices

Let C be a regular category. Extending the previous point, we want to construct an involutive ordered category $\mathsf{R} = \mathrm{Rel}\,(\mathsf{C})$ of relations on C, in the 'cartesian form'.

For this goal, we assume that C is equipped with a choice of binary products and subobjects. (Let us note that, in many concrete cases, these selections can be made without any use of the axiom of choice, by cartesian products and subsets in Set, via a forgetful functor $U\colon \mathsf{C} \rightarrow \mathsf{Set}$.)

As a consequence, every morphism $f\colon X \to Y$ has *precisely one* factorisation $f = mp$ where p is a regular epi and m is a subobject of Y, and we shall write

$$f = mp, \qquad m = \mathrm{im}\,(f)\colon \mathrm{Im}\,(f) \to Y. \qquad (4.30)$$

(We are using a *strict* factorisation system in C, see 2.5.3(f).)

4.5.3 Constructing the category of relations

For every pair of objects X, Y we let $\mathsf{R}(X, Y)$ be the ordered set of subobjects of the product $X \times Y$.

Given two consecutive relations $r = (r', r'')\colon R \rightarrowtail X \times Y$ and $s = (s', s'')\colon S \rightarrowtail Y \times Z$, their composite $s.r = w$ is a subobject of $X \times Z$, determined by a construction based on *a* pullback P of (r'', s'), and the strict factorisation (4.30), where p is a regular epi and w is a subobject

$$ (4.31) $$

$$ P \xrightarrow{\;p\;} W \xrightarrow{\;w\;} X \times Z $$
$$ \underset{(r'u', s''u'')}{} $$

$$ s.r = \mathrm{im}\,(r'u', s''u'')\colon W \rightarrowtail X \times Z. \qquad (4.32) $$

The reader will note that any choice of the pullback gives the same subobject of $X \times Z$. The identity 1_X is the subobject associated to the diagonal monomorphism

$$ 1_X = \mathrm{im}\,((1,1)\colon X \rightarrowtail X \times X) = (u, u)\colon \Delta_X \rightarrowtail X \times X, \qquad (4.33) $$

where u is an isomorphism. The involution sends a subobject $r\colon R \rightarrowtail X \times Y$ to the 'opposite' relation

$$ r^{\sharp} = \mathrm{im}\,(ir\colon R \rightarrowtail X \times Y \to Y \times X), \qquad (4.34) $$

where $i = (p_2, p_1)\colon X \times Y \to Y \times X$ is the canonical symmetry of the cartesian product.

For two parallel relations $r, r'\colon X \to Y$, the order relation

$$ r \leqslant r' \qquad (4.35) $$

has already been defined, as the canonical order of subobjects of $X \times Y$.

4.5.4 Theorem and Definition

These definitions produce an involutive ordered category $R = \mathrm{Rel}\,(C)$, *called the category of relations over the regular category* C.

The embedding $C \to \mathrm{Rel}\,(C)$ *is the identity on the objects and sends the morphism* $f\colon X \to Y$ *to its graph, namely the subobject associated to the monomorphism* $(1, f)\colon X \to X \times Y$.

The category R *has small hom-sets if and only if every object of* C *has a small set of subobjects.*

Proof First, the associativity of the composition law, for three relations $r\colon X \to Y$, $s\colon Y \to Z$, $t\colon Z \to U$, is proved by the following diagram

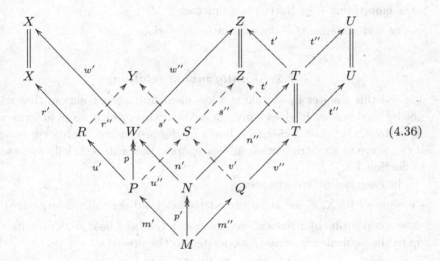

$$(4.36)$$

At the lower level we have a symmetric construction, based on three pullbacks (P, u', u''), (Q, v', v'') and (M, m', m''). At the upper level we first compute $sr = \mathrm{im}\,(r'u', s''u'') = (w', w'')\colon W \rightarrowtail X \times Z$, then the pullback (N, n', n'') of (w'', t'), so that $t(sr) = \mathrm{im}\,(N \to X \times U)$.

Now, the square $pm' = n'p'$ is a pullback, by Lemma 2.3.7, and p is a regular epi, whence p' is also. But then $\mathrm{im}\,(N \to X \times U)$ coincides with $\mathrm{im}\,(M \to X \times U)$, and this subobject of $X \times U$ is determined by the *symmetric* construction of the lower level.

By symmetry, $(ts)r$ too coincides with this subobject.

Second, the identity $1_X\colon \Delta_X \rightarrowtail X \times X$ defined above gives $r1_X = r$

$$
\begin{array}{ccccccc}
X & & X & & & Y \\
& \nwarrow\!u & \nearrow\!u & \nwarrow\!r' & & \nearrow\!r'' & \\
& \Delta_X & & & R & & \\
& \searrow\!v & & \nearrow\!1 & & & \\
& & R & & & &
\end{array}
\tag{4.37}
$$

because u is an isomorphism and we can take R itself as the pullback of (u, r'); therefore the image of $(uv, r''.1) = (r', r'')\colon R \rightarrowtail X \times Y$ is precisely the relation r. Symmetrically, $1_Y s = s$.

We leave to the interested leader the verification of the remaining points:

- the involution defined in (4.34) is consistent with the composition (in a contravariant way),

- the order defined in (4.35) is consistent with composition and involution,

- the embedding $\mathsf{C} \to \mathrm{Rel}\,(\mathsf{C})$ is a functor.

The last statement, about smallness, is obvious. $\qquad\qquad\square$

4.5.5 *Jointly monic relations

We end this chapter by sketching a less elementary construction, that will not be used below. It gives a preordered structure, which we shall write as $\mathsf{R}' = \mathrm{Rel}\,'\mathsf{C}$. It is not a category, but a *locally preordered bicategory*; again, it can serve as an introduction to bicategories, that will be briefly analysed in Section 7.1.

The essential differences are:

- a hom-set $\mathsf{R}'(X, Y)$ is now a *preordered* set (and generally a large one),

- the composition of arrows is 'weakly associative' and has 'weak identities', *up* to the equivalence relation associated to the preorder.

A *jointly monic relation* $r = (r', r'')\colon X \to Y$ is a jointly monic pair of morphisms of C, as in the left diagram below, with an arbitrary domain R (it is also called a jointly monic *span*, from X to Y)

$$
\begin{array}{ccccccccc}
X & & & & Y & \qquad & X & \xleftarrow{\;s'\;}\;\bullet\;\xrightarrow{\;s''\;} & Y \\
& \nwarrow\!r' & & \nearrow\!r'' & & & & \nwarrow\!r'\;\uparrow\!u\;\nearrow\!r'' & \\
& & R & & & & & \bullet &
\end{array}
\tag{4.38}
$$

$\mathsf{R}'(X, Y)$ is the preordered set of such spans, with a 2-cell $r \prec s$ if there exists a (unique) mapping u that makes the right diagram above commute.

Equivalently, $\mathsf{R}'(X, Y)$ is the set of all monomorphisms $r\colon \bullet \rightarrowtail X \times Y$ with values in $X \times Y$, with the canonical preordering $r \prec s$ of monomorphisms with the same codomain (see 2.4.1). Let us note that the previous $R(X, Y)$ can be identified with the associated ordered set $R'(X, Y)/\sim$.

For a consecutive jointly monic relation $s \colon Y \rightarrowtail Z$, the composite $w = sr$ is obtained by:

- *choosing* a pullback (P, u', u'') of the pair (r'', s'),

- then replacing the mapping $(r'u', s''u'') \colon P \to X \times Z$ with an associated jointly monic span, i.e. *any* monomorphism $w \colon W \rightarrowtail X \times Z$ of the factorisation regular epi - mono $(r'u', s''u'') = wp$, in C

$$
\begin{array}{ccc}
X & Y & Z
\end{array}
$$

(4.39)

$$
P \xrightarrow{\;p\;} W \xrightarrow[(r'u',\, s''u'')]{\;w\;} X \times Z
$$

Then one defines the vertical composition and the whisker composition of 2-cells. We only remark that the vertical composition is associated to a preorder, and is therefore strictly associative, with strict identities; on the other hand, the composition of morphisms is weakly associative and has weak identities, up to the equivalence relation associated to the preorder (as follows readily from the first approach). In this way one gets a *locally preordered bicategory* $R' = \mathrm{Rel}'C$ (see 7.1.3).

The latter gives back the locally ordered 2-category R, by identifying its arrows up to the equivalence relation associated to the preorder. Moreover, *defining* R *as this quotient gives an ordered category that is independent of any choice*.

We also note that an equivalence relation *in* the object A has been defined in 4.2.4 following this approach.

5

Applications in Topology and Algebraic Topology

Various categorical topics have interesting applications in Topology and Algebraic Topology. One of the first interactions of this kind can be found in Eilenberg–Steenrod's foundations of (topological) homology theories [EiS], whose axioms deal with categories of 'pairs' of topological spaces, homology functors and natural transformations. (Something about this will be said in 6.1.5.)

Here we begin by studying various adjunctions between categories of topological spaces. Then, in Section 5.2, the cylinder-cocylinder adjunction is used to define homotopies in Top, and set up their structure. Simplicial sets and cubical sets are dealt with in Sections 5.3 and 5.4, as combinatorial data that represent elementary topological spaces and allow us to define the singular homology of spaces. We end with some hints at Directed Algebraic Topology, in Section 5.5.

For a reader who would like to explore the fascinating domain of Algebraic Topology there are excellent texts, like the classical Hilton–Wylie [HiW] or the recent book by Hatcher [Ha], which is freely downloadable. For an easier approach, one could begin from an elementary text on singular homology, like Vick [Vi] or Massey [Mas] (for the simplicial or cubical construction, respectively).

As to our notation, a functor will be defined by its action on objects whenever its extension to morphisms is evident (more frequently than above). A component $\varphi X \colon FX \to GX$ of a natural transformation will often be written as $\varphi \colon FX \to GX$.

If X is a topological space and A is a subset, X/A denotes the quotient space of X where the subset A is collapsed to a point. When X is a pointed space and A is a pointed subset, the quotient X/A is pointed at the equivalence class $[a]$ of any $a \in A$ (formally, the subset A itself).

Intervals of the real line are always denoted by square brackets, as $[a, b]$, $[a, b[,]a, b[$, etc. The standard euclidean interval is written as $\mathbb{I} = [0, 1]$, the standard euclidean n-sphere as $\mathbb{S}^n$. The two-valued index α (or β) varies in the set $2 = \{0, 1\}$, also written as $\{-, +\}$.

5.1 Adjoints and limits in Topology

We already know that the forgetful functor $U\colon \mathsf{Top} \to \mathsf{Set}$ has both adjoints $D \dashv U \dashv C$, where DX (resp. CX) is the set X with the discrete (resp. codiscrete) topology (see 3.1.1(a)).

Here we examine various adjunctions of interest in Topology, starting from exponentiation. Some of them are used to compute colimits. Many other similar facts can be found in the literature, under the heading of 'Categorical Topology'.

5.1.1 Exponentiable spaces

We have seen in 3.4.3 that every set A is exponentiable in Set: this means that the endofunctor $F(X) = X \times A$ has a right adjoint $G(Y) = Y^A = \mathsf{Set}(A, Y)$, linked to the former by the exponential law.

In Top the situation is more complex: not every space is exponentiable, but it is not difficult to see that every *locally compact Hausdorff space* is.

Indeed, let A be such a space and consider the functor

$$F\colon \mathsf{Top} \to \mathsf{Top},$$
$$F(X) = X \times A, \qquad F(f) = f \times A = f \times \mathrm{id}A. \tag{5.1}$$

F has a right adjoint

$$G\colon \mathsf{Top} \to \mathsf{Top},$$
$$G(Y) = Y^A, \quad G(g) = \mathsf{Top}(g, Y)\colon h \mapsto gh \qquad (\text{for } g\colon Y \to Y'), \tag{5.2}$$

where the space Y^A is the set $\mathsf{Top}(A, Y)$ endowed with the *compact-open topology*, also called the *topology of uniform convergence on compact subspaces*.

The open subsets of the latter are generated by the sets

$$W(K, V) = \{h \in \mathsf{Top}(A, Y) \mid h(K) \subset V\}, \tag{5.3}$$

where K is any compact subspace of X and V is any open subset of Y. A general open subset is a union of finite intersections of 'distinguished' open sets, of the previous form.

The proof is left to the reader, who can follow the outline given in the exercise below.

In particular the standard euclidean interval $[0, 1]$ is exponentiable in

Top, with all its cartesian powers. This is a crucial fact in homotopy theory, as we shall see in Section 5.2.

(As a partial converse to the previous result, every exponentiable space which is Hausdorff must be locally compact, because the product $- \times A$ preserves coequalisers precisely in this case: see [Mic], Theorem 2.1 and footnote (5)).

5.1.2 Exercise

This is an outline of the proof of the thesis of 5.1.1, whose details are left to the reader. We let ε_Y be the evaluation mapping

$$\varepsilon_Y \colon FG(Y) = Y^A \times A \to Y, \qquad \varepsilon_Y(h,a) = h(a), \qquad (5.4)$$

where A is a locally compact Hausdorff space and Y^A has the compact-open topology. We want to prove that (GY, ε_Y) is a universal arrow from the functor F to the object Y. (Note that the action of G on the morphisms is then a consequence: we do not have to prove the continuity of $G(g)$ in (5.2).)

The reader will prove the following two points.

(a) For every space Y, the mapping ε_Y is continuous at any point $(h_0, a_0) \in Y^A \times A$.

(b) For every map $f \colon X \times A \to Y$, the mapping

$$g \colon X \to Y^A, \qquad g(x) = f(x, -) \colon A \to Y, \qquad (5.5)$$

is well defined (i.e. $f(x, -)$ is continuous on A, for every $x \in X$) and a continuous mapping on X. (It is then the unique map such that $f = \varepsilon_Y . (g \times \mathrm{id}A) \colon X \times A \to Y^A \times A \to Y$.)

The only point requiring some care is the last, namely the continuity of g.

5.1.3 Proposition (The reflector of Hausdorff spaces)

The full subcategory Hsd $\subset$ Top *of all Hausdorff spaces is reflective in* Top, *with a unit whose components* $p_X \colon X \to X/R$ *are projections on a quotient.*

Proof Let X be a space and R the equivalence relation in X where xRx' means that every map $f \colon X \to Y$ with values in a Hausdorff space has $f(x) = f(x')$.

Let $H(X) = X/R$ be the quotient space and $p\colon X \to X/R$ the projection. We prove that the latter is the universal arrow from the space X to the inclusion $U\colon \mathsf{Hsd} \subset \mathsf{Top}$.

Plainly, every map $f\colon X \to Y$ with values in a Hausdorff space factorises as $gp\colon X \to X/R \to Y$, by a unique map g. It is thus sufficient to prove that the quotient space X/R is Hausdorff.

For two equivalence classes $[x] \neq [x']$ there exist some map $f\colon X \to Y$, with values in a Hausdorff space, such that $f(x) \neq f(x')$. If V, V' are disjoint open neighbourhoods of these two points in Y, their preimages U, U' are disjoint open neighbourhoods of x, x' in X. Moreover the sets U, U' are saturated for the equivalence relation R_f associated to f, whence they also are for the (finer) equivalence relation R. It follows that $p(U)$ and $p(U')$ are disjoint open neighbourhoods of $[x], [x']$ in X/R. $\qquad\square$

5.1.4 *Exercises and complements* (Separation axioms)

(a) The 'constructive' character of the previous definition $H(X) = X/R$ is limited, because it may be difficult to determine the equivalence relation R of a given space.

However, let us consider – for a given space X – the equivalence relation R' formed by all pairs of points x, x' that do not have a pair of disjoint neighbourhoods in X. Then $R' \subset R$. If it happens that the quotient X/R' is Hausdorff, these equivalence relations coincide and our goal is reached. On the contrary, one might reiterate the same procedure on $X' = X/R'$, and hope of getting a Hausdorff space; and so on.

(b) The functor $U\colon \mathsf{Hsd} \subset \mathsf{Top}$ creates all limits and all sums, which thus exist in Hsd. It does not create general colimits, but it is easy to prove that they exist: for a functor $S \to \mathsf{Hsd}$, one can compute the colimit in Top and apply the left adjoint $H\colon \mathsf{Top} \to \mathsf{Hsd}$ (following the same pattern used for colimits of ordered sets, in 3.2.3(b)).

(c) These results can be easily adapted to T_0-spaces (and many other separation axioms):

- the full embedding $U\colon T_0\mathsf{Top} \subset \mathsf{Top}$ of T_0-spaces has a left adjoint $H\colon \mathsf{Top} \to T_0\mathsf{Top}$,

- this functor U creates all limits and all sums. To compute the colimit of a functor $S\,\text{`}\!\to T_0\mathsf{Top}$, one can compute the colimit in Top and then apply the left adjoint H.

5.1.5 Topological spaces and preorder relations

There is an interesting idempotent adjunction between Top and the category pOrd of preordered sets

$$T: \text{pOrd} \rightleftarrows \text{Top}: W, \qquad\qquad T \dashv W. \qquad\qquad (5.6)$$

For a space X, the preordered set $W(X)$ has the same underlying set and the *specialisation preorder* $x \prec y$, defined by the following equivalent conditions (where $\overline{y}$ is the closure of $\{y\}$ in X)

$$x \in \overline{y}, \qquad\qquad \overline{x} \subset \overline{y}. \qquad\qquad (5.7)$$

Plainly, a continuous mapping $f: X \to Y$ becomes a monotone mapping $f: W(X) \to W(Y)$.

The other way round, for a preordered set X, the space $T(X)$ has the same underlying set and the topology whose closed subsets $C \subset X$ are the downward closed ones

$$\text{if} \quad x \prec y \text{ in } X \text{ and } y \in C, \text{ then } x \in C. \qquad\qquad (5.8)$$

One verifies easily that:

- this defines an *Alexandrov topology* on X (which means that the family of closed subsets is closed under arbitrary intersections *and* unions),

- a monotone mapping $f: X \to Y$ becomes a continuous mapping $f: T(X) \to T(Y)$.

For a preordered set X, the unit $\eta_X: X \to WT(X)$ is the identity (because the closure of $x \in X$ in the space $T(X)$ is $\downarrow x$). For a space X, the counit is

$$\varepsilon_X: TW(X) \to X, \qquad\qquad |\varepsilon_X| = 1_X, \qquad\qquad (5.9)$$

because $TW(X)$ has a finer topology than X: every closed subset C of X is downward closed for the specialisation preorder. The triangular identities are trivially satisfied.

Therefore T embeds pOrd in Top, as the full coreflective subcategory of Alexandrov spaces.

5.1.6 Exercises and complements (Topology and order)

(a) The previous adjunction can be restricted to an adjunction

$$T: \text{Ord} \rightleftarrows T_0\text{Top}: W, \qquad\qquad T \dashv W, \qquad\qquad (5.10)$$

between ordered sets and T_0-spaces. Now T embeds Ord in T_0Top, as the full coreflective subcategory of Alexandrov spaces which are T_0.

(b) This adjunction and the previous one (in 5.1.5) can be inserted in a commutative square of adjunctions (i.e. a commutative square in AdjCat, leaving apart questions of size)

$$
\begin{array}{ccc}
\text{Ord} & \underset{W}{\overset{T}{\rightleftarrows}} & T_0\text{Top} \\
F \mathbin{|} U & & H \mathbin{|} U \\
\text{pOrd} & \underset{W}{\overset{T}{\rightleftarrows}} & \text{Top}
\end{array}
\tag{5.11}
$$

where the right adjoints are dashed. The adjunction $F \dashv U$ between ordered and preordered sets can be found in 3.2.3(b), while the adjunction $H \dashv U$ between T_0-spaces and spaces is in 5.1.4(c).

(c) Consider the specialisation order of a T_1-space.

(d) Describe the topology of the space $T(\mathbb{R}, \leqslant)$ associated to the ordered real line.

5.1.7 Exercises and complements (Metric spaces)

(a) Let us recall that a metric space is a set X equipped with a mapping d (or d_X), its distance (or metric), satisfying the following conditions, for all x, y, z in X

(i) $d \colon X \times X \to [0, +\infty[$,

(ii) $d(x, y) = 0 \;\Leftrightarrow\; x = y$ (*separation*),

(iii) $d(x, y) + d(y, z) \geqslant d(x, z)$ (*triangular inequality*),

(iv) $d(x, y) = d(y, x)$ (*symmetry*).

A mapping $f \colon X \to Y$ is said to be *Lipschitz* if it admits a Lipschitz constant $L \in \mathbb{R}$, such that

$$
d_Y(f(x), f(x')) \leqslant L.d_X(x, x'), \qquad \text{for } x, x' \in X,
\tag{5.12}
$$

and a *weak contraction* if it admits 1 as a Lipschitz constant, so that

$$
d_Y(f(x), f(x')) \leqslant d_X(x, x'), \qquad \text{for } x, x' \in X,
$$

We have the categories

- Mtr *of metric spaces and Lipschitz mappings,*
- Mtr$_1$ *of metric spaces and weak contractions,*

with obvious forgetful functors

$$
\text{Mtr}_1 \subset \text{Mtr} \to \text{Hsd} \subset \text{Top},
$$
$$
\text{Ban} \to \text{Mtr}, \qquad \text{Ban}_1 \to \text{Mtr}_1.
\tag{5.13}
$$

(b) A metric space is *complete* if every Cauchy sequence converges in it. The full subcategory CplMtr $\subset$ Mtr of complete metric spaces is reflective, with reflector given by the completion functor.

The same holds for the full subcategory CplMtr$_1$ $\subset$ Mtr$_1$.

(c) The classical notion recalled in (a) has been generalised in various way. For an *extended semimetric* (called an *écart* in Bourbaki [Bor2]), the conditions (i), (ii) are replaced by the following more general ones (while (iii) and (iv) are kept as above)

(i') $d\colon X \times X \to [0, +\infty]$,

(ii') $d(x, x) = 0$.

We shall see that the resulting categories

- esMtr *of extended semimetric spaces and Lipschitz mappings,*

- esMtr$_1$ *of extended semimetric spaces and weak contractions,*

have much better categorical properties than the classical ones.

(Dropping also the symmetry axiom gives a further, important extension, which will be considered in 7.3.4.)

(d) Prove that the forgetful functors

$$U\colon \text{esMtr} \to \text{Set}, \qquad\qquad U\colon \text{esMtr}_1 \to \text{Set}, \qquad\qquad (5.14)$$

have both adjoints $D \dashv U \dashv C$, which give to any set the *discrete extended metric* (the largest possible one) or the *codiscrete semimetric* (the least possible one). Therefore these functors U preserve (the existing) limits and colimits.

(e) Prove that esMtr$_1$ has all limits and colimits. Some work is required to define the extended semimetric of a coequaliser.

(f) For an *extended metric* we only replace (i) with (i'). Prove that the embedding $U\colon \text{esMtr}_1 \subset \text{eMtr}_1$, with values in the category of extended metric spaces and weak contractions, has a left adjoint $F(X) = X/\sim$, by means of the obvious equivalence relation that turns an extended semimetric into an extended metric.

(g) Use this left adjoint to construct all colimits in eMtr$_1$.

5.1.8 Frames, locales and pointless topology

There is a natural functor

$$\mathcal{O}\colon \text{Top} \to \text{Lth}^{\text{op}}, \qquad\qquad (5.15)$$

that assigns to a space X the lattice $\mathcal{O}(X)$ of its open subsets, and to a continuous mapping $f\colon X \to Y$ the pre-image homomorphism $f^*\colon \mathcal{O}(Y) \to \mathcal{O}(X)$.

In fact, $\mathcal{O}(X)$ is a *frame*, i.e. a complete lattice where arbitrary joins distribute over binary meets

$$(\vee U_i) \wedge V = \vee (U_i \wedge V),$$

since these operations are unions and intersections in $\mathcal{P}(X)$. In the same way $f^*\colon \mathcal{O}(Y) \to \mathcal{O}(X)$ is a *homomorphism of frames*, which means that it preserves arbitrary joins and binary meets.

As a surrogate of Top, we now introduce the formal *category of locales* Loc, as the opposite of the category of frames and their homomorphisms. The previous functor (5.15) is thus reinterpreted as

$$\mathcal{O}\colon \mathsf{Top} \to \mathsf{Loc}. \tag{5.16}$$

Most topological notions can be studied in Loc, even though some results can be different. This is the domain of 'pointless topology', which is also important in Topos Theory. The interested reader can begin by [Jo1, Jo2, Bo3].

5.1.9 *Complements (Compact Hausdorff spaces as algebras)

The full subcategory CmpHsd $\subset$ Top of *compact Hausdorff spaces* has peculiar properties among the categories of topological spaces.

To begin with, CmpHsd is balanced (see 1.3.1), as a consequence of a classical theorem: any map $f\colon X \to Y$ from a compact space to a Hausdorff space is closed, and any bijective map of this kind is a homeomorphism. Moreover, an epimorphism has a dense image (by the same argument used for Hsd, in 2.3.6(c)), and must be surjective.

We now review some important properties of CmpHsd, without proofs (referred to [M4]).

The embedding $V\colon$ CmpHsd $\subset$ Top has a left adjoint

$$\beta\colon \mathsf{Top} \to \mathsf{CmpHsd}, \tag{5.17}$$

called the *Stone-Čech compactification* (see [M4], Section V.8). The unit $\eta X\colon X \to V\beta(X)$ is a topological embedding if and only if X is a completely regular space. (Classically a 'compactification' was meant to be an embedding, and this procedure was only considered for completely regular spaces.)

The composed adjunction (where D equips a set with the discrete topology)

$$\mathsf{Set} \underset{W}{\overset{D}{\rightrightarrows}} \mathsf{Top} \underset{V}{\overset{\beta}{\rightrightarrows}} \mathsf{CmpHsd} \qquad\qquad \beta D \dashv WV, \qquad (5.18)$$

gives the left adjoint $\beta D \colon \mathsf{Set} \to \mathsf{CmpHsd}$ of the forgetful functor $U = WV \colon \mathsf{CmpHsd} \to \mathsf{Set}$.

The latter is monadic, so that a compact Hausdorff space can be viewed as an algebra $(X, a \colon TX \to X)$, *where* $T = U\beta D \colon \mathsf{Set} \to \mathsf{Set}$ *is the monad* associated to the adjunction (5.18) (see [M4], Section VI.9).

The existence of the left adjoint β can be proved by the Special Adjoint Functor Theorem (as in [M4]) or by some 'constructive' procedure which can be found in many books on General Topology (for instance in [Dg, Ke]).

Here we only sketch, without proofs, one of the most frequently used – based on the compact interval $[0, 1]$ of the euclidean line.

We consider the set $C = \mathsf{Top}(X, [0, 1])$ and the set

$$[0, 1]^C = \mathsf{Set}(C, [0, 1]).$$

The latter is given the product topology, which is Hausdorff compact, by Tychonoff's theorem. Its projections are written as $p_f \colon [0, 1]^C \to [0, 1]$, for $f \in C$.

Then one takes the mapping of *evaluation*

$$\mathrm{ev}_X \colon X \to [0, 1]^C,$$
$$\mathrm{ev}_X(x) \colon C \to [0, 1], \qquad\qquad \mathrm{ev}_X(x)(f) = f(x), \qquad (5.19)$$

which is continuous, because all its components are:

$$p_f . \mathrm{ev}_X = f \colon X \to [0, 1].$$

Now we let βX be the closure of the subset $\mathrm{ev}_X(X)$ in the space $[0, 1]^C$, and take the codomain-restriction of ev_X

$$\eta_X \colon X \to V\beta X, \qquad\qquad \eta_X(x) = \mathrm{ev}_X(x). \qquad (5.20)$$

The delicate point is proving the universal property of this arrow, from the space X to the functor V.

The interval $[0, 1]$ works in this construction because it is a *cogenerator* of the category CmpHsd: this means that if A and B are compact Hausdorff spaces, and $f, g \colon A \to B$ are distinct maps, there is a map $h \colon B \to [0, 1]$ that distinguishes them: $hf \neq hg$.

5.2 Cylinder, cocylinder and homotopies

In Top the standard interval $\mathbb{I} = [0,1]$ gives the cylinder endofunctor $I = -\times\mathbb{I}$ and its right adjoint, the cocylinder or path endofunctor $P = (-)^{\mathbb{I}}$.

Homotopies between maps $X \to Y$ are equivalently defined as maps $IX \to Y$ or $X \to PY$. Their structure is produced by a rich structure on the standard interval, covariantly extended to the cylinder functor and contravariantly extended to the path functor.

Similar constructions can be made in the category Top. of pointed spaces.

(Many other situations where homotopies can be considered, classical or non classical, have a similar framework, organised by a cylinder or cocylinder endofunctor, or both. Many of them are studied in [G8]; this includes cases where we should rather speak of 'directed homotopies', as Cat itself, or the category of preordered topological spaces that will be sketched in Section 5.5.)

In Top and Top. we also have the cone and suspension endofunctors; but their right adjoints, the cocone and loop endofunctors, only exist in the pointed case.

We recall that the two-valued index α (or β) varies in the set $2 = \{0,1\}$, also written as $\{-,+\}$.

5.2.1 The structure of the euclidean interval

A path in the space X is a map $a\colon \mathbb{I} \to X$ defined on the standard interval $\mathbb{I} = [0,1]$, with euclidean topology.

The basic, 'first degree' structure of $\mathbb{I}$ consists of four maps, linking $\mathbb{I}$ to its 0-th cartesian power, the singleton $\mathbb{I}^0 = \{*\}$

$$\partial^\alpha\colon \{*\} \to \mathbb{I}, \qquad \partial^-(*) = 0,\ \partial^+(*) = 1 \qquad (faces),$$

$$e\colon \mathbb{I} \to \{*\}, \qquad e(t) = * \qquad\qquad (degeneracy), \qquad (5.21)$$

$$r\colon \mathbb{I} \to \mathbb{I}, \qquad r(t) = 1 - t \qquad\qquad (reversion).$$

We shall say that $(\mathbb{I}, \partial^\alpha, e)$ is an *internal interval* in Top, and $(\mathbb{I}, \partial^\alpha, e, r)$ an *internal interval with reversion*. (We shall use the same terminology replacing Top with any monoidal category, provided that $e\partial^\alpha = \mathrm{id}$ and moreover, in the reversible case: $rr = \mathrm{id},\ er = e,\ \partial^- r = \partial^+$.)

Identifying a point x of the space X with the corresponding map $x\colon \{*\} \to X$, this basic structure determines:

(a) the endpoints of a path $a\colon \mathbb{I} \to X$, $a\partial^- = a(0)$ and $a\partial^+ = a(1)$,

(b) the trivial path at the point x, which will be written as $0_x = xe$,

(c) the reversed path of a, written as $-a = ar$.

Two consecutive paths $a, b \colon \mathbb{I} \to X$ (with $a\partial^+ = b\partial^-$, i.e. $a(1) = b(0)$) have a *concatenated path* $a + b$

$$(a+b)(t) = \begin{cases} a(2t), & \text{for } 0 \leqslant t \leqslant 1/2, \\ b(2t-1), & \text{for } 1/2 \leqslant t \leqslant 1. \end{cases} \tag{5.22}$$

(This operation is not associative, and only works well up to homotopy with fixed endpoints, see 5.2.9). Formally, this can be expressed saying that the *standard concatenation pushout* – pasting two copies of the interval, one after the other – is homeomorphic to $\mathbb{I}$ and can be realised as $\mathbb{I}$ itself

$$
\begin{array}{ccc}
\{*\} & \xrightarrow{\partial^+} & \mathbb{I} \\
{\scriptstyle \partial^-} \downarrow & & \downarrow {\scriptstyle c^-} \\
\mathbb{I} & \xrightarrow{c^+} & \mathbb{I}
\end{array}
\qquad
\begin{aligned}
c^-(t) &= t/2, \\[2ex]
c^+(t) &= (t+1)/2.
\end{aligned}
\tag{5.23}
$$

Indeed the concatenated path $a + b \colon \mathbb{I} \to X$ comes from the universal property of the pushout, and is characterised by the conditions

$$(a+b).c^- = a, \qquad (a+b).c^+ = b. \tag{5.24}$$

Finally, there is a 'second degree' structure which involves the standard square $\mathbb{I}^2 = [0,1] \times [0,1]$ and is used to construct homotopies of paths

$$
\begin{aligned}
g^- &\colon \mathbb{I}^2 \to \mathbb{I}, & g^-(t,t') &= \max(t,t') & &\text{(\textit{lower connection})}, \\
g^+ &\colon \mathbb{I}^2 \to \mathbb{I}, & g^+(t,t') &= \min(t,t') & &\text{(\textit{upper connection})}, \\
s &\colon \mathbb{I}^2 \to \mathbb{I}^2, & s(t,t') &= (t',t) & &\text{(\textit{transposition})}.
\end{aligned}
\tag{5.25}
$$

Together with (5.21), these maps complete the structure of $\mathbb{I}$ as an *involutive lattice* in Top. The choice of the superscripts of g^-, g^+ comes from the fact that the unit of g^α is $\partial^\alpha(*)$. Within homotopy theory, the importance of these binary operations has been highlighted by R. Brown and P.J. Higgins [BroH1, BroH2], which introduced the term of *connection*, or *higher degeneracy*.

5.2.2 Cylinder and homotopies

Given two continuous mappings $f, g \colon X \to Y$ (in Top), a homotopy $\varphi \colon f \to g$ 'is' a map $\varphi \colon X \times \mathbb{I} \to Y$ defined on the cylinder $I(X) = X \times \mathbb{I}$, which coincides with f on the lower basis and with g on the upper one

$$\varphi(x,0) = f(x), \quad \varphi(x,1) = g(x) \qquad \text{for } x \in X. \tag{5.26}$$

This map will be written as $\hat{\varphi} \colon X \times \mathbb{I} \to Y$ when we want to distinguish it from the homotopy $\varphi \colon f \to g$ which it represents.

All this is based on the *cylinder* endofunctor:

$$I: \mathsf{Top} \to \mathsf{Top}, \qquad I(X) = X \times \mathbb{I}, \tag{5.27}$$

and on the four natural transformations which it inherits from the structural maps of the standard interval, and which will be written as the latter

$$
\begin{aligned}
&\partial^\alpha: X \to IX, && \partial^\alpha(x) = (x, \alpha), && (faces), \\
&e: IX \to X, && e(x, t) = x && (degeneracy), \\
&r: IX \to IX, && r(x, t) = (x, 1 - t) && (reversion).
\end{aligned}
\tag{5.28}
$$

Identifying $I\{*\} = \{*\} \times \mathbb{I} = \mathbb{I}$, the structural maps of the standard interval coincide with the components of the transformations (5.28) on the singleton.

The transformations ∂^α, e, r give rise to the *faces* $\varphi\partial^\alpha$ of a homotopy, the *trivial homotopy* $0_f = fe$ of a map and the *reversed homotopy* $-\varphi = \varphi r$. (More precisely, the homotopy $-\varphi: g \to f$ is represented by the map $(-\varphi)\hat{} = \hat{\varphi}r: IX \to Y$.) A path $a: \mathbb{I} \to X$ is the same as a homotopy $a: x \to x'$ between its endpoints (always by identifying $I\{*\} = \mathbb{I}$).

Two consecutive homotopies $\varphi: f \to g$ and $\psi: g \to h$ have a *concatenation*, or *vertical composition*

$$\varphi + \psi: f \to g, \tag{5.29}$$

that extends the concatenation of paths (in (5.22)).

Again, this can be formally expressed noting that the *concatenation pushout of the cylinder* – pasting two copies of a cylinder IX, 'one on top of the other' – can be realised as the cylinder itself

$$
\begin{array}{ccc}
X & \xrightarrow{\;\partial^+\;} & IX \\
{\scriptstyle \partial^-}\downarrow & \nearrow & \downarrow{\scriptstyle c^-} \\
IX & \xrightarrow{\;c^+\;} & IX
\end{array}
\qquad
\begin{array}{l}
c^-(x, t) = (x, t/2), \\[2ex]
c^+(x, t) = (x, (t + 1)/2).
\end{array}
\tag{5.30}
$$

Indeed, the subspaces $c^-(IX) = X \times [0, 1/2]$ and $c^+(IX) = X \times [1/2, 1]$ form a finite closed cover of IX, so that a mapping defined on IX is continuous if and only if this holds for its restrictions to these subspaces.

The concatenated homotopy $\varphi + \psi: f \to h$ is represented by the map $(\varphi + \psi)\hat{}: IX \to Y$ which reduces to φ on c^-, and to ψ on c^+.

Note that the fact that 'pasting two copies of the cylinder gives back the cylinder' is rather *peculiar of spaces*; e.g. it does not hold for chain complexes, where the concatenation of homotopies is based on a more general procedure: an assigned concatenation morphism $c: IX \to IX +_X IX$.

Here also, there is a 'second degree' structure, with three natural transformations which involve the iterated cylinder $I^2(X) = I(I(X)) = X \times \mathbb{I}^2$ and are used to construct higher homotopies (i.e. homotopies of homotopies)

$$
\begin{aligned}
g^\alpha \colon I^2 X \to IX, \quad & g^\alpha(x, t, t') = (x, g^\alpha(t, t')) \quad (connections), \\
s \colon I^2 X \to I^2 X, \quad & s(x, t, t') = (x, t', t) \quad (transposition).
\end{aligned}
\tag{5.31}
$$

For instance, given a homotopy $\varphi \colon f^- \to f^+ \colon X \to Y$, we need the transposition to construct a homotopy $I\varphi \colon If^- \to If^+$

$$
\begin{aligned}
(I\varphi)\hat{\ } = I(\hat{\varphi}).sX \colon I^2 X \to IY, \\
I(\hat{\varphi}).sX.\partial^\alpha(IX) = I(\hat{\varphi}).I(\partial^\alpha X) = If^\alpha,
\end{aligned}
\tag{5.32}
$$

by modifying the mapping $I(\hat{\varphi})$, *which does not have the correct faces.*

5.2.3 Cocylinder and homotopies

We conclude this brief review of the formal bases of classical homotopy recalling that a homotopy $\varphi \colon f \to g$, represented by a map $\hat{\varphi} \colon IX \to Y$, also has a dual representation as a map $\check{\varphi} \colon X \to PY$ with values in the *path-space* $PY = Y^{\mathbb{I}}$.

In fact, as we know from 5.1.1, the cylinder functor $I = - \times \mathbb{I}$ has a right adjoint

$$
P \colon \mathsf{Top} \to \mathsf{Top}, \qquad P(Y) = Y^{\mathbb{I}}, \tag{5.33}
$$

called the *cocylinder* or *path functor*: $P(Y)$ is the space of paths $\mathbb{I} \to Y$, with the compact-open topology (also called the topology of uniform convergence on $\mathbb{I}$).

The counit

$$
\mathrm{ev} \colon P(Y) \times \mathbb{I} \to Y, \qquad (a, t) \mapsto a(t), \tag{5.34}
$$

is also called *evaluation* (of paths).

The functor P inherits from the interval $\mathbb{I}$, contravariantly, a *dual* structure, which we write with the same symbols. It consists of a basic, first degree part:

$$
\begin{aligned}
\partial^\alpha \colon PY \to Y, \quad & \partial^\alpha(a) = a(\alpha), \quad & (faces), \\
e \colon Y \to PY, \quad & e(y)(t) = y \quad & (degeneracy), \\
r \colon PY \to PY, \quad & r(a)(t) = a(1 - t) \quad & (reversion),
\end{aligned}
\tag{5.35}
$$

and a second degree structure: two *connections* and a *transposition*

$$g^\alpha \colon PY \to P^2Y, \qquad g^\alpha(a)(t, t') = a(g^\alpha(t, t')),$$
$$s \colon P^2Y \to P^2Y, \qquad s(a)(t, t') = a(t', t). \tag{5.36}$$

In this description, the faces of a homotopy $\check\varphi \colon X \to PY$ are defined as $\partial^\alpha\check\varphi \colon X \to Y$.

Concatenation of homotopies can now be performed with the *concatenation pullback* QY (which can be realised as *the object of pairs of consecutive paths*) and the concatenation map c

$$\begin{array}{ccc} QY & \xrightarrow{\ c^+\ } & PY \\ {\scriptstyle c^-}\downarrow & \diagup\ \bullet & \downarrow{\scriptstyle \partial^-} \\ PY & \xrightarrow[\ \partial^+\ ]{} & Y \end{array} \tag{5.37}$$

$$Q(Y) = \{(a, b) \in PY \times PY \mid \partial^+(a) = \partial^-(b)\},$$

$$c \colon QY \to PY, \qquad c(a, b) = a + b \qquad (\textit{concatenation map}).$$

Again, as a property of topological spaces, the natural transformation c is invertible (by splitting a path into its two halves), and we can also realise QY as PY. (For chain complexes one has a non-invertible concatenation map.)

5.2.4 The homotopy category

Two parallel maps $f, g \colon X \to Y$ in Top are said to be *homotopic*, written as $f \simeq g$, if there exists a homotopy $\varphi \colon f \to g$. This is an equivalence relation in $\mathsf{Top}(X, Y)$, because of trivial homotopies, reversed homotopies and their concatenation.

Furthermore, homotopies have a *whisker composition* with maps

$$\mathsf{X}' \xrightarrow{\ h\ } \mathsf{X} \underset{g}{\overset{f}{\underset{\Downarrow\varphi}{\rightrightarrows}}} \mathsf{Y} \xrightarrow{\ k\ } \mathsf{Y}' \tag{5.38}$$

$$k\varphi k \colon kfh \to kgh \colon X' \to Y',$$

where $k\varphi h$ is represented on the cylinder by $k.\check\varphi.I(h) \colon IX' \to Y'$ (or on the cocylinder by $P(k).\check\varphi.h \colon X' \to PY'$). As a consequence, the homotopy relation is a congruence of categories in Top.

The quotient category $\mathsf{hoTop} = \mathsf{Top}/{\simeq}$ is called the *homotopy category of topological spaces*, and is important in Algebraic Topology. Its objects are the topological spaces, and a morphism $[f] \colon X \to Y$ is a homotopy class of continuous mappings $X \to Y$.

An isomorphism in hoTop is the same as a *homotopy equivalence*: namely a pair of maps $f \colon X \rightleftarrows Y \colon g$ such that $gf \simeq \mathrm{id}X$ and $fg \simeq \mathrm{id}Y$.

As a basic notion of Algebraic Topology, a functor $F \colon \mathsf{Top} \to \mathsf{C}$ with values in an arbitrary category is said to be *homotopy invariant* if $F(f) = F(g)$ whenever $f \simeq g$ in Top, so that F induces a functor hoTop $\to \mathsf{C}$. Typically, this is the case of the homology and homotopy functors with values in Ab, Gp, etc.

5.2.5 Cones and suspension for spaces

Let X be a topological space. Its *upper cone* C^+X is defined by the following pushout, where $\top = \{*\}$ is the terminal object

$$
\begin{array}{ccc}
X & \xrightarrow{\ p\ } & \{*\} \\
{\scriptstyle \partial^+}\downarrow & {\diagup} & \downarrow{\scriptstyle v^+} \\
IX & \xrightarrow[\gamma]{} & C^+X
\end{array}
\qquad C^+X = (IX + \{*\})/(\partial^+X + \{*\}). \qquad (5.39)
$$

Note that $C^+(\emptyset) = \{*\}$: *only when X is not empty*, the cone C^+X is a quotient of the cylinder IX (by collapsing its upper bases to a point).

The cone comes equipped with an *upper vertex* $v^+ \colon \{*\} \to X$ and a homotopy

$$
\gamma \colon u^- \to v^+p \colon X \to C^+X, \qquad u^- = \gamma\partial^- \colon X \to C^+X, \qquad (5.40)
$$

between the embedding u^- of X as the (lower) basis of the cone and the constant mapping to the (upper) vertex. The points of C^+X will be written as equivalence classes $[x, s]$, with $[x, 1] = v^+ = v^+(*)$, for $x \in X$ and $s \in \mathbb{I}$ (adding the vertex $v^+ = v^+(*)$ when $X = \emptyset$).

The cone C^+X is always a *contractible space*, homotopy equivalent to the point, by the following homotopy between the identity and the constant endomap ending in its vertex

$$
h \colon C^+X \times \mathbb{I} \to C^+X, \qquad h([x, s], t) = [x, s \vee t] \quad (h(v^+, t) = v^+). \qquad (5.41)
$$

The higher structure of the cylinder I gives rise to a monad (C, u, g) on the upper cone functor $C = C^+$. The unit is the lower basis $u \colon 1 \to C$ and the multiplication $g \colon C^2 \to C$ is induced by $g^- \colon \mathbb{I}^2 \to \mathbb{I}$. (A proof can be found in [G8], Theorem 4.8.3.)

Symmetrically, one has the *lower cone*

$$
C^-X = (IX + \{*\})/(\partial^-X + \{*\}),
$$

with a lower vertex v^- and an upper basis u^+.

The *suspension* ΣX is defined by the following colimit

$$
\begin{array}{ccc}
X & \longrightarrow & \{*\} \\
{\scriptstyle \partial^+}\downarrow & & \downarrow{\scriptstyle v^+} \\
X \xrightarrow{\ \partial^-\ } & IX & \\
\downarrow & \searrow{\scriptstyle \sigma} & \\
\{*\} & \xrightarrow[\ v^-\ ]{} & \Sigma X
\end{array}
\tag{5.42}
$$

which collapses, independently, the bases of IX to a lower and an upper vertex, v^- and v^+. Note that $\Sigma(\emptyset) \cong \mathbb{S}^0$, and $\Sigma\mathbb{S}^n \cong \mathbb{S}^{n+1}$ for all $n \geqslant 0$.

The suspension ΣX can also be obtained by the following pushout: the pasting of both cones $C^\alpha X$ of X, along their bases

$$
\begin{array}{ccc}
X & \xrightarrow{\ u^-\ } & C^+ X \\
{\scriptstyle u^+}\downarrow & & \downarrow \\
C^- X & \longrightarrow & \Sigma X
\end{array}
\tag{5.43}
$$

(Again, this coincidence is *not* a 'general' fact, in homotopy theory, but rests on the 'pasting property' of the topological cylinder.)

As we have seen, the cone and suspension functors of Top do not preserve the initial object $\bot = \emptyset$ (nor sums, since the result is always a connected space); therefore they do not have right adjoints.

But in the category Top. of pointed spaces, the cylinder, cone and suspension functors have each a right adjoint, described below.

5.2.6 Cylinder, cone and suspension for pointed spaces

In Top. one works with *pointed homotopies*, which preserve the base point. They are represented by the *pointed cylinder* endofunctor

$$
I\colon \mathsf{Top}_{\bullet} \to \mathsf{Top}_{\bullet}, \qquad I(X, x_0) = (IX/I\{x_0\}, [x_0, t]),
\tag{5.44}
$$

defined as the quotient of the unpointed cylinder IX which collapses the fibre of the base point x_0 (providing thus the new base point).

The upper cone and suspension are defined by the same colimits as in (5.39) and (5.42), in the category Top.. They can also be obtained from the corresponding unpointed constructions, by collapsing the fibre of the base point

$$
\begin{aligned}
C^+(X, x_0) &= I(X, x_0)/\partial^+(X, x_0) = (C^+(X)/I\{x_0\}, [x_0, t]), \\
\Sigma(X, x_0) &= ((\Sigma X)/I\{x_0\}, [x_0, t]).
\end{aligned}
\tag{5.45}
$$

Also here, the pointed suspension of a pointed sphere gives a higher sphere

$$\Sigma(\mathbb{S}^n, x_0) \cong (\mathbb{S}^{n+1}, y_0), \qquad \text{for } n \geqslant 0.$$

5.2.7 *Exercises* (The pointed interval)

Let us recall that the important monoidal structure of Top. is the smash product $- \wedge -$, introduced in 3.4.2(c).

(a) Define a *pointed interval* $\mathbb{I}_{\bullet}$, so that the cylinder endofunctor of pointed spaces can be obtained as $I(X, x_0) = (X, x_0) \wedge \mathbb{I}_{\bullet}$.

(b) Define on $\mathbb{I}_{\bullet}$ a structure of internal interval for the monoidal structure, as in 5.2.1.

5.2.8 *Cocylinder, cocone and loop objects, for pointed spaces*

All the previous endofunctors have a right adjoint

$$
\begin{aligned}
&P \colon \mathsf{Top}_{\bullet} \to \mathsf{Top}_{\bullet}, &&I \dashv P &&(path\,functor), \\
&E^+ \colon \mathsf{Top}_{\bullet} \to \mathsf{Top}_{\bullet}, &&C^+ \dashv E^+ &&(upper\ cocone), &&(5.46)\\
&\Omega \colon \mathsf{Top}_{\bullet} \to \mathsf{Top}_{\bullet}, &&\Sigma \dashv \Omega &&(loop\ functor).
\end{aligned}
$$

The pointed path object $P(Y, y_0)$ is simply the unpointed path space $P(Y)$, pointed at the constant path $0_{y_0}(t) = y_0$

$$P(Y, y_0) = (P(Y), 0_{y_0}). \qquad (5.47)$$

The adjunction $I \dashv P$ is induced by the corresponding adjunction of topological spaces

$$
\begin{aligned}
&\varphi \colon \mathsf{Top}_{\bullet}(I(X, x_0), (Y, y_0)) \to \mathsf{Top}_{\bullet}((X, x_0), P(Y, y_0)), \\
&\varphi(f) \colon (X, x_0) \to P(Y, y_0), \qquad \varphi(f)(x)(t) = f[x, t],
\end{aligned} \qquad (5.48)
$$

since a map $f \colon I(X, x_0) \to (Y, y_0)$ is the same as a map $f' \colon IX \to Y$ that sends the fibre $I\{x_0\}$ to y_0.

Finally, the upper cocone and the loop object in Top. are constructed as limits

$$E^+(Y, y_0) = (\{(y, b) \in Y \times PY \mid b(0) = y, b(1) = y_0\}, (y_0, 0_{y_0})),$$

$$\Omega(Y, y_0) = (\{b \in PY \mid b(0) = y_0 = b(1)\}, 0_{y_0}),$$

with the topology of subspace of $Y \times PY$ and PY, respectively.

$\Omega(Y, y_0)$ is indeed the space of loops at the base point, with compact-open topology.

5.2.9 *The fundamental groupoid

The reader has likely seen the construction of the fundamental group $\pi_1(X, x)$ of a pointed space, where an element is the class $[a]$ of a loop $a \in \Omega(X, x)$, *up to homotopy with fixed endpoints.*

The operation is induced by path-concatenation:

$$[a] * [b] = [a + b],$$

which becomes a group law in the quotient: the unit is the class $[0_x]$ of the constant loop at x, and the inverse is obtained by reversing a representative path. (We do not review here the easy but tiresome construction of the homotopies required to prove all this.)

This gives a functor

$$\pi_1 \colon \mathsf{Top_\bullet} \to \mathsf{Gp}, \tag{5.50}$$

where a pointed map $f \colon (X, x) \to (Y, y)$ is sent to the homomorphism

$$f_* \colon \pi_1(X, x) \to \pi_1(Y, y), \qquad f_*([a]) = [fa].$$

Working on an unpointed space X, one gets a small groupoid $\Pi_1(X)$ whose objects (or vertices) are the points of X, while an arrow $[a] \colon x \to y$ is a class of paths $a \colon \mathbb{I} \to X$ from x to y, *up to homotopy with fixed endpoints.* Composition with a consecutive arrow $[b] \colon y \to z$ is again by concatenation: $[a] * [b] = [a + b]$.

This becomes indeed a groupoid law in the quotient; the proof (for associativity, identities and inverses) is a straightforward extension of the corresponding proof for the fundamental group.

We have now a functor with values in the category Gpd of *small groupoids and functors*

$$\Pi_1 \colon \mathsf{Top} \to \mathsf{Gpd}. \tag{5.51}$$

A map $f \colon X \to Y$ is sent to the functor

$$f_* \colon \Pi_1(X) \to \Pi_1(Y), \qquad f_*(x) = f(x), \quad f_*([a]) = [fa].$$

For every $x \in X$, the fundamental group $\pi_1(X, x)$ is the group of automorphisms $\Pi_1(X)(x, x)$ at the vertex x of $\Pi_1(X)$.

5.3 Simplicial sets

After describing simplicial objects by faces and degeneracies, we show that every topological space S has an associated *singular simplicial set* Smp(S) (see 53.4), which is used to construct singular homology.

On the other hand, a simplicial set can be viewed as a combinatorial description of a topological space of a rather simple type. The latter is produced by the geometric realisation functor $\mathcal{R} \colon$ SmpSet $\to$ Top (in 5.3.5), by pasting tetrahedra of any dimension, according to the rules 'written' in X. This functor is left adjoint to the singular simplicial functor Smp$\colon$ Top $\to$ SmpSet.

5.3.1 Simplicial objects

As already said in 1.6.3, a *simplicial object* in the category C is a functor $X \colon \Delta^{\mathrm{op}} \to$ C defined on the category Δ of finite positive ordinals (and increasing maps). A *morphism of simplicial objects* $f \colon X \to Y$ is a natural transformation of such functors; they form the category Smp(C) $=$ C$^{\Delta^{\mathrm{op}}}$ of simplicial objects in C.

In particular we have the categories SmpSet of *simplicial sets*, SmpGp of *simplicial groups*, etc. The category Δ, as the 'domain' of simplicial objects, is also called the *simplicial site*.

We also consider the *augmented simplicial site* $\Delta^{\sim}$ of all finite ordinals; an *augmented simplicial object* is a functor $X \colon (\Delta^{\sim})^{\mathrm{op}} \to$ C.

(Let us note that in Mac Lane [M4], Section VII.5, the main subject is the category $\Delta^{\sim}$, written as Δ and called the *simplicial category*, while its subcategory Δ is written as Δ^{+}. Here we prefer to follow the notation used in Topology. The terminology about simplicial objects in [M4] is the classical one, the same as here.)

5.3.2 Faces and degeneracies

The simplicial site Δ can be described by simple generators.

Δ has objects $[n] = \{0, ..., n\}$, starting with the singleton $[0] = \{0\}$. We are interested in the following morphisms. The *face* δ_i (or δ_{ni})

$$\delta_i \colon [n-1] \to [n],$$

$$0, ..., i-1, i, ..., n-1 \mapsto 0, ..., i-1, i+1, ..., n \quad (0 \leqslant i \leqslant n),$$

(5.52)

is the injective map which misses i, while the *degeneracy* ε_i (or ε_{ni})

$$\varepsilon_i: [n] \to [n-1],$$
$$0, ..., i, i+1, ..., n \mapsto 0, ..., i, i, ..., n-1 \quad (0 \leqslant i \leqslant n-1), \tag{5.53}$$

is the surjective map which repeats i. (These maps are also called *cofaces* and *codegeneracies*.)

These mappings satisfy the *cosimplicial relations*

$$\delta_j \delta_i = \delta_{i+1} \delta_j, \qquad \text{for } j \leqslant i,$$

$$\varepsilon_i \varepsilon_j = \varepsilon_j \varepsilon_{i+1}, \qquad \text{for } j \leqslant i,$$

$$\varepsilon_j \delta_i = \begin{cases} \delta_{i-1} \varepsilon_j, & \text{for } j < i-1, \\ \text{id}, & \text{for } j = i-1, i, \\ \delta_i \varepsilon_{j-1}, & \text{for } j > i. \end{cases} \tag{5.54}$$

Plainly, every injective increasing mapping $\delta: [m] \to [n]$ can be uniquely factorised as a composite of faces, indexed by $\{i_1, ..., i_k\} = [n] \setminus \varphi([m])$

$$\delta = \delta_{i_k} ... \delta_{i_1}: [m] \to [n] \quad (0 \leqslant i_1 < i_2 < ... < i_k \leqslant n), \tag{5.55}$$

while every surjective increasing mapping $\varepsilon: [m] \to [n]$ can be uniquely factorised as a composite of degeneracies indexed by the 'stationary points' $j \leqslant m-1$, where $\varepsilon(j) = \varepsilon(j+1)$

$$\varepsilon = \varepsilon_{j_1} ... \varepsilon_{j_h}: [m] \to [n] \quad (0 \leqslant j_1 < j_2 < ... < j_h \leqslant m-1). \tag{5.56}$$

Finally an increasing mapping $\varphi: [m] \to [n]$ can be uniquely factorised epi-mono $\varphi = \delta \varepsilon$, and as a composite

$$\varphi = \delta_{i_k} ... \delta_{i_1} \varepsilon_{j_1} ... \varepsilon_{j_h}$$
$$(0 \leqslant j_1 < j_2 < ... < j_h < m, \ 0 \leqslant i_1 < i_2 < ... < i_h \leqslant n), \tag{5.57}$$

with $m - h = n - k$ (the cardinal of the image of φ).

All this shows that the category $\boldsymbol{\Delta}$ is generated by faces and degeneracies, under the cosimplicial relations (applying the Special Form Lemma 5.3.8).

Similar facts hold for the augmented simplicial site $\boldsymbol{\Delta}^\sim$, adding $\delta_0: \emptyset \to [0]$ to the previous generators.

5.3.3 Simplicial objects by faces and degeneracies

As a consequence, a simplicial object $X: \boldsymbol{\Delta}^{\text{op}} \to \mathsf{C}$ can be equivalently assigned as a family $(X_n)_{n \geqslant 0}$ of objects of C, linked by *faces* and *degeneracies*

$$\partial_i: X_n \to X_{n-1} \qquad (0 \leqslant i \leqslant n),$$
$$e_i: X_{n-1} \to X_n \qquad (0 \leqslant i \leqslant n-1), \tag{5.58}$$

$$X_0 \underset{e_0}{\overset{\partial_i}{\rightleftarrows}} X_1 \underset{e_i}{\overset{\partial_i}{\rightleftarrows}} X_2 \underset{e_i}{\overset{\partial_i}{\rightleftarrows}} X_3 \ ...$$

satisfying the *simplicial relations* (dual to the cosimplicial ones of (5.54))

$$
\begin{aligned}
\partial_i \partial_j &= \partial_j \partial_{i+1}, && \text{for } j \leqslant i, \\
e_j e_i &= e_{i+1} e_j, && \text{for } j \leqslant i, \\
\partial_i e_j &= \begin{cases} e_j \partial_{i-1}, & \text{for } j < i-1, \\ \text{id}, & \text{for } j = i-1, i, \\ e_{j-1} \partial_i, & \text{for } j > i. \end{cases}
\end{aligned}
\qquad (5.59)
$$

5.3.4 Singular simplices

The category Δ can be embedded in Top, forming the cosimplicial object of the *standard simplices* Δ^n (with euclidean topology)

$$J \colon \Delta \to \mathsf{Top},$$
$$[n] \mapsto \Delta^n = \{(t_0, ..., t_n) \in \mathbb{R}^{n+1} \mid t_i \geqslant 0, \ \Sigma\, t_i = 1\},$$

$$
\begin{aligned}
\delta_i &\colon \Delta^{n-1} \to \Delta^n \qquad (0 \leqslant i \leqslant n), \\
\delta_i(t_0, ..., t_{n-1}) &= (t_0, ..., t_{i-1}, 0, t_i, ..., t_{n-1}),
\end{aligned}
\qquad (5.60)
$$

$$
\begin{aligned}
\varepsilon_i &\colon \Delta^n \to \Delta^{n-1} \qquad (0 \leqslant i \leqslant n-1), \\
\varepsilon_i(t_0, ..., t_i, ..., t_n) &= (t_0, ..., t_i + t_{i+1}, ..., t_n).
\end{aligned}
$$

For every topological space S, there is an associated *singular simplicial set* Smp(S), obtained by composing this functor with the contravariant functor $\mathsf{Top}(-, S)\colon \mathsf{Top}^{\mathrm{op}} \to \mathsf{Set}$

$$
\begin{aligned}
\mathrm{Smp(S)} &\colon \Delta^{\mathrm{op}} \to \mathsf{Set}, & (\mathrm{Smp(S)})_n &= \mathsf{Top}(\Delta^n, S), \\
\partial_i(a) &= a\delta_i \colon \Delta^{n-1} \to S & (a &\in \mathsf{Top}(\Delta^n, S)), \\
e_i(a) &= a\varepsilon_i \colon \Delta^n \to S & (a &\in \mathsf{Top}(\Delta^{n-1}, S)).
\end{aligned}
\qquad (5.61)
$$

Globally we have the *singular simplicial* functor

$$\mathrm{Smp} \colon \mathsf{Top} \to \mathsf{SmpSet}, \qquad (5.62)$$

which is used to define the singular homology of a space (see 5.3.7).

5.3.5 The geometric realisation

The *geometric realisation* of a simplicial set $X \colon \mathbf{\Delta}^{\mathrm{op}} \to \mathsf{Set}$ is a topological space $\mathcal{R}(X)$ constructed by pasting a standard simplex $\Delta^n(x)$ for every $x \in X_n$, under identifications induced by the faces and degeneracies of X.

More precisely, we form the topological sum

$$S = \Sigma_n \, DX_n \times \Delta^n$$

(where $D \colon \mathsf{Set} \to \mathsf{Top}$ equips a set with the discrete topology), modulo the equivalence relation R spanned by the following identifications

$$
\begin{aligned}
[x, \delta_i t] &= [\partial_i x, t] & (x \in X_n, \ t \in \Delta^{n-1}), \\
[x, \varepsilon_i t] &= [e_i x, t] & (x \in X_{n-1}, \ t \in \Delta^n).
\end{aligned}
\tag{5.63}
$$

The space $\mathcal{R}(X) = S/R$ is thus the coend (see 2.8.6)

$$\mathcal{R}(X) = \int^n DX_n \times \Delta^n, \qquad u_n \colon DX_n \times \Delta^n \to \mathcal{R}(X), \tag{5.64}$$

of the functor $\mathbf{\Delta}^{\mathrm{op}} \times \mathbf{\Delta} \to \mathsf{Top}$, produced by the composite $DX \colon \mathbf{\Delta}^{\mathrm{op}} \to \mathsf{Top}$ and the embedding $J \colon \mathbf{\Delta} \to \mathsf{Top}$. (The latter, defined in (5.60), realises $[n]$ as the standard simplex $\Delta^n \subset \mathbb{R}^{n+1}$.) The map u_n is the composite $DX_n \times \Delta^n \subset S \to S/R$.

We have now, for any simplicial set X, a canonical morphism of simplicial sets

$$\eta X \colon X \to \mathrm{Smp}\mathcal{R}(X), \qquad (\eta X)_n(x) = u_n(x, -) \colon \Delta^n \to \mathcal{R}(X). \tag{5.65}$$

Finally, $(\mathcal{R}(X), \eta X)$ is a universal arrow from X to the functor $\mathrm{Smp} \colon \mathsf{Top} \to \mathsf{SmpSet}$, and the geometric realisation is its left adjoint

$$\mathcal{R} \colon \mathsf{SmpSet} \to \mathsf{Top}, \qquad \mathcal{R} \dashv \mathrm{Smp}. \tag{5.66}$$

5.3.6 Exercises and complements

(a) *Faces and degeneracies as adjoints.* The unbounded chain of adjunctions of the ordered integral line, in 1.7.7(e), can be restricted to finite ordinals, giving a finite chain (involving $n + 1$ faces, n degeneracies and $2n$ adjunctions)

$$\delta_i \colon [n-1] \rightleftarrows [n] \colon \varepsilon_j \qquad (0 \leqslant i \leqslant n, \ 0 \leqslant j \leqslant n-1),$$

$$\delta_n \dashv \varepsilon_{n-1} \dashv \ \dots \ \delta_{i+1} \dashv \varepsilon_i \dashv \delta_i \dashv \ \dots \ \varepsilon_0 \dashv \delta_0, \tag{5.67}$$

$$\varepsilon_i \delta_i = \mathrm{id} = \varepsilon_{i-1} \delta_i, \qquad \delta_{i+1} \varepsilon_i \leqslant \mathrm{id}, \qquad \delta_i \varepsilon_i \geqslant \mathrm{id}.$$

(b) The category $\mathbf{\Delta}^\sim$ of finite ordinals $n = \{0, \dots, n-1\}$ has a strict monoidal structure (see Section 3.4) given by the ordinal sum $m + n$, with identity 0.

The interested reader can see in [M4], Section VII.5, that the terminal object 1 has a unique structure of internal monoid in $\Delta^\sim$ (in the sense of 3.4.8). The latter can be characterised as *the universal strict monoidal category equipped with an internal monoid*.

(c) The augmented *symmetric simplicial site* $!\Delta^\sim$ is the category *of finite cardinals* (equivalent to the category of finite sets). It also has a description by generators and relations, and can be characterised as *the universal strict monoidal category equipped with an internal symmetric monoid*: see [G5].

5.3.7 *Exercises and complements* (Singular homology)

These brief hints are addressed to a reader with some interest in the argument and no real knowledge of it. To go on, Vick [Vi] is a very clear elementary book on singular homology; other texts are cited in the Introduction to this chapter.

(a) A (positive) *chain complex* of abelian groups is a diagram in Ab

$$A = ((A_n)_{n \geqslant 0}, (\partial_n)_{n \geqslant 1})$$

$$\ldots A_{n+1} \xrightarrow{\partial_{n+1}} A_n \xrightarrow{\partial_n} A_{n-1} \ldots A_1 \xrightarrow{\partial_1} A_0 \dashrightarrow 0 \ldots \qquad (5.68)$$

where $\partial_n \partial_{n+1} = 0$ for $n \geqslant 1$.

The homomorphisms ∂_n are called *boundaries* or *differentials* (and prolonged at the right with null homomorphisms ∂_n for $n \leqslant 0$, when useful).

The condition $\partial_n \partial_{n+1} = 0$ is equivalent to saying that

$$\operatorname{Im} \partial_{n+1} \subset \operatorname{Ker} \partial_n.$$

A *morphism of chain complexes* $f : A \to B$ has components $f_n : A_n \to B_n$ in Ab, that commute with the boundaries of the two complexes: $\partial_n f_n = f_{n-1} \partial_n$ for $n \geqslant 0$. With the componentwise composition we get the category $\mathrm{Ch}_+(\mathsf{Ab})$ of *chain complexes of abelian groups*.

(The reader may know that this construction can be applied to every abelian category A, giving an abelian category $\mathrm{Ch}_+(\mathsf{A})$. We are here in the domain of Homological Algebra, at its beginning.)

(b) The reader should now construct a functor

$$\mathrm{Ch}_+ : \mathsf{SmpSet} \to \mathrm{Ch}_+(\mathsf{Ab}), \qquad \mathrm{Ch}_+(K) = ((\mathrm{F}(K_n)), (\partial_n)), \qquad (5.69)$$

where, for a simplicial set K, the component $\mathrm{Ch}_n(K)$ is the free abelian group $F(K_n) = \mathbb{Z}K_n$. The differential is defined on the basis K_n as

$$\partial_n(a) = \Sigma_i \, (-1)^i \, \partial_{ni}(a) \in F(K_{n-1}), \tag{5.70}$$

where $i = 0, ..., n$. The only non trivial point, and a nice exercise, is proving that $\partial_n \partial_{n+1} = 0$, by applying the simplicial relations of the faces.

(c) Construct the (algebraic) *homology functor*

$$H_n \colon \mathsf{Ch}_+(\mathsf{Ab}) \to \mathsf{Ab}, \qquad H_n(A) = (\operatorname{Ker} \partial_n)/(\operatorname{Im} \partial_{n+1}), \tag{5.71}$$

where $H_n(A)$ is called the n-th *homology group* of the chain complex $A = ((A_n), (\partial_n))$.

(d) Finally we have the n-th *singular homology functor*

$$H_n \colon \mathsf{Top} \to \mathsf{Ab}, \qquad H_n(X) = H_n(\mathsf{Ch}_+(\mathsf{Smp}(\mathsf{X}))), \tag{5.72}$$

by composing the singular simplicial functor $\mathsf{Smp} \colon \mathsf{Top} \to \mathsf{SmpSet}$ of (5.62) with the functor $\mathsf{Ch}_+ \colon \mathsf{SmpSet} \to \mathsf{Ch}_+(\mathsf{Ab})$, above, and the algebraic homology functor $H_n \colon \mathsf{Ch}_+(\mathsf{Ab}) \to \mathsf{Ab}$.

For a topological space X, a (formal) linear combination of singular simplices $a_i \colon \Delta^n \to X$ (with coefficients in $\mathbb{Z}$)

$$c = \Sigma_i \, \lambda_i a_i \in F(\mathsf{Top}(\Delta^n, X)) \tag{5.73}$$

is called a *singular (simplicial) chain* of dimension n. It is an *n-cycle* if $\partial_n(c) = 0$, and an *n-boundary* if $c \in \operatorname{Im} \partial_{n+1}$. Every cycle has a homology class $[c] \in H_n(S)$, which annihilates if and only if c is a boundary.

(e) Marginally, we note that the differential (5.70) can be viewed as a morphism $K_n \to K_{n-1}$ of the Kleisli category Set_T of sets and $\mathbb{Z}$-weighted mappings, considered in 3.7.1(b).

(f) A comonad on a category C can be used to construct simplicial objects on C, and then homology or cohomology theories on this category. Mac Lane gives a brief, clear presentation of group (co)homology in this way, in [M4], Section VII.6. This is an excellent introduction to a field also called 'triple cohomology' or 'cohomology on standard constructions' [Bc, BarB, In].

5.3.8 Lemma (Special Form Lemma)

Let S be a category generated by a subgraph G, whose maps satisfy in S a system of relations Φ.

Then S is freely generated by G under such relations if and only if every S-map can be expressed in a unique special form *$f = g_m...g_1$, as a composite* of G-maps, and every G-factorisation $f = g'_m...g'_1$ in S can be made special by applying the relations Φ, finitely many times.

Proof The necessity of this condition is easily proved by *choosing*, arbitrarily, one special form in each equivalence class of the quotient of the free category generated by G, under the congruence generated by Φ.

Conversely, we take a graph-morphism $F_0 \colon G \to C$, with values in an arbitrary category. If F_0 satisfies the system of relations Φ, it extends to at most one functor $F \colon S \to C$, letting it operate on special forms $F(g_m...g_1) = F_0(g_m)...F_0(g_1)$. This construction defines indeed a functor, since any composite $f'f$ in S can be rewritten in special form using relations which are preserved in C. □

5.3.9 Further comments on horror vacui

Finally, let us remark that the categories $\Delta^\sim$ and $!\Delta^\sim$, of *all* finite ordinals or cardinals, have substantial categorical properties and simple characterisations (recalled in 5.3.6).

Taking out the initial object (the empty ordinal or cardinal) destroys these properties, but gives two categories – namely Δ and $!\Delta$ – which have an important *auxiliary role* in Combinatorial and Algebraic Topology, as the sites of Smp(C) and its symmetric version !Smp(C).

We can therefore refine our comments on *horror vacui*, in 1.1.6: artificial exclusions *can* be of interest, for a category S which is used in an 'auxiliary role' – typically as a domain of functors $X \colon S \to C$ with values in a (generally 'well behaved') category C.

In the same perspective we can also note that, when discussing limits, *we do not expect the domain of a functor $S \to C$ to have good categorical properties*; this can even be a hindrance: for instance, if S has an initial object the limit of each functor defined on S is trivial (as we have observed in 2.8.1).

5.4 Cubical sets

After simplicial sets, cubical sets have also been studied in Algebraic Topology, in a parallel way. They are 'modelled' on the standard euclidean cubes $\mathbb{I}^n$, much in the same way as simplicial sets are modelled on the euclidean tetrahedra Δ^n.

The main advantage with respect to the simplicial framework comes from the fact that the cubes $\mathbb{I}^n$ are closed under cartesian products (in Top), while the products of tetrahedra have to be covered with tetrahedra.

Here we describe cubical objects as functors $X \colon \mathsf{I}^{\mathrm{op}} \to \mathsf{C}$ defined on the basic cubical site, that was introduced and characterised in [GM]. Various extensions, studied in this paper, are mentioned at the end of this section.

Also here the two-valued index α (or β) varies in the cardinal $2 = \{0, 1\} = \{-, +\}$. The singleton $1 = 2^0$ is also written as $\{*\}$.

5.4.1 The basic cubical site

Let I be the subcategory of Set consisting of the *elementary cubes* $2^n = \{0, 1\}^n$ (for $n \geqslant 0$) together with the mappings $f \colon 2^m \to 2^n$ which delete some coordinates and insert some 0's and 1's (without modifying the order of the remaining coordinates).

I is a strict symmetric monoidal category; its tensor product $2^m \, \square \, 2^n = 2^{m+n}$ is induced by the cartesian product of Set, but is no longer a cartesian product in the subcategory (where the diagonal $2 \to 2 \times 2$ of Set is not present); the exponent of 2^n denotes now a tensor power.

The object $2 = \{0, 1\}$ is an *internal interval* (both in $(\mathsf{Set}, \times)$ and $(\mathsf{I}, \square)$), with (elementary) *faces* δ^α and *degeneracy* ε (as in 5.2.1, with slightly different notation)

$$\delta^\alpha \colon 1 \to 2, \qquad \varepsilon \colon 2 \to 1, \qquad \varepsilon . \delta^\alpha = \mathrm{id},$$
$$\delta^\alpha(*) = \alpha \qquad (\alpha = 0, 1). \tag{5.74}$$

Higher faces and degeneracies are derived from the previous ones, via the monoidal structure, for $1 \leqslant i \leqslant n$ and $\alpha = \pm$

$$\delta^\alpha_{ni} = 2^{i-1} \, \square \, \delta^\alpha \, \square \, 2^{n-i} \colon 2^{n-1} \to 2^n,$$
$$\delta^\alpha_{ni}(t_1, ..., t_{n-1}) = (t_1, ..., t_{i-1}, \alpha, t_i, ..., t_{n-1}),$$
$$\varepsilon_{ni} = 2^{i-1} \, \square \, \varepsilon \, \square \, 2^{n-i} \colon 2^n \to 2^{n-1}, \tag{5.75}$$
$$\varepsilon_{ni}(t_1, ..., t_n) = (t_1, ..., t_{i-1}, t_{i+1}, ..., t_n).$$

The *cocubical relations* follow easily from the formulas (5.74) (omitting the index n):

$$\delta^\beta_j . \delta^\alpha_i = \delta^\alpha_{i+1} . \delta^\beta_j, \qquad \text{for } j \leqslant i,$$
$$\varepsilon_i . \varepsilon_j = \varepsilon_j . \varepsilon_{i+1}, \qquad \text{for } j \leqslant i,$$
$$\varepsilon_j . \delta^\alpha_i = \begin{cases} \delta^\alpha_{i-1} . \varepsilon_j, & \text{for } j < i, \\ \mathrm{id}, & \text{for } j = i, \\ \delta^\alpha_i . \varepsilon_{j-1}, & \text{for } j > i. \end{cases} \tag{5.76}$$

5.4.2 *Lemma* (The canonical form of the cubical site)

Using the cocubical relations (5.76) *as* rewriting rules, *each composite in* Set *of the cubical faces and degeneracies can be rewritten in a unique canonical factorisation (empty for an identity)*

$$\delta_{j_s}^{\alpha_s} \ldots \delta_{j_1}^{\alpha_1} \, \varepsilon_{i_1} \ldots \varepsilon_{i_r} \colon 2^m \to 2^{m-r} \to 2^n,$$

$$1 \leqslant i_1 < \ldots < i_r \leqslant m, \qquad 1 \leqslant j_1 < \ldots < j_s \leqslant n, \qquad (5.77)$$

$$m - r = n - s \geqslant 0,$$

consisting of a surjective composed degeneracy $\varepsilon_{i_1} \ldots \varepsilon_{i_r}$ (that deletes the co-ordinates specified by the indices) and an injective composed face $\delta_{j_s}^{\alpha_s} \ldots \delta_{j_1}^{\alpha_1}$ (that inserts 0's and 1's in the specified positions).

Proof Obvious.   □

5.4.3 *Theorem* (The basic cubical site)

The category I *can be characterised in the following equivalent ways:*

(a) the subcategory of Set *with objects 2^n, generated by all faces and de-generacies* (5.75),

(b) the subcategory of Set *with objects 2^n, closed under the binary-product functor (realised as $2^m \times 2^n = 2^{m+n}$), and generated by the basic faces ($\delta^\alpha \colon 1 \to 2$) and degeneracy ($\varepsilon \colon 2 \to 1$),*

(c) the category generated by the graph (5.75), *subject to the cocubical relations* (5.76),

(d) the free strict monoidal category with an assigned internal interval, namely $(2, \delta^\alpha, \varepsilon)$.

Proof The characterisation (a) is obvious: every map of I can clearly be factorised as in (5.77), in a unique way. Therefore, (b) follows from the construction of higher faces and degeneracies as tensor products, in (5.75), while (c) follows from the Special Form Lemma, in 5.3.8.

For (d), let $A = (A, \otimes, E)$ be a strict monoidal category with an assigned internal interval $(A, \delta^\alpha, \varepsilon)$. We write a tensor power as $(-)^n$.

Defining the higher faces and degeneracies of A as above, in (5.75)

$$\delta_{ni}^\alpha = A^{i-1} \otimes \delta^\alpha \otimes A^{n-i} \colon A^{n-1} \to A^n,$$

$$\varepsilon_{ni} = A^{i-1} \otimes \varepsilon \otimes A^{n-i} \colon A^n \to A^{n-1}, \qquad (5.78)$$

the cocubical relations are satisfied, and there is a unique functor $F \colon I \to A$ sending 2^n to A^n and preserving higher faces and degeneracies.

It is now sufficient to prove that this F is strictly monoidal: then, it will be the unique such functor sending 2 to A and preserving $\delta^\alpha, \varepsilon$. As we already know that F is a functor, our result follows from the following formulas

$$F(2^p \square 2^q) = F(2^{p+q}) = A^{p+q} = A^p \otimes A^q,$$

$$F(2^{i-1} \square \delta^\alpha \square 2^{n-i}) = F(\delta_{ni}^\alpha) = \delta_{ni}^\alpha = A^{i-1} \otimes \delta^\alpha \otimes A^{n-i}, \qquad (5.79)$$

$$F(2^{i-1} \square \varepsilon \square 2^{n-i}) = F(\varepsilon_{ni}) = \varepsilon_{ni} = A^{i-1} \otimes \varepsilon \otimes A^{n-i},$$

since the tensor product of two arbitrary I-maps $f = f_r...f_1$ and $g = g_s...g_1$, both in canonical form, can be decomposed as

$$f \square g = (f_r \square 1)...(f_1 \square 1).(1 \square g_s)...(1 \square g_1).$$

$\square$

5.4.4 Cubical objects

A *cubical object* in the category C is a presheaf $X \colon \mathsf{I}^{\mathrm{op}} \to \mathsf{C}$ on the cubical site. A *morphism of cubical objects* $f \colon X \to Y$ is a natural transformation of such functors; they form the category $\mathrm{Cub}(\mathsf{C}) = \mathsf{C}^{\mathsf{I}^{\mathrm{op}}}$ of cubical objects in C.

As a consequence, a cubical object $X \colon \mathsf{I}^{\mathrm{op}} \to \mathsf{C}$ can be equivalently assigned as a family $(X_n)_{n \geqslant 0}$ of objects of C, linked by *faces* and *degeneracies* (for $1 \leqslant i \leqslant n$ and $\alpha = \pm$)

$$\partial_{ni}^\alpha \colon X_n \to X_{n-1} \qquad e_{ni} \colon X_{n-1} \to X_n, \qquad (5.80)$$

that satisfy the *cubical relations* (dual to the cocubical ones of (5.76))

$$\partial_i^\alpha . \partial_j^\beta = \partial_j^\beta . \partial_{i+1}^\alpha, \qquad \text{for } j \leqslant i,$$

$$e_j . e_i = e_{i+1} . e_j, \qquad \text{for } j \leqslant i,$$

$$\partial_i e_j = \begin{cases} e_j . \partial_{i-1}^\alpha, & \text{for } j < i, \\ \mathrm{id}, & \text{for } j = i, \\ e_{j-1} . \partial_i^\alpha, & \text{for } j > i. \end{cases} \qquad (5.81)$$

5.4.5 The singular cubical set of a space

In Top (with the cartesian monoidal structure) the standard interval $\mathbb{I} = [0, 1]$ of the euclidean line has an obvious structure of internal interval

(already considered in 5.2.1)

$$\delta^\alpha \colon \{0\} \to \mathbb{I}, \quad \varepsilon \colon \mathbb{I} \to \{0\}, \qquad \varepsilon.\delta^\alpha = \mathrm{id},$$
$$\delta^\alpha(0) = \alpha \qquad\qquad (\alpha = 0, 1). \tag{5.82}$$

Higher faces and degeneracies are constructed as above, for $1 \leqslant i \leqslant n$ and $\alpha = \pm$

$$\delta^\alpha_{ni} = \mathbb{I}^{i-1} \times \delta^\alpha \times \mathbb{I}^{n-i} \colon \mathbb{I}^{n-1} \to \mathbb{I}^n,$$
$$\varepsilon_{ni} = \mathbb{I}^{i-1} \times \varepsilon \times \mathbb{I}^{n-i} \colon \mathbb{I}^n \to \mathbb{I}^{n-1}, \tag{5.83}$$

and satisfy the cocubical relations.

We have thus an embedding

$$J \colon \mathsf{I} \to \mathsf{Top}, \qquad\qquad 2^n \mapsto \mathbb{I}^n, \tag{5.84}$$

or – in other words – a cocubical space.

For every space S the contravariant functor $\mathsf{Top}(-, S) \colon \mathsf{Top}^{\mathrm{op}} \to \mathsf{Set}$ produces the *singular cubical set*

$$\mathrm{Cub}(S) = \mathsf{Top}(J, S) \colon \mathsf{I}^{\mathrm{op}} \to \mathsf{Set},$$
$$\mathrm{Cub}(S) = ((\mathrm{Cub}_n(S)), (\partial^\alpha_i), (e_i)), \qquad \mathrm{Cub}_n(S) = \mathsf{Top}(\mathbb{I}^n, S),$$
$$\partial^\alpha_i(a \colon \mathbb{I}^n \to S) = (a.\delta^\alpha_i \colon \mathbb{I}^{n-1} \to S), \tag{5.85}$$
$$e_i(a \colon \mathbb{I}^{n-1} \to S) = (a.\varepsilon_i \colon \mathbb{I}^n \to S).$$

Acting similarly on each map $f \colon S \to T$ of topological spaces

$$\mathrm{Cub}(f) \colon \mathrm{Cub}(S) \to \mathrm{Cub}(T),$$
$$(\mathrm{Cub}(f))(a \colon \mathbb{I}^n \to S) = (fa \colon \mathbb{I}^n \to T), \tag{5.86}$$

we have a functor:

$$\mathrm{Cub} \colon \mathsf{Top} \to \mathsf{CubSet}. \tag{5.87}$$

5.4.6 The geometric realisation

Proceeding in the same way as for simplicial sets, in 5.3.5, the *geometric realisation* of a cubical set $X \colon \mathsf{I}^{\mathrm{op}} \to \mathsf{Set}$ is a topological space $\mathcal{R}(X)$ constructed by pasting a standard cube $\mathbb{I}^n(x)$ for every $x \in X_n$, under the identifications induced by the faces and degeneracies of X.

This is the coend (see 2.8.6)

$$\mathcal{R}(X) = \int^n DX_n \times \mathbb{I}^n, \qquad u_n \colon DX_n \times \mathbb{I}^n \to \mathcal{R}(X), \tag{5.88}$$

of the functor $\mathsf{I}^{\mathrm{op}} \times \mathsf{I} \to \mathsf{Top}$ produced by the composite $DX \colon \mathsf{I}^{\mathrm{op}} \to \mathsf{Top}$

and the embedding $J\colon \mathsf{I} \to \mathsf{Top}$ defined in (5.84) (that realises the vertex 2^n of I as the standard euclidean cube $\mathbb{I}^n$).

We have now, for any cubical set X, a canonical morphism of cubical sets

$$\eta X\colon X \to \mathrm{Cub}(\mathcal{R}(\mathrm{X})), \qquad (\eta \mathrm{X})_n(\mathrm{x}) = \mathrm{u}_n(\mathrm{x}, -)\colon \mathbb{I}^n \to \mathcal{R}(\mathrm{X}), \qquad (5.89)$$

and the geometric realisation is left adjoint to the singular-cubical-set functor $\mathrm{Cub}\colon \mathsf{Top} \to \mathsf{CubSet}$

$$\mathcal{R}\colon \mathsf{CubSet} \to \mathsf{Top}, \qquad \mathcal{R} \dashv \mathrm{Cub}. \qquad (5.90)$$

5.4.7 Extended cubical sites

As already remarked in Section 5.2, the formal study of homotopies requires to add further structure to cubical faces and degeneracies: namely the connections (introduced in Brown–Higgins [BroH1]), together with reversions and transpositions (cf. [G8]).

These structures are embodied in the *intermediate cubical site* J, the *extended cubical site* K and its *symmetric* version !K, that have been introduced in [GM].

5.4.8 *The cubical construction of singular homology

Singular homology can be equivalently constructed from the cubical models $\mathbb{I}^n$ instead of the simplicial models Δ^n (as in 5.3.7). The interested reader is referred to the excellent, introductory text by Massey [Mas]. (A proof of the equivalence of these constructions is given, at a more advanced level, in Hilton–Wylie [HiW], using the 'acyclic model' technique.)

The singular homology of the space X is defined here as the homology of the (normalised) chain complex associated to the cubical set $\mathrm{Cub}(X)$

$$H_n\colon \mathsf{Top} \to \mathsf{Ab}, \qquad H_n(X) = H_n(\mathrm{Ch}_+(\mathrm{Cub}(\mathrm{X}))). \qquad (5.91)$$

To make sense of this, we must construct the chain complex $\mathrm{Ch}_+(K)$ associated to a cubical set K; this is more complicated than the simplicial analogue: here we have to quotient the free abelian group $F(K_n) = \mathbb{Z}K_n$ modulo the subgroup spanned by the 'degenerate elements' (otherwise we would not get the correct homology groups).

First, the *degenerate cubes* of a cubical set K are all the elements of type $e_i(a)$

$$\mathrm{Deg}_n K = \bigcup_i \mathrm{Im}\,(e_i\colon K_{n-1} \to K_n), \qquad \mathrm{Deg}_0 K = \emptyset. \qquad (5.92)$$

Because of the cubical relations, we have (for $i = 1, ..., n$)

$$a \in \mathrm{Deg}_n K \;\Rightarrow\; (\partial_i^\alpha a \in \mathrm{Deg}_{n-1} K \text{ or } \partial_i^- a = \partial_i^+ a). \tag{5.93}$$

Now $\mathrm{Ch}_+(K)$ is defined as follows:

$$\mathrm{Ch}_n(K) = (\mathbb{Z} K_n)/(\mathbb{Z}\mathrm{Deg}_n K),$$
$$\partial_n : \mathrm{Ch}_n(K) \to \mathrm{Ch}_{n-1}(K), \tag{5.94}$$
$$\partial_n(\hat{a}) = \Sigma_{i,\alpha} \, (-1)^{i+\alpha} (\partial_i^\alpha a)\hat{} \qquad (a \in K_n).$$

Here $\hat{a}$ is the class of the n-cube a, up to degenerate cubes; the differential is legitimate because of (5.93). Again, the only non trivial point is proving that $\partial_n \partial_{n+1} = 0$, by applying the cubical relations of the faces.

The heavier definition of $\mathrm{Ch}_+(X)$ is compensated by the advantages in all the arguments related with cartesian products of spaces, starting from the Homotopy Invariance Theorem of singular homology, as one can see exploring Massey's book.

5.5 Hints at Directed Algebraic Topology

Directed Algebraic Topology is a recent subject which arose in the 1990's, on the one hand in abstract settings for homotopy theory, like [G3], and on the other hand in investigations in the theory of concurrent processes, like [FGR1, FGR2]. Its general aim can be stated as 'modelling non-reversible phenomena'. The subject has a deep relationship with category theory.

The domain of Directed Algebraic Topology is distinguished from the domain of classical Algebraic Topology by the principle that *directed spaces have privileged directions*, and their (directed) paths need not be reversible.

While Topology and Algebraic Topology deal with reversible worlds, where a path can always be travelled backwards, the study of non-reversible phenomena requires broader worlds, where a directed space *can* have non-reversible paths. (It is important to develop the directed theory so that it *extends* the reversible one.)

The classical tools, consisting of:

- ordinary homotopies, fundamental groups and fundamental n-groupoids,

are now replaced by (possibly) non-reversible versions:

- *directed homotopies, fundamental monoids* and *fundamental n-categories*.

Similarly, the homological theories of the directed setting take values in 'directed' algebraic structures, like *preordered* abelian groups or abelian *monoids*.

Here we give a few hints at the starting points of this subject, taken from the book [G8].

The prefixes ↑ and d- are used to distinguish a directed notion from the corresponding 'reversible' one.

5.5.1 Preordered topological spaces

The simplest topological setting where one can study directed paths and directed homotopies is likely the category pTop of *preordered topological spaces* and *preorder-preserving continuous mappings*.

(We are not requiring any particular relationship between topology and preorder; this would only complicate the category without any advantage.)

Here, the *standard directed interval* $\uparrow \mathbb{I} = \uparrow[0,1]$ has the euclidean topology and the natural order. A (directed) *path* in a preordered space X is – by definition – a map $a\colon \uparrow \mathbb{I} \to X$ (continuous and monotone).

The category pTop has all limits and colimits, constructed as for topological spaces and equipped with the initial or final preorder for their structural maps; for instance, in a product $X = \Pi X_j$, we have the product preorder: $(x_j) \prec_X (x'_j)$ if and only if, for each index j, $x_j \prec x'_j$ in X_j.

The forgetful functor $U\colon$ pTop $\to$ Top has both a left and a right adjoint, $D \dashv U \dashv C$ where DS (resp. CS) is the space S equipped with the *discrete* order (resp. the *chaotic*, or *codiscrete* preorder).

The standard embedding of Top in pTop will be the one given by the *codiscrete preorder*, so that all (ordinary) paths in S are directed in CS.

Our category is not cartesian closed, of course; but it is easy to transfer here the classical result for topological spaces, recalled above (in 5.1.1).

Thus, every preordered space A that has a *locally compact Hausdorff topology* and an *arbitrary preorder* is exponentiable in pTop, with Y^A consisting of the set pTop$(A, Y) \subset$ Top(UA, UY) of *monotone continuous mappings*, equipped with the (induced) compact-open topology and the *pointwise preorder*

$$f \prec g \quad \text{if} \quad (\forall x \in A,\ f(x) \prec_Y g(x)). \tag{5.95}$$

The main setting studied in [G8] is in fact a richer one, the category dTop of d-spaces, or *spaces with distinguished paths*, where an object can have *vortices* (i.e. non-reversible loops); this framework will not be recalled here.

5.5.2 Comments

As to our choice of the category pTop, we note two points – in the line of the remarks in 1.1.5(b).

First, *if we replace preordered spaces with the ordered ones we miss the embedding* C: Top $\to$ pTop by the chaotic preorder, and can no longer view classical Algebraic Topology within the directed one. Moreover, colimits of ordered spaces are 'different' from the topological ones (in the same way as the colimits of ordered sets do not agree with those of sets).

Second, if we require that in a preordered space X the graph of the pre-order is closed in $X \times X$ (as is often done), then the discrete (pre)order would only be allowed in Hausdorff spaces (which are precisely characterised by having a closed diagonal), and the previous embedding D: Top $\to$ pTop would fail. Now we would complicate limits in pTop.

5.5.3 The basic structure of the directed interval

The standard directed interval $\uparrow \mathbb{I} = \uparrow [0, 1]$, in the category pTop of pre-ordered topological spaces, has a structure *partially* similar to the ordinary one, in (5.21).

Faces and *degeneracy* are as in the classical case

$$\partial^\alpha : \{*\} \rightleftarrows \uparrow \mathbb{I} : e, \quad \partial^-(*) = 0, \quad \partial^+(*) = 1, \quad e(t) = *, \qquad (5.96)$$

where $\uparrow \mathbb{I}^0 = \{*\}$ is now an ordered space, with the unique order-relation on the singleton.

On the other hand, the classical reversion $r(t) = 1 - t$ is not an *endomap* of $\uparrow \mathbb{I}$, but becomes a map which we prefer to call *reflection*

$$r : \uparrow \mathbb{I} \to \uparrow \mathbb{I}^{\mathrm{op}}, \qquad\qquad r(t) = 1 - t, \qquad (5.97)$$

as it takes values in the *opposite* preordered space $\uparrow \mathbb{I}^{\mathrm{op}}$ (with the opposite preorder).

Again, the *standard concatenation pushout* can be realised as $\uparrow \mathbb{I}$ itself

$$
\begin{array}{ccc}
\{*\} & \xrightarrow{\ \partial^+\ } & \uparrow \mathbb{I} \\
{\scriptstyle \partial^-}\downarrow & \nearrow & \downarrow{\scriptstyle c^-} \\
\uparrow \mathbb{I} & \xrightarrow{\ c^+\ } & \uparrow \mathbb{I}
\end{array}
\qquad
\begin{array}{l}
c^-(t) = t/2, \\
\\
c^+(t) = (t+1)/2,
\end{array}
\qquad (5.98)
$$

since a mapping a: $\uparrow \mathbb{I} \to X$ with values in a preordered space is a *map* (continuous and monotone) if and only if its two restrictions ac^α (to the first or second half of the interval) are maps.

5.5.4 *Cylinder, cocylinder and homotopies*

The directed interval produces the cylinder functor of preordered topological spaces, where the product $I(X) = X \times {\uparrow}\mathbb{I}$ has the product topology and the product preorder:

$$I: \mathsf{pTop} \to \mathsf{pTop}, \qquad I(X) = X \times {\uparrow}\mathbb{I},$$

$$(x, t) \prec (x', t') \quad \text{if} \quad (x \prec x' \text{ in } X \text{ and } t \leqslant t' \text{ in } {\uparrow}\mathbb{I}). \tag{5.99}$$

The cylinder functor has a first-degree structure, formed of four natural transformations. Faces and degeneracy are as in the classical case (see (5.28)), while the reflection r, again, has to be expressed via the *reversor*, i.e. the involutive endofunctor R which reverses the preorder-relation

$$\partial^\alpha: 1 \to I, \qquad e: I \to 1, \qquad r: IR \to RI,$$

$$R: \mathsf{pTop} \to \mathsf{pTop}, \qquad R(X) = X^{\mathrm{op}} \qquad (\textit{reversor}), \qquad (5.100)$$

$$r: I(X^{\mathrm{op}}) \to (IX)^{\mathrm{op}}, \qquad r(x, t) = (x, 1 - t) \qquad (\textit{reflection}).$$

Since ${\uparrow}\mathbb{I}$ is exponentiable in pTop (by 5.5.1), we also have a path functor, right adjoint to the cylinder functor, where the preordered space $P(Y)$ has the compact-open topology and the pointwise preorder (for $a, b: {\uparrow}\mathbb{I} \to Y$)

$$P: \mathsf{pTop} \to \mathsf{pTop}, \qquad P(Y) = Y^{{\uparrow}\mathbb{I}},$$

$$a \prec b \quad \text{if} \quad (\forall t \in [0, 1], \ a(t) \prec_Y b(t)). \tag{5.101}$$

The path functor is equipped with a dual (first-degree) structure, formed of four natural transformations (with the same reversor R as above)

$$\partial^\alpha: P \to 1, \qquad e: 1 \to P, \qquad r: RP \to PR, \tag{5.102}$$

A (directed) *homotopy* $\varphi: f^- \to f^+: X \to Y$ is defined by a map

$$\hat{\varphi}: X \times {\uparrow}\mathbb{I} \to Y, \qquad \hat{\varphi}\partial^\alpha = f^\alpha,$$

or, equivalently, by a map

$$\check{\varphi}: X \to Y^{{\uparrow}\mathbb{I}}, \qquad \partial^\alpha \check{\varphi} = f^\alpha.$$

It yields a *reflected* homotopy between the opposite spaces

$$\varphi^{\mathrm{op}}: Rf^+ \to Rf^-: X^{\mathrm{op}} \to Y^{\mathrm{op}},$$

$$(\varphi^{\mathrm{op}})\hat{\ } = R(\hat{\varphi}).rX: IRX \to RIX \to RY. \tag{5.103}$$

Extending the construction of the fundamental groupoid of a space, in 5.2.8, we get here the *fundamental category* ${\uparrow}\Pi_1(X)$ of a preordered space, and a functor

$${\uparrow}\Pi_1: \mathsf{pTop} \to \mathsf{Cat}, \tag{5.104}$$

which sends the opposite space to the opposite category.

The fundamental monoids $\uparrow\pi_1(X,x) = \uparrow\Pi_1(X)(x,x)$ are often of little importance, even in richer settings where vortices are possible.

5.5.5 Some examples

We end with some hints at notions and applications dealt with in [G8], without any proof and still in the elementary framework of preordered spaces. Consider the following order relation in the plane

$$(x,y) \leqslant (x',y') \Leftrightarrow |y'-y| \leqslant x'-x. \qquad (5.105)$$

The picture shows the 'cone of the future' at a point p (i.e. the set of points which follow it) and a *directed path* from p' to p'', i.e. a continuous mapping $a\colon [0,1] \to \mathbb{R}^2$ which is (weakly) *increasing*, with respect to the natural order of the standard interval and the previous order of the plane: if $t \leqslant t'$ in $[0,1]$, then $a(t) \leqslant a(t')$ in the plane.

Take now the following (compact) subspaces X, Y of the plane, with the induced order (the cross-marked open rectangles are taken out). A directed path in X or Y satisfies the same conditions as above

Then – as displayed in the figures above and easy to guess – there are, respectively, 3 or 4 'homotopy classes' of directed paths from the point p' to the point p'', in the fundamental categories $\uparrow\Pi_1(X)$, $\uparrow\Pi_1(Y)$; in both cases there are none from p'' to p', and every loop is constant.

First, we can view each of these 'directed spaces' as a stream with two islands, and the induced order as an upper bound for the relative velocity feasible in the stream. Secondly, one can interpret the horizontal coordinate as (a measure of) time, the vertical coordinate as position in a

1-dimensional physical medium, and the order as the possibility of going from (x, y) to (x', y') with velocity $\leqslant 1$ (with respect to a 'rest frame' of the medium). The two forbidden rectangles are now linear obstacles in the medium, with a bounded duration in time. Thirdly, our figures can be viewed as execution paths of concurrent automata subject to some conflict of resources, as in [FGR2], fig. 14.

In all these cases, the fundamental category distinguishes between obstructions (islands, temporary obstacles, conflict of resources) which intervene essentially together (in the earlier diagram on the left) or one after the other, *in a discernible way* (on the right). On the other hand, the underlying topological spaces are homeomorphic, and topology, or algebraic topology, cannot distinguish these two situations.

Again, all the fundamental monoids $\uparrow\pi_1(X, x_0)$ are trivial: as a striking difference to the classical case, the fundamental monoids can carry a very minor part of the information of the fundamental category $\uparrow\Pi_1(X)$.

6

Applications in Homological Algebra

In this chapter we present some hints at a particular approach to Homological Algebra, exposed in the book [G9] and further extended in [G10]. This presentation cannot be self-contained, and various proofs will be referred to [G9].

After the classical text on Homological Algebra by Cartan–Eilenberg [CE], in 1956, a crucial contribution to this discipline came from the theory of abelian categories, introduced by Buchsbaum and Grothendieck [Bu1, Bu2, Gt]. This extension reaches new domains, and allows one to 'duplicate' theorems by duality. Yet some aspects of the theory are better viewed in the more general context of Puppe-exact categories [Pu, Mit], or p-exact categories for short.

In fact, as shown in [G9] and briefly suggested in Section 6.6, below, the problem of coherence of induction (particularly relevant in complicated systems like those that generate spectral sequences) can only be solved by working in auxiliary 'distributive' p-exact categories. This cannot be done in the setting of abelian categories, which are only distributive when trivial (i.e. equivalent to the category $\mathbf{1}$).

Here, after a general definition of kernels and cokernels, we show in Section 6.2 how suitable categories of lattices are of help in studying exactness properties in Ab and R Mod. The natural framework of this approach is p-exact categories, since we need from the very beginning a *non-abelian* p-exact category Mlc, of modular lattices and modular connections. Puppe-exact and abelian categories are presented in Sections 6.3–6.5. We end by introducing the category of relations of a p-exact category, as the basis for studying its subquotients, their induced relations and the coherence problem of the latter (Section 6.6).

Two-dimensional categories can also be useful in Homological Algebra, as we shall see in Sections 7.4 and 7.5.

Other perspectives on the use of category theory in Homological Algebra can be found in [BoB, In].

6.1 Kernels and cokernels

Kernels and cokernels are the basic notions in classical Homological Algebra, the starting point to define exact sequences and exact functors. They can be defined in an arbitrary pointed category C.

Some hints at a more general framework are given in 6.1.5.

6.1.1 Pointed categories

We assume that we are working in a *pointed* category C. As already said in 2.1.9, this means that there is a *zero object* 0, that is both initial and terminal in C: for every object A there is precisely one morphism $0 \to A$ and precisely one morphism $A \to 0$. *The* zero object is determined up to isomorphism; as usual, one of them is chosen.

For instance Ab, R Mod, Gp, Set., gRng, Ban and Ban$_1$ are pointed categories, with zero object given by the (adequate) singleton.

Given two objects A, B in C, the composite $A \to 0 \to B$ is called the *zero morphism* from A to B, and written as $0_{AB} \colon A \to B$, or also as 0 when the context identifies it.

The morphism $0 \to A$ is plainly a monomorphism, also written as $0_A \colon 0 \rightarrowtail A$ and called the *zero subobject* of A; dually the morphism $A \to 0$ is necessarily an epimorphism, written as $0^A \colon A \twoheadrightarrow 0$ and called the *zero quotient* of A.

6.1.2 Kernels and cokernels

In the pointed category C the *kernel* of a morphism $f \colon A \to B$ is defined by a universal property, as an object Ker f equipped with a morphism ker $f \colon \text{Ker}\, f \to A$, such that $f.(\ker f) = 0$ and

(i) every map h such that $fh = 0$ factorises uniquely through ker f

$$\text{Ker}\, f \xrightarrow{\ker f} A \xrightarrow{f} B$$

$$\nearrow^{u} \quad \uparrow^{h}$$

$$\bullet \tag{6.1}$$

which means that there exists a unique morphism u such that $h = (\ker f)u$.

Equivalently ker f is the equaliser of f and the zero morphism $A \to B$. A *normal monomorphism* is any kernel of a morphism, and is always a regular mono. Note that the existence of kernels does not require the existence of all equalisers, as many examples will show, in 6.3.3.

Dually, the *cokernel* of f is the coequaliser of f and the zero morphism $A \to B$, i.e. an object $\mathrm{Cok}\, f$ equipped with an (epi)morphism $\mathrm{cok}\, f \colon B \to \mathrm{Cok}\, f$, such that $(\mathrm{cok}\, f).f = 0$ and

(i*) every map h such that $hf = 0$ factorises uniquely through $\mathrm{cok}\, f$

$$A \xrightarrow{\;f\;} B \xrightarrow{\;\mathrm{cok}\, f\;} \mathrm{Cok}\, f$$

(6.2)

i.e. there is a unique morphism u such that $h = u(\mathrm{cok}\, f)$.

A *normal epimorphism* is any cokernel of a morphism, and is always a regular epi.

6.1.3 Exercises and complements, I

(a) For a morphism $f \colon A \to B$ in a pointed category, $\ker f$ is the same as the preimage $f^*(0)$ (see 2.4.6) of the zero subobject $0 \rightarrowtail B$, while $\mathrm{cok}\, f$ amounts to the image of the zero quotient $A \twoheadrightarrow 0$.

(b) Prove that normal monos (resp. epis) are closed under preimages (resp. direct images).

(c) In $R\,\mathsf{Mod}$ or $\mathsf{Set}_{\bullet}$, or $\mathsf{Top}_{\bullet}$, or Gp, or gRng, or Ban the natural 'kernel-object' is the usual subobject $\mathrm{Ker}\, f$ (the preimage of the zero subobject) and $\ker f$ is its embedding in the domain of f. However, if $u \colon K \to \mathrm{Ker}\, f$ is an isomorphism, also the composite $(\ker f)u \colon K \to A$ is *a* kernel of f. This is important, because the preservation of kernels by functors holds in this sense, generally.

(d) In $R\,\mathsf{Mod}$ and $\mathsf{Set}_{\bullet}$ every subobject is normal, while in $\mathsf{Top}_{\bullet}$ (resp. Gp, gRng, Ban) the normal subobjects correspond to the pointed subspaces (resp. invariant subgroups, bilateral ideals, closed linear subspaces). Note that *normal monomorphisms are not closed under composition* in Gp, nor in gRng, but they are in Ban.

(e) In $R\,\mathsf{Mod}$ the natural cokernel-object is the quotient $B/f(A)$, and $\mathrm{cok}\, f$ is the canonical projection $B \to B/f(A)$. In Gp (resp. gRng) the cokernel-object is the quotient of B modulo the invariant subgroup (resp. the bilateral ideal) generated by $f(A)$. In these categories all epimorphisms are normal.

(f) Show that in $\mathsf{Set}_{\bullet}$ and $\mathsf{Top}_{\bullet}$ cokernels exist, and there are non-normal epimorphisms. *The same holds in Ban.*

(g) Let us note that in R Mod the image $f(A)$ *is determined as* $\mathrm{Ker}\,(\mathrm{cok}\,f)$, i.e. *the kernel-object of the morphism cokf*, while $\mathrm{Cok}\,(\mathrm{ker}\,f) = A/\mathrm{Ker}\,f$ gives the *coimage* of f, an isomorphic object. The other categories we have been considering above do not behave in this 'simple' way, which holds in all abelian categories, and – loosely speaking – characterises the Puppe-exact ones (see Section 6.3).

6.1.4 Exercises and complements, II

Let C be a pointed category.

(a) Define a functor $\mathsf{C} \to \mathsf{C}^2$ which has a right adjoint $K\colon \mathsf{C}^2 \to \mathsf{C}$ if and only if C has kernels; describe the action of K on objects and arrows.

(a*) Define a functor $\mathsf{C} \to \mathsf{C}^2$ dually related to cokernels.

6.1.5 *Categories with null maps

This point is addressed to readers with some knowledge of homology theories, or interest in them.

Kernels, cokernels and all related notions of exactness can be defined, more generally, in a category C equipped with an ideal N of (so called) *null morphisms*, that extends the zero-morphisms of the pointed case. Homological Algebra in this framework is studied in [G10], where the hints below are developed.

The reader likely knows that homology theories of topological spaces are generally defined (and axiomatised) on the category Top_2 of *pairs of topological spaces*, as in the foundational text of Eilenberg–Steenrod [EiS]: an object is a 'pair' (X, A) formed of a space X and a subspace A, while a morphism $f\colon (X, A) \to (Y, B)$ is a continuous mapping $f\colon X \to Y$ such that $f(A) \subset B$. Composition is as in Top, and the latter is embedded in Top_2 by identifying the space X with the pair $(X, \emptyset)$.

The object (X, A) is usually read as 'X modulo A' and viewed as a formal quotient. Now, as a natural choice, we define an object (X, A) to be *null* when $X = A$, and we say that the morphism $f\colon (X, A) \to (Y, B)$ is *null* if it factorises through a null object, or equivalently if $f(X) \subset B$.

Kernels and cokernels (with respect to this ideal) exist and a general short exact sequence is – up to isomorphism – of the following type, with morphisms induced by the inclusions $A \to X$ and $X \to X$

$$(A, B) \rightarrowtail (X, B) \twoheadrightarrow (X, A). \tag{6.3}$$

If we take $B = \emptyset$ we get that (X, A) is indeed the quotient X/A in Top_2.

Loosely speaking, Eilenberg–Steenrod's exactness axiom for a (relative) homology theory defined on Top_2 amounts to saying that the theory takes any short exact sequence of pairs of spaces to a long homology sequence which is exact (in the codomain of the theory, usually an abelian category).

Top_2, equipped with this ideal, is a *homological category*, in the sense of [G10].

A similar analysis holds for the category Set_2 of *pairs of sets* (where A is a subset of X), or Gp_2 of *pairs of groups* (where A is a subgroup of X), etc.

The categories $\mathsf{Set}_\bullet$, $\mathsf{Top}_\bullet$ and Ban are *pointed homological*, i.e. homological with respect to the ideal of zero morphisms.

6.2 Categories of lattices in Homological Algebra

We show, *in an informal way*, how the category Mlc of modular lattices and modular (Galois) connections can be of use while dealing with subobjects. For simplicity, we work in an elementary context, the category $R\,\mathsf{Mod}$ of (left) modules on a (unitary) ring R.

This section is also a preparation for the next: the introduction of the non-abelian category Mlc, together with the present analysis of its exactness properties, will prepare the definition of Puppe-exact categories in Section 6.3.

6.2.1 An outline

Direct and inverse images of *subsets* have been dealt with in 1.7.5. Images and preimages of *submodules* will be analysed in this section, according to the following layout.

We already know that, for every object A in $R\,\mathsf{Mod}$, the ordered set $\mathrm{Sub}A$ of its subobjects (i.e. submodules) is a modular lattice (see 1.2.6), with maximum $1_A : A \to A$ and minimum $0_A : 0 \to A$. For submodules $X, X' \subset A$ the meet and the join are

$$X \wedge X' = X \cap X',$$

$$X \vee X' = X + X' = \{x + x' \mid x \in X,\, x' \in X'\}.$$

A homomorphism $f \colon A \to B$ gives two increasing mappings, called *direct* and *inverse image*, or image and preimage

$$f_* \colon \mathrm{Sub}A \rightleftarrows \mathrm{Sub}B : f^*,$$

$$f_*(X) = f(X), \qquad f^*(Y) = f^{-1}(Y) \qquad (X \subset A,\, Y \subset B),$$

(6.4)

with $\operatorname{Ker} f = f^*(0_B)$ and $\operatorname{Im} f = f_*(1_A)$.

These mappings form a Galois connection $f_* \dashv f^*$ (as in 1.7.1), but here we have a stronger property

$$
\begin{aligned}
f^* f_*(X) &= X \vee f^*(0_B) \supset X, \\
f_* f^*(Y) &= Y \wedge f_*(1_A) \subset Y,
\end{aligned}
\tag{6.5}
$$

which we describe saying that the pair $(f_*, f^*) \colon \operatorname{Sub} A \to \operatorname{Sub} B$ is a *modular connection*.

All this will be used to define a *transfer functor* for subobjects of $R\operatorname{Mod}$

$$
\operatorname{Sub}_R \colon R\operatorname{Mod} \to \operatorname{Mlc},
\tag{6.6}
$$

with values in the category Mlc of modular lattices and modular connections – a subcategory of the category AdjOrd of ordered sets and Galois connections (see Section 1.7). This functor will be seen to be exact in 6.3.8, i.e. a functor which preserves kernels and cokernels (or equivalently exact sequences, as defined in 6.3.4).

6.2.2 Modular lattices and modular connections

We begin now to study the category Mlc of *modular lattices and modular connections*, that we have seen to abstract the properties of direct and inverse images for $R\operatorname{Mod}$ (and will similarly work for a wide class of categories).

An object is a modular lattice (with 0 and 1). A morphism, called a *modular connection*

$$
f = (f_\bullet, f^\bullet) \colon X \to Y,
$$

is a pair where

(i) $f_\bullet \colon X \to Y$ and $f^\bullet \colon Y \to X$ are increasing mappings,

(ii) $f^\bullet f_\bullet(x) = x \vee f^\bullet 0, \quad f_\bullet f^\bullet(y) = y \wedge f_\bullet 1 \qquad$ for $x \in X,\ y \in Y$.

As a consequence $f^\bullet f_\bullet \geqslant \operatorname{id} X$ and $f_\bullet f^\bullet \leqslant \operatorname{id} Y$, and we have an adjunction $f_\bullet \dashv f^\bullet$ between ordered sets, i.e. a covariant Galois connection (see 1.7.1). As we have seen $f_\bullet$ determines $f^\bullet$, and conversely

$$
\begin{aligned}
f^\bullet(y) &= \max\{x \in X \mid f_\bullet(x) \leqslant y\}, \\
f_\bullet(x) &= \min\{y \in Y \mid f^\bullet(y) \geqslant x\}.
\end{aligned}
\tag{6.7}
$$

Moreover $f_\bullet$ preserves all the existing joins (including $0 = \vee \emptyset$), $f^\bullet$ preserves all the existing meets (including $1 = \wedge \emptyset$), and

$$
f_\bullet f^\bullet f_\bullet = f_\bullet, \qquad f^\bullet f_\bullet f^\bullet = f^\bullet.
\tag{6.8}
$$

Condition (ii) can be equivalently rewritten in a seemingly stronger form:

(ii') $\quad f^\bullet(f_\bullet x \vee y) = x \vee f^\bullet y, \qquad f_\bullet(f^\bullet y \wedge x) = y \wedge f_\bullet x \qquad (x \in X, y \in Y).$

Indeed, from (ii) *and* the modularity of lattices, we have

$$
\begin{aligned}
x \vee f^\bullet y &= (x \vee f^\bullet y) \vee f^\bullet 0 = f^\bullet f_\bullet(x \vee f^\bullet y) = f^\bullet(f_\bullet x \vee f_\bullet f^\bullet y) \\
&= f^\bullet(f_\bullet x \vee (y \wedge f_\bullet 1)) = f^\bullet((f_\bullet x \vee y) \wedge f_\bullet 1) \\
&= f^\bullet(f_\bullet f^\bullet(f_\bullet x \vee y)) = f^\bullet(f_\bullet x \vee y).
\end{aligned}
$$

This makes evident that modular connections *are closed under composition* in the category AdjOrd of ordered sets and Galois connections (introduced in 1.7.4),

$$(g_\bullet, g^\bullet) \cdot (f_\bullet, f^\bullet) = (g_\bullet f_\bullet, f^\bullet g^\bullet). \tag{6.9}$$

and form a subcategory Mlc of the latter.

Mlc inherits a (contravariant) involution from AdjOrd, and an order consistent with composition and involution (see (1.7.4))

$$
\begin{aligned}
X \mapsto X^{\mathrm{op}}, \quad ((f_\bullet, f^\bullet) \colon X \to Y) &\mapsto ((f^\bullet, f_\bullet) \colon Y^{\mathrm{op}} \to X^{\mathrm{op}}), \\
f \leqslant g \;&\Leftrightarrow\; f_\bullet \leqslant g_\bullet \;\Leftrightarrow\; f^\bullet \geqslant g^\bullet.
\end{aligned}
\tag{6.10}
$$

Mlc is thus selfdual.

We are also interested in the category Dlc of *distributive lattices and modular connections*, a full subcategory of Mlc.

6.2.3 Exercises and complements

(a) Prove that the morphism f of Mlc is an isomorphism (i.e. it has an inverse *in* the category) if and only if the following equivalent conditions hold:

(i) $\quad f^\bullet f_\bullet = 1_X, \qquad f_\bullet f^\bullet = 1_Y,$

(ii) $\quad f^\bullet 0 = 0, \qquad f_\bullet 1 = 1,$

(iii) $\quad f_\bullet$ is a bijective mapping,

(iii*) $\quad f^\bullet$ is a bijective mapping,

(iv) $\quad f_\bullet$ is an isomorphism of ordered sets (hence of lattices, in the usual sense),

(iv*) $\quad f^\bullet$ is an isomorphism of ordered sets (hence of lattices).

(b) Therefore *no distinction is needed between the isomorphisms of the category* Mlc *and the usual isomorphisms of* Mlh, i.e. the bijective homomorphisms of modular lattices.

(c) Both of the following faithful functors reflect the isomorphisms (where $|X|$ denotes the underlying set of X)

$$U: \mathsf{Mlc} \to \mathsf{Set}, \qquad X \mapsto |X|, \qquad (f_\bullet, f^\bullet) \mapsto f_\bullet, \qquad (6.11)$$

$$V: \mathsf{Mlc} \to \mathsf{Set}^{\mathrm{op}}, \qquad X \mapsto |X|, \qquad (f_\bullet, f^\bullet) \mapsto f^\bullet. \qquad (6.12)$$

U will be called the *forgetful functor* of Mlc.

(d) Let $f = (f_\bullet, f^\bullet): X \to Y$ be a modular connection between distributive lattices. Prove that the mappings $f_\bullet$ and $f^\bullet$ are homomorphisms of *quasi* lattices (in the sense of 1.2.4): in other words, $f_\bullet$ also preserves binary meets (but need not preserve the maximum, of course), while – dually – $f^\bullet$ also preserves binary joins (but need not preserve the minimum).

Hint: one can begin by writing $f_\bullet(x \wedge y) = f_\bullet f^\bullet f_\bullet(x \wedge y)$.

6.2.4 Exactness properties

We prove now that the category Mlc of modular lattices and modular connections satisfies exactness conditions that will make it Puppe-exact, or p-exact for short (according to a definition made precise below, in 6.3.2).

The reader is warned that Mlc is not abelian, and cannot even be exactly embedded in an abelian category (this will be proved in Theorem 6.5.5).

First there is a *zero object* (both terminal and initial, see 6.1.1), namely the one-point lattice $0 = \{*\}$, since every object X has unique morphisms t and s:

$$X \xrightarrow{\;t\;} \{*\} \xrightarrow{\;s\;} X \qquad (t^\bullet(*) = 1, \; s_\bullet(*) = 0). \qquad (6.13)$$

The zero morphism $0_{XY}: X \to \{*\} \to Y$ is defined by $x \mapsto 0$ and $y \mapsto 1$.

Every morphism $f = (f_\bullet, f^\bullet): X \to Y$ has a *kernel* $\ker f$ and a *cokernel* $\operatorname{cok} f$ (see 6.1.2)

$$\begin{aligned} m = \ker f: \; \downarrow f^\bullet 0 \to X, \qquad m_\bullet(x') = x', \qquad m^\bullet(x) = x \wedge f^\bullet 0, \\ p = \operatorname{cok} f: Y \to \uparrow f_\bullet 1, \qquad p_\bullet(y) = y \vee f_\bullet 1, \qquad p^\bullet(y') = y'. \end{aligned} \qquad (6.14)$$

The verification of the universal property of m is straightforward: essentially, if $h = (h_\bullet, h^\bullet): Z \to X$ is annihilated by f, then for every $z \in Z$ we have:

$$f^\bullet 0 = f^\bullet(f_\bullet h_\bullet(z)) = h_\bullet(z) \vee f^\bullet 0 \geqslant h_\bullet(z),$$

so that $h_\bullet$ takes values in $\downarrow f^\bullet 0$. Dually for p.

As a consequence of these properties, the morphism f has a canonical factorisation $f = ngq$ through $q = \operatorname{cok}(\ker f)$ (the *coimage* of f) and $n = \ker(\operatorname{cok} f)$ (the *image* of f), that determines a precise morphism g

$$
\begin{array}{ccccc}
\downarrow f^\bullet 0 & \xrightarrow{\ \ m\ \ } & X & \xrightarrow{\ \ f\ \ } & B & \xrightarrow{\ \ p\ \ } & \uparrow f_\bullet 1 \\
& & {\scriptstyle q}\downarrow & & \uparrow{\scriptstyle n} & & \\
& & \uparrow f^\bullet 0 & \xrightarrow{\ \ g\ \ } & \downarrow f_\bullet 1 & &
\end{array}
\tag{6.15}
$$

$$
q_\bullet(x) = x \vee f^\bullet 0, \quad q^\bullet(x) = x, \quad n_\bullet(y) = y, \quad n^\bullet(y) = y \wedge f_\bullet 1,
$$
$$
g_\bullet(x) = f_\bullet(x), \quad g^\bullet(y) = f^\bullet(y).
$$

Moreover g is an *isomorphism* of Mlc (as characterised above), because

$$
g^\bullet g_\bullet(x) = f^\bullet f_\bullet(x) = x, \quad g_\bullet g^\bullet(y) = f_\bullet f^\bullet(y) = y,
$$

for $x \geqslant f^\bullet 0$ and $y \leqslant f_\bullet 1$.

This will mean that the category Mlc is Puppe-exact.

As an easy consequence of the canonical factorisation, each monomorphism is normal (i.e. a kernel of some arrow) and each epimorphism is normal (i.e. a cokernel).

6.2.5 Subobjects and quotients

We also note that every element $a \in X$ determines a *subobject* and a *quotient* of X

$$
m\colon \downarrow a \to X, \qquad m_\bullet(x') = x', \quad m^\bullet(x) = x \wedge a, \tag{6.16}
$$

$$
p\colon X \to \uparrow a, \qquad p_\bullet(x) = x \vee a, \quad p^\bullet(x') = x', \tag{6.17}
$$

where $m = \ker p$ and $p = \operatorname{cok} m$.

The correspondence $a \mapsto m$ establishes an isomorphism between the *lattice* X and the ordered set Sub$X$ of *subobjects* of X in Mlc, while the correspondence $a \mapsto p$ gives an anti-isomorphism of X with the ordered set of *quotients* of X.

The subobject m and the quotient p determined by the element $a \in X$ form a *short exact sequence* (defined in 6.3.4)

$$
\downarrow a \to X \to \uparrow a, \qquad m = \ker p, \quad p = \operatorname{cok} m. \tag{6.18}
$$

Conversely, every short exact sequence (m, p) in Mlc with central object X is isomorphic to a unique sequence of this type, with $a = m_\bullet 1 = p^\bullet 0$.

6.2.6 Exercises and complements (The transfer functor)

(a) Coming back to 6.2.1, the reader can prove that we have a functor

$$\text{Sub}_R \colon R\,\text{Mod} \to \text{Mlc}, \qquad A \mapsto \text{Sub}(A), \qquad f \mapsto (f_*, f^*). \qquad (6.19)$$

Hints. It is sufficient to show that $f \mapsto f^*$ preserves composition (contravariantly) and identities.

(b) Prove that this functor is *exact*, i.e. preserves kernels and cokernels.

Hints. Take an R-homomorphism $h \colon A \to B$ with kernel $m \colon h^{-1}\{0\} \subset A$ and cokernel $p \colon B \to B/h(A)$.

Then the embedding $m_* \colon \text{Sub}(h^{-1}\{0\}) \to \text{Sub}A$ is the same as the inclusion $\downarrow h^*(0) \subset \text{Sub}A$, while the embedding $p^* \colon \text{Sub}(B/h(A)) \to \text{Sub}B$ can be identified with the inclusion $\uparrow h_*(1) \subset \text{Sub}B$.

6.3 Puppe-exact categories

We review the main definitions about p-exact categories, exact sequences and exact functors. Then we deal with the category Mlc of modular lattices and modular connections, introduced in the previous section.

We show now that this category, which is p-exact and not abelian, abstracts the behaviour of direct and inverse images of subobjects *for all p-exact categories* (including the abelian ones); p-exact categories form thus the natural setting of this analysis.

Subquotients, a crucial tool of Homological Algebra, are introduced in 6.3.9. Some proofs are only referred to the book [G9].

6.3.1 Three forms of exactness

The term 'exact category' has assumed different meanings in category theory, *and at least three of them are still in use* (listed as (b), (c), (d) below).

(a) (*Abelian categories*). The term 'exact category' was first used in 1955-56 by Buchsbaum [Bu1, Bu2], essentially meaning what is now called an abelian category (even if the existence of finite biproducts was seemingly deferred to an additional axiom). The name subsisted – in this sense – in various papers of the 1950's and 1960's (by Atiyah, Hilton, Heller, etc.), together with the term 'abelian category' that had been introduced by Grothendieck in his Tôhoku paper of 1957 [Gt]. Gradually the last term was universally accepted in the literature, including the first two books on category theory, by Freyd [Fr1] (1964) and Mitchell [Mit] (1965) – both focused on the embedding of abelian categories in categories of modules.

(b) (*Puppe-exact categories*). Meanwhile, in 1962, Puppe [Pu] had introduced a more general notion, called a 'quasi exact category', still selfdual but not additive (and based on the axioms AB1, AB2 of Grothendieck). This framework was investigated by Tsalenko [T1, T2] (also transliterated as 'Calenko') in 1964 and 1967, for the construction of the category of relations, and by many researchers for diagram lemmas. It became an 'exact category' in Mitchell's book [Mit] (1965), where abelian categories are defined as additive exact categories. The name is kept, in this sense, in subsequent works by Brinkmann and Puppe [Bri, BriP], in 1969, and in the text by Herrlich and Strecker [HeS], in 1973. The books [AHS, FrS] still use in 1990 the name of exact category in the sense of Puppe and Mitchell, as a first step towards abelian categories.

(c) (*Barr-exact categories*). Later, in 1971, Barr [Bar] used the term 'exact category' for a different generalisation of abelian categories, based on regular categories and not selfdual (see Chapter 4). This setting became popular in category theory and has been extended in various forms (see [Bou1, BoB, JaMT, Bo4, Bou2]). As we have seen, it contains all varieties of algebras, and can give a general categorical presentation of Universal Algebra.

(d) (*Quillen-exact categories*). In 1973 a paper by Quillen [Qu] on higher K-theory introduced another notion of 'exact category', as a full additive subcategory E of an abelian category, closed under extensions; this means that, for every short exact sequence $A \rightarrowtail C \twoheadrightarrow B$ of the abelian category, if A and B are in E, so is C. This notion can be defined intrinsically, without reference to an abelian environment, as an additive category equipped with a class of 'short exact sequences' satisfying some axioms.

We must thus distinguish the cases (b), (c), (d) as: *Puppe-exact, Barr-exact* and *Quillen-exact* categories. It is often remarked that the first two notions satisfy the 'equation': exact + additive = abelian; on the other hand, Quillen-exact categories *are* additive. A Puppe-exact category is abelian if and only if it has finite products, or equivalently finite sums (by Theorem 6.4.6). Therefore the intersection of each pair of these three frameworks only contains the abelian categories.

6.3.2 Exact and abelian categories

A *Puppe-exact*, or *p-exact*, category E is a category satisfying the following two self-dual axioms (essentially corresponding to AB1, AB2 in [Gt]):

(pex.1) E is pointed and every morphism $f \colon A \to B$ has a kernel and a cokernel, (pex.2) in the *canonical factorisation* of a morphism f through its *coimage* and its *image*

$$\text{Ker} f \xrightarrow{\ \ k\ \ } A \xrightarrow{\ \ f\ \ } B \xrightarrow{\ \ c\ \ } \text{Cok} f$$

(6.20)

$$\text{Coim} f \xrightarrow{\ \ g\ \ } \text{Im} f$$

$$\text{Coim} f = \text{Cok}(\ker f), \qquad \text{Im} f = \text{Ker}(\text{cok} f),$$
$$q = \text{coim} f = \text{cok}(\ker f), \qquad n = \text{im} f = \ker(\text{cok} f),$$

the unique morphism g such that $f = ngq$ is an *isomorphism*.

This formulation is redundant, but clearer and often simpler to check than the more concise ones. For instance – as is easy to verify – it is sufficient to say that E is pointed and every morphism factorises as a normal epi followed by a normal mono.

As an easy consequence of the axioms, each monomorphism is normal (i.e. a kernel of some arrow) and each epimorphism is normal (i.e. a coker-nel). Every morphism has an essentially unique epi-mono factorisation, the canonical one, as in diagram (6.20) – which *motivates* our names of coimage and image, above. E is a *balanced* category, i.e. epi and mono implies iso.

Following the notation of 2.4.1, we write $\text{Sub}_{\mathsf{E}} A$ and $\text{Quo}_{\mathsf{E}} A$ for the (possibly large) ordered sets of subobjects and quotients of an object A in E.

Cokernels and kernels give two *decreasing* mappings

$$\text{cok} \colon \text{Sub}_{\mathsf{E}} A \rightleftarrows \text{Quo}_{\mathsf{E}} A \colon \ker \qquad \text{(kernel duality)}, \qquad (6.21)$$

that are easily proved to be inverse to each other, using the fact that all monos and epis are normal

$$\ker(\text{cok}\, m) = m, \qquad \text{cok}(\ker p) = p. \qquad (6.22)$$

(On the other hand, *categorical* duality links $\text{Sub}_{\mathsf{E}} A$ and $\text{Quo}_{\mathsf{E}^{op}} A$, pre-serving the ordering. Working with both dualities simplifies many proofs.)

From now on we assume that a p-exact category is well powered (see 2.4.1), i.e. that all sets $\text{Sub} A$ (and $\text{Quo} A$, as a consequence) are small. We shall see in 6.3.8 that $\text{Sub} A$ and $\text{Quo} A$ are modular lattices, anti-isomorphic by (6.21).

A p-exact category is said to be *trivial* if all its objects are zero objects; or, in other words, if it is equivalent to the singleton category **1**.

An *abelian category* can be defined as a p-exact category having finite products, or equivalently finite sums, or finite limits and colimits; this equivalence will be proved in Theorem 6.4.6, but we shall use from now the notion of abelian category, much better known than that of a p-exact category.

6.3.3 Examples and complements

The pointed categories Gp and gRng are not p-exact, since there are non-normal monomorphisms; Set. and Top. are not p-exact, since there are non-normal epimorphisms. *(But we have mentioned in 6.1.5 that Set., Top., Set₂, Top₂, Gp₂, Ban are homological, in the sense of [G10].)*

The following examples of p-exact categories are taken from [G9], Subsection 1.5.6.

(a) Every abelian category.

(b) Every non-empty full subcategory of a p-exact category that is closed under subobjects and quotients; for instance:

- cyclic groups; finite cyclic groups,

- abelian groups of cardinality (or rank) lower than a fixed integer.

- vector spaces on a fixed field, of dimension lower than a fixed integer,

(Replacing a 'fixed integer' with a 'fixed infinite cardinal' just gives an abelian subcategory.)

(c) The category Mlc of modular lattices and modular connections, as proved in 6.2.2–6.2.4. Mlc abstracts the behaviour of direct and inverse images of subobjects in (abelian or) p-exact categories (as we shall see in 6.3.8). It is *not* abelian and cannot even be exactly embedded in an abelian category (as will be proved in Theorem 6.5.5).

Its full subcategory Dlc of *distributive lattices* (also defined in 6.2.2) will play the same role for *distributive p-exact categories*, i.e. those p-exact categories whose lattices of subobjects are distributive. As a first example, Exercise 1.2.7(c) proves that the category of cyclic groups is distributive p-exact.

(d) The category $\mathcal{I}$ of sets and partial bijections, introduced in 1.8.7. A morphism $f\colon X \to Y$ in $\mathcal{I}$ is a bijection between a subset (Def f) of X and a subset (Val f) of Y, or equivalently a single-valued, injective relation; they compose as relations in Rel (Set). $\mathcal{I}$ is self-dual; the zero object is $\emptyset$, and

$$\operatorname{Ker} f = X \setminus \operatorname{Def} f, \qquad \operatorname{Cok} f = Y \setminus \operatorname{Val} f.$$

This category is distributive p-exact (with Sub$X = \mathcal{P}X$), and 'universal' in this domain, in the sense that every small category of this kind has an exact embedding in $\mathcal{I}$ (cf. [G9], 4.6.7(b)).

(e) The *projective category* $\mathsf{Pr}\,\mathsf{E}$ associated to any p-exact category E, that will be studied in Section 6.5. In particular, if $\mathsf{E} = K\,\mathsf{Vct}$ is the abelian category of vector spaces over the commutative field K, $\mathsf{Pr}\,\mathsf{E}$ 'is' the (p-exact) category of projective spaces and projective linear maps over K.

*(f) The *distributive expansion* $\mathsf{Dst}\,\mathsf{E}$ of any p-exact category E, defined in [G9], Section 2.8. It is a distributive p-exact category, and is never abelian – unless E (and therefore $\mathsf{Dst}\,\mathsf{E}$) is trivial.

Its objects are the pairs (A, X) consisting of an object A of E and a distributive sublattice $X \subset \mathrm{Sub}(A)$. A morphism $f\colon (A, X) \to (B, Y)$ comes from an E-morphism $f\colon A \to B$ which carries the subobjects of X into Y and those of Y into X (by direct and inverse images, see 6.3.7).

(g) Every category of functors E^S, where E is a p-exact category and S is a small category. This includes all cartesian powers of E (when S is a discrete category), the category of morphisms of E (when $S = \mathbf{2}$ is the 'arrow category' $0 \to 1$), the category of commutative squares of E (when $S = \mathbf{2} \times \mathbf{2}$), etc. The proof that E^S is also p-exact is straightforward.

6.3.4 Exact functors and exact sequences

A functor $F\colon \mathsf{E} \to \mathsf{E}'$ between p-exact categories is said to be *exact* if it preserves kernels and cokernels, in the usual sense of preserving limits, that is – in the present case – up to equivalence of monos and epis, respectively.

As a consequence it also preserves the zero object (that is the kernel and cokernel of any identity), canonical factorisations and exact sequences.

The latter are defined in the usual way: in the p-exact category E the sequence

$$A \xrightarrow{\ f\ } B \xrightarrow{\ g\ } C \tag{6.23}$$

is said to be *exact* (in B) if $\mathrm{im}\,f = \ker g$, or equivalently $\mathrm{cok}\,f = \mathrm{coim}\,g$. A sequence of consecutive morphisms is said to be *exact* if it is in all locations where the condition makes sense.

Short exact sequences are of particular interest (see 6.3.5). By definition, this is a sequence of the following form, that is exact (in A, B, C)

$$0 \longrightarrow A \xrightarrow{\ m\ } B \xrightarrow{\ p\ } C \longrightarrow 0 \tag{6.24}$$

Plainly, this means that the sequence satisfies the following equivalent conditions:

(i) m is a monomorphism, p is an epimorphism and $\mathrm{im}\,m = \ker p$,

(ii) $m \sim \ker p$ (as a mono in B) and $p \sim \operatorname{cok} m$ (as an epi from B).

This sequence will also be written in the form $A \rightarrowtail B \twoheadrightarrow C$.

A functor F between p-exact categories is *left exact* (resp. *right* exact) if it preserves kernels (resp. cokernels), or equivalently exact sequences of the form $0 \to A \to B \to C$ (resp. $A \to B \to C \to 0$). More generally, F is said to be *half exact* if it takes any sort exact sequence $0 \to A \to B \to C \to 0$ to an exact sequence $FA \to FB \to FC$; note that this property is not closed under composition.

A functor is exact if and only if it is left and right exact. A left adjoint functor between p-exact categories is necessarily right exact. We shall see in 6.4.7 that any exact (or even half exact) functor between abelian categories preserves finite products and sums, and the additive structure.

6.3.5 *Exercises and complements

(a) Prove that, in a p-exact category, a sequence (f, g) of consecutive morphisms is exact if and only if it can be inserted in the commutative diagram below, with a slanting short exact sequences

$$A \ {\overset{f}{\relbar\joinrel\longrightarrow}}\ B \ {\overset{g}{\relbar\joinrel\longrightarrow}}\ C \tag{6.25}$$

(b) An object X is a zero object if and only if the pair $(1_X, 1_X)$ is an exact sequence, if and only if it forms a short exact one.

(c) Prove that a functor between p-exact categories is exact if and only if it preserves short exact sequences.

6.3.6 *Lemma (Pullbacks and pushouts in p-exact categories)*

A p-exact category E *has* pullbacks of monos *and* pushouts of epis *along arbitrary maps. The following points are a more detailed formulation.*

(a) Along a map f, the pullback (or preimage) $m = f^*(n)$ of a monomorphism n and the pushout (or direct image) $q = f_\circ(p)$ of an epimorphism p always exist, and are computed as follows

$$f^*(n) = \ker\left((\operatorname{cok} n).f\right), \qquad f_\circ(p) = \operatorname{cok}\left(f.(\ker p)\right), \tag{6.26}$$

(b) In the same situation, if in the left diagram f is epi, then so is g and the square is also a pushout (a bicartesian square*). Dually, if in the right diagram f is mono, then so is g and the square is also a pullback.*

(c) Given the following commutative diagram with (short) exact rows

$$
\begin{array}{ccccc}
A & \xrightarrow{\;m\;} & B & \xrightarrow{\;p\;} & C \\
\downarrow{\scriptstyle u} & & \downarrow{\scriptstyle v} & & \downarrow{\scriptstyle w} \\
A' & \xrightarrow[\;m'\;]{} & B' & \xrightarrow[\;p'\;]{} & C'
\end{array}
\tag{6.27}
$$

the left square is a pullback if and only if w is mono. The right square is a pushout if and only if u is epi.

Proof The reader can prove the statement for a category of modules, or try the general case. The latter can be found in [G9], Lemma 2.2.4. □

6.3.7 Direct and inverse images

In a p-exact category E every morphism $f\colon A \to B$ gives rise to two increasing mappings between ordered sets, called – respectively – *direct and inverse image* of subobjects along f

$$
\begin{aligned}
f_*\colon \mathrm{Sub}A \to \mathrm{Sub}B, &\qquad f_*(m) = \mathrm{im}\,(fm) = \ker\,(\mathrm{cok}\,(fm)), \\
f^*\colon \mathrm{Sub}B \to \mathrm{Sub}A, &\qquad f^*(n) = \ker\,((\mathrm{cok}\,n).f),
\end{aligned}
\tag{6.28}
$$

$$
\begin{array}{ccc}
A & \xrightarrow{\;f\;} & B \\
\uparrow{\scriptstyle m} & & \uparrow{\scriptstyle f_*m} \\
\bullet & \longrightarrow & \bullet
\end{array}
\qquad\qquad
\begin{array}{ccc}
A & \xrightarrow{\;f\;} & B \\
\uparrow{\scriptstyle f^*n} & & \uparrow{\scriptstyle n} \\
\bullet & \dashrightarrow & \bullet
\end{array}
$$

The left square is commutative and the right one is a pullback (by the previous lemma). The theorem below proves that the pair (f_*, f^*) is a *modular connection between modular lattices*, i.e. a morphism of the category Mlc studied in Section 6.2.

In particular:

$$
f_*0 = 0, \quad f_*1 = \mathrm{im}\,f; \qquad f^*0 = \ker f, \quad f^*1 = 1.
\tag{6.29}
$$

6.3.8 Theorem and Definition (The transfer functor)

(a) Every (well-powered) p-exact category E *has an exact functor with values in the p-exact category* Mlc *of modular lattices and modular connections*

$$\mathrm{Sub}_E \colon E \to \mathsf{Mlc},$$

$$A \mapsto \mathrm{Sub}A, \qquad f \mapsto \mathrm{Sub}(f) = (f_*, f^*). \tag{6.30}$$

This functor also reflects exactness: a sequence of E *is exact if and only if its image in* Mlc *is exact.*

Sub_E *will be called the* transfer functor *(for subobjects), or also the* projective functor *of* E. (The name is motivated by the projective category $\mathrm{Pr}\,E$, that will be studied in Section 6.5.)

(b) The lattice operations and the ordering of $\mathrm{Sub}A$ *can be described as follows, for* $m, n \in \mathrm{Sub}A$ *and* $p = \mathrm{cok}\,m$, $q = \mathrm{cok}\,n$ *(the corresponding quotients)*

$$m \wedge n = m_* m^*(n) = n_* n^*(m),$$
$$m \vee n = p^* p_*(n) = q^* q_*(m), \tag{6.31}$$
$$n \leqslant m \iff n^*(m) = 1.$$

(c) An exact functor $F \colon E \to E'$ *preserves direct and inverse images of subobjects, as well as their finite joins and meets. Therefore it induces a homomorphism of lattices,* for every A in E

$$(\mathrm{Sub}_F)A \colon \mathrm{Sub}_E A \to \mathrm{Sub}_{E'}(FA), \qquad \mathrm{Sub}_F(m) = \mathrm{im}\,(Fm), \tag{6.32}$$

(where $\mathrm{im}\,(Fm)$ *is the subobject associated to the monomorphism* Fm).

Proof The particular case $E = R\,\mathsf{Mod}$ has been studied in 6.2.1, 6.2.6. The general case is proved in [G9], Theorem 2.2.6. ☐

6.3.9 Subquotients

As a prime example, the homology group

$$H_n(A) = \mathrm{Ker}\,\partial_n / \mathrm{Im}\,\partial_{n+1}$$

of a chain complex A of abelian groups (see 5.3.7) is a subquotient of the component A_n. More complex situations arise in the study of spectral sequences (see [M2, G9]).

In general, a *subquotient* $S = M/N$ of an object A in a p-exact category E is a quotient of a subobject (M) of A, or equivalently a subobject of a quotient (A/N) of A.

It is determined by a decreasing pair $m\colon M \rightarrowtail A$, $n\colon N \rightarrowtail A$ $(n \prec m)$ of subobjects of A, via the commutative square produced by the epi-mono factorisation of $pm\colon M \to A \to A/N$, where $p = \operatorname{cok} n$

$$
\begin{array}{ccc}
M & \xrightarrow{\ m\ } & A \\
\scriptstyle{p'}\downarrow & & \downarrow\scriptstyle{p} \\
S & \xrightarrow[\ m'\ ]{} & A/N
\end{array}
\qquad
\begin{array}{l}
p = \operatorname{cok} n, \\[2em]
n = \ker p \prec m.
\end{array}
\tag{6.33}
$$

This square is bicartesian, i.e. pullback and pushout at the same time. (In fact, it is a pullback by (6.26), because $p^*(m') = p^* p_*(m) = m \vee \ker p = m$; and a pushout by 6.3.6(b), or by duality).

Conversely, every bicartesian square in E formed of two 'parallel' monos and two 'parallel' epis is of this type, up to isomorphism.

6.4 Additive and abelian categories

We now review additive and abelian categories, with a more complete analysis of biproducts – after 2.1.9 and 3.2.8.

The reader can find further results in [M4], Chapter VIII, and in [Gt]. The embedding of abelian categories in categories of modules was studied in the books [Mit, Fr1], right at the beginning of category theory.

6.4.1 Preadditive categories and biproducts

A *preadditive*, or $\mathbb{Z}$-*linear*, category is a category C where every hom-set $\mathsf{C}(A, B)$ is equipped with a structure of abelian group, so that composition is bilinear over $\mathbb{Z}$.

More generally, we are also interested in $\mathbb{N}$-*linear categories*, where every hom-set $\mathsf{C}(A, B)$ is equipped with a structure of abelian monoid, and composition is bilinear over $\mathbb{N}$. The zero element of $\mathsf{C}(A, B)$ is generally written as $0_{AB}\colon A \to B$, or simply 0. (Both these structures are related with 'enrichment', see 7.3.3.)

By definition, the *biproduct* of two objects A, B in an arbitrary *pointed* category is an object C equipped with four maps u, v, p, q, so that

$$
\begin{array}{ccc}
A & & B \\
\Big\| \quad {\scriptstyle u}\searrow & \swarrow{\scriptstyle v} \quad & \Big\| \\
& C & \\
\Big\| \quad {\scriptstyle p}\nwarrow & \nearrow{\scriptstyle q} \quad & \Big\| \\
A & & B
\end{array}
\tag{6.34}
$$

$$
pu = \operatorname{id} A, \quad qv = \operatorname{id} B, \qquad qu = 0, \quad pv = 0,
$$

- $(C; p, q)$ is the product of A, B and $(C; u, v)$ is their sum.

The object C is often written as $A \oplus B$; the maps p, q are called *projections* while u, v are called *injections*.

Similarly one defines the biproduct $\bigoplus_i A_i$ of any family of objects; the biproduct of the empty family is the zero object. A morphism $C \to \bigoplus_i A_i$ will often be written as (f_i), by its *components* $f_i : C \to A_i$, while a morphism $\bigoplus_i A_i \to D$ can be written as $[f_i]$, by its *co-components* $f_i : A_i \to D$ (as in our notation for products and coproducts, in 2.1.1 and 2.1.5).

As we know, all categories of modules, Abm and Ban have finite biproducts; other examples will be examined in 6.4.5, including some cases of arbitrary biproducts.

6.4.2 Lemma

Let C *be an* $\mathbb{N}$-*linear category.*

The following conditions on the object Z *are equivalent:*

(a) Z *is terminal,*

(a*) Z *is initial,*

(b) Z *is the zero object,*

(c) $C(Z, Z)$ *is the trivial group,*

(d) $\mathrm{id} Z = 0_{ZZ}$.

Secondly, the following conditions on the diagram (6.34) *are equivalent (using the notation by components and co-components, recalled above at the end of 6.4.1):*

(i) $(C; p, q)$ *is the product of* A, B *and* $u = (\mathrm{id} A, 0)$, $v = (0, \mathrm{id} B)$,

(i*) $(C; u, v)$ *is the sum of* A, B *and* $p = [\mathrm{id} A, 0]$, $q = [0, \mathrm{id} B]$,

(ii) $pu = \mathrm{id} A$, $qv = \mathrm{id} B$, $up + vq = \mathrm{id} C$,

(iii) p, q *are jointly mono,* $pu = \mathrm{id} A$, $qv = \mathrm{id} B$, $pv = 0$, $qu = 0$,

(iii*) u, v *are jointly epi,* $pu = \mathrm{id} A$, $qv = \mathrm{id} B$, $pv = 0$, $qu = 0$.

Therefore in an $\mathbb{N}$-*linear category (a fortiori in a preadditive category), the existence of binary (or finite) products is equivalent to the existence of binary (or finite) sums, which are then biproducts.*

Proof Well-known and straightforward. We write down the proof of the second part, taking advantage of categorical duality.

(i) $\Rightarrow$ (iii) Obvious. (iii) $\Rightarrow$ (ii) We have $p(up + vq) = p$ and $q(up + vq) = q$, whence $up + vq = \mathrm{id} C$. (ii) $\Rightarrow$ (i) The maps $f : X \to A$ and

$g: X \to B$ are given. If $h: X \to C$ satisfies $ph = f$ and $qh = g$, then $h = (up + vq)h = uf + vg$; conversely, the morphism $h = uf + vg: X \to C$ does have f, g as components. $\qquad\square$

6.4.3 Theorem and Definition (Semiadditive categories)

Let C be a category.

(a) C is said to be semiadditive if it satisfies the following equivalent conditions:

(i) C is an ℕ-linear category with finite products,

(i*) C is an ℕ-linear category with finite sums,

(ii) C is a pointed category with finite biproducts.

When these conditions hold, the sum $f + g$ of two maps $f, g: A \to B$ is determined by the categorical structure

$$A \xrightarrow{d} A \oplus A \xrightarrow{f \oplus g} B \oplus B \xrightarrow{\partial} B \qquad f + g = \partial(f \oplus g)d, \qquad (6.35)$$

where $d = (1, 1): A \to A \oplus A$ is the diagonal of the product and $\partial = [1, 1]: B \oplus B \to B$ is the codiagonal of the coproduct.

(b) A functor $F: C \to C'$ between semiadditive categories is said to be additive if it satisfies the following equivalent conditions:

(i) F preserves finite products,

(i*) F preserves finite sums,

(ii) F preserves finite biproducts,

(iii) F preserves (finite) sums of parallel maps.

Proof (a) We already know, from the previous lemma, that (i) and (i*) are equivalent. Furthermore, if they hold, C has finite biproducts and the binary ones satisfy the conditions of 6.4.2. Therefore, for $f, g: A \to B$, the composition (6.35) is computed as follows (letting p', q' be the projections of $B \oplus B$):

$$\partial(f \oplus g)d = (p' + q')(u'fp + v'gq)(u + v) = f + g.$$

Now let us suppose that C is pointed and has finite biproducts (as defined in 6.4.1), and define the sum of parallel maps as in (6.35). One verifies that this is indeed an enrichment over Abm (see [M4], Section VIII.2, exercise 4a); moreover, in diagram (6.34), the map $up + vq: Z \to Z$ must be the identity, because:

$$p(up + vq) = pup + pvq = p, \qquad q(up + vq) = qup + qvq = q.$$

(b) It is a straightforward consequence.   □

6.4.4 Definition (Additive categories)

As a consequence of the previous theorem, we can define an *additive category* C as a preadditive category (i.e. a $\mathbb{Z}$-linear category) with finite products, or equivalently with finite sums.

Equivalently, we can say that C is a pointed category with finite biproducts, such that the $\mathbb{N}$-linear structure defined in (6.35) (and determined by the categorical structure) is actually $\mathbb{Z}$-linear, i.e. all abelian monoids $C(A, B)$ are groups.

It is easy to see that a $\mathbb{Z}$-linear category is finitely complete if and only if it is additive and has kernels. In fact, in any category the existence of finite limits is equivalent to the existence of finite products and equalisers (by Theorem 2.2.4). Furthermore, if C is $\mathbb{Z}$-linear, the equaliser of two maps $f, g: A \to B$ amounts to the kernel of the morphism $f - g$, while the kernel of f amounts to the equaliser of f and 0_{AB}.

6.4.5 Exercises and complements

(a) We know that all categories $R\,\mathsf{Mod}$ and Ban are additive.

As a less obvious example, the reader can prove that the category Set_T of sets and $\mathbb{Z}$-weighted maps (see 3.7.1(b)) is additive, with the obvious sum of maps; a finite biproduct is a disjoint union $X = \sum X_i$ (a sum in Set), with the usual injections $u_i: X_i \to X$ and suitable projections $p_i: X \to X_i$ (that are not mappings).

But also this example becomes obvious, if we replace Set_T with the equivalent category of free abelian groups – a full subcategory of Ab closed under finite biproducts (see 3.7.1(c)).

(b) Abm is obviously a semiadditive, not additive, category. The examples below consist of selfdual categories with finite (or arbitrary) biproducts.

(c) The category Ltc of *lattices and Galois connections* is a full subcategory of AdjOrd. It has finite biproducts (while Mlc lacks products and sums), realised as cartesian products ΠX_i of lattices

$$\Pi X_i \overset{p_i}{\dashrightarrow} X_i \overset{m_i}{\dashrightarrow} \Pi X_i \qquad (6.36)$$

$$m_{i\bullet} \dashv m_i{}^\bullet = \mathrm{pr}_i = p_{i\bullet} \dashv p_i{}^\bullet,$$

where pr_i is a cartesian projection; for an element $x \in \Pi X_i$ and $j \neq i$

$$(m_{i\bullet}(x))_i = x, \quad (m_{i\bullet}(x))_j = 0, \qquad (p_i{}^\bullet(x))_i = x, \quad (p_i{}^\bullet(x))_j = 1.$$

(For an infinite family (X_i) the universal properties of product and sum would fail, unless we restrict to complete lattices.)

Ltc is thus semiadditive, and the sum $f + g$ of two maps $f, g \colon X \to Y$ is defined as in (6.35)

$$X \xrightarrow{\ d\ } X \times X \xrightarrow{\ f \times g\ } Y \times Y \xrightarrow{\ \partial\ } Y \tag{6.37}$$

$$d_{\bullet}(x) = (x, x), \qquad d^{\bullet}(x, x') = x \wedge x',$$
$$\partial_{\bullet}(y, y') = y \vee y', \qquad \partial^{\bullet}(y) = (y, y),$$
$$f + g = \partial.(f \times g).d = (f_{\bullet} \vee g_{\bullet}, f^{\bullet} \wedge g^{\bullet}).$$

In other words, $f + g$ *is the join* $f \vee g$ with respect to the order relation of Galois connections, defined in (6.10). *The sum is idempotent* $(f \vee f = f)$, so that the cancellation law does not hold and *our structure is not additive.*

Restricting to the category of *complete* lattices and Galois connections we even have arbitrary biproducts of objects, and arbitrary sums of parallel morphisms $\vee f_i$.

(d) Rel Set is also semiadditive, and has arbitrary biproducts realised as disjoint unions (see 2.2.6(b)). The sum of a family of parallel relations $a_i \colon X \to Y$ is their join $\cup a_i$, a union of subsets of $X \times Y$.

6.4.6 Theorem and definition (Abelian categories)

Let E *be a Puppe-exact category. We say that* E *is* abelian *if it satisfies the following equivalent properties:*

(i) E *has finite products,*

(i*) E *has finite sums,*

(ii) E *has finite biproducts (in the sense of 6.4.1),*

(iii) E *has pullbacks,*

(iii*) E *has pushouts,*

(iv) E *has finite limits,*

(iv*) E *has finite colimits,*

(v) E *is additive.*

The sum of parallel maps is then determined by the categorical structure, as in (6.35).

Note. A preadditive Puppe-exact category need not be abelian: see the examples of 6.3.3(b).

Proof See [G9], Theorem 2.1.5. The essential part of this result was proved in Freyd–Scedrov [FrS], but previously stated in [HeS]. A weaker, easier result based on the existence of biproducts was already in [Mit]. □

6.4.7 Biproducts in abelian categories

We already know that, in any $\mathbb{N}$-linear category C, the biproduct $A \oplus B$ can be characterised as a diagram (6.34) satisfying the condition 6.4.2(iii):

(iii) p, q are jointly mono, $pu = \mathrm{id}A$, $qv = \mathrm{id}B$, $pv = 0$, $qu = 0$,

It is now easy to see that, if C is *preadditive* (i.e. $\mathbb{Z}$-linear) and *p-exact*, the biproduct $A \oplus B$ can also be characterised as a diagram (6.34) satisfying:

(iv) the diagram is commutative, with (short) exact 'diagonals'.

Indeed, if (iv) holds and a morphism $f \colon X \to C$ gives $pf = 0$, $qf = 0$, it must factorise through $v = \ker p$, as $f = vg$. It follows that $g = qvg = qf = 0$ and $f = 0$. For two parallel morphisms $f_i \colon A \to Z$ on which p, q coincide, take $f = f_1 - f_2$. Conversely, if (iii) is satisfied, let us prove that $v = \ker p$. Take $f \colon X \to C$ such that $pf = 0$; since we know that $\mathrm{id}C = up + vq$, it follows that $f = (up + vq)f = vqf$ factorises through v (that is mono).

Let $F \colon \mathsf{E} \to \mathsf{E}'$ be a functor *between abelian categories*. As an immediate consequence of the previous point, if F is exact (in the previous sense, i.e. it preserves kernels and cokernels), then it preserves biproducts, the sum of parallel maps, all finite limits and finite colimits.

In fact, the preservation of all biproducts, including the sum of maps, *also holds if F is just half exact* (as defined in 6.3.4) because, in the diagonals of (6.34), all monos and epis are split, and preserved by any functor.

6.4.8 Pullbacks, pushouts and exact squares

Let us start from a commutative square in the abelian category E

$$
\begin{array}{ccc}
A & \xrightarrow{\ f\ } & X \\
{\scriptstyle g}\downarrow & & \downarrow{\scriptstyle h} \\
Y & \xrightarrow[\ k\ ]{} & B
\end{array}
\qquad (6.38)
$$

and form the sequence

$$
0 \to A \xrightarrow{\ (f,g)\ } X \oplus Y \xrightarrow{\ [h,-k]\ } B \to 0 \qquad (6.39)
$$

which is *of order two* (i.e. its composed morphisms are null), since $hf - kg = 0$.

It is easy to see that:

(a) the square (6.38) is a pullback if and only if the sequence (6.39) is exact in A and $X \oplus Y$,

(a*) the square (6.38) is a pushout if and only if the sequence (6.39) is exact in B and $X \oplus Y$.

It is sufficient to prove (a), by duality. We know (from 2.3.2) that the pullback-object of the cospan (h, k) is the equaliser of the morphisms $hp_1, kp_2 \colon X \oplus Y \to B$, that is the subobject

$$\operatorname{Ker}(hp_1 - kp_2) = \operatorname{Ker}[h, -k] \rightarrowtail X \oplus Y.$$

Therefore the square (6.38) is a pullback if and only if $(f, g) \colon A \to X \oplus Y$ is a monomorphism equivalent to the subobject $\ker[h, -k]$ of $X \oplus Y$.

More generally the square (6.38) is called *exact* in the sense of Hilton [Hi] if the sequence is exact in the central object $X \oplus Y$, which means that

$$\operatorname{im}(f, g) = \ker[h, -k]. \tag{6.40}$$

If $\mathsf{E} = R\operatorname{Mod}$ this is the same as condition 2.3.8(ii) and is equivalent to saying that *the square* (6.38) *becomes bicommutative in* $\operatorname{Rel}\mathsf{E}$. In fact this equivalence holds for all abelian categories E, as stated in [Hi]. More general characterisations of this fact in p-exact categories can be found in 6.6.4.

It will also be useful to know that the exactness of square (6.38) is equivalent to each of the following (dual) conditions:

(i) the pair (h, k) and the pushout of (f, g) have the same pullback,

(i*) the pair (f, g) and the pullback of (h, k) have the same pushout.

It is sufficient to prove the equivalence of (i), because the exactness condition is self-dual. We have already noted that the pullback-object of (h, k) is the subobject $\operatorname{Ker}[h, -k] \rightarrowtail X \oplus Y$; now, by (a*), the pushout of (f, g) is a pair (h', k') with $\operatorname{Ker}[h', -k'] = \operatorname{Im}(f, g)$, and its pullback is this subobject of $X \oplus Y$. Thus the exactness condition is the same as saying that the two pullbacks coincide.

6.5 Projective p-exact categories and projective spaces

Categories of projective spaces are rarely considered in category theory; the reason might be that one needs the (generally overlooked) theory of p-exact categories. The present analysis first appeared in the article [G1], and later in the book [G9].

We show here that every p-exact category E has an associated *projective* p-exact category $\Pr E$, with a faithful projective functor. If E is the abelian category $K\,\mathsf{Vct}$ of vector spaces on the field K, $\Pr E$ can be identified with the category $K\,\mathsf{Prj}$ of projective spaces and projective linear maps on K (see 6.5.4(a)). This p-exact category is *not* abelian, generally, and cannot even be exactly embedded in an abelian category (see Theorem 6.5.5). But it inherits from $K\,\mathsf{Vct}$ a symmetric monoidal closed structure (see 6.5.6).

A characterisation of the projective categories associated to abelian categories was given in [CaG].

K is always a (commutative) field and K^* the multiplicative group of its non-zero elements.

6.5.1 The associated projective category

We say that a p-exact category is *projective* if its transfer functor $\mathrm{Sub}_E \colon E \to \mathsf{Mlc}$ is faithful. Let us recall that Sub_E, defined in 6.3.8, is also called the projective functor of E.

We construct the projective p-exact category *associated* to an arbitrary p-exact category E as the quotient of E modulo the *projective congruence* $\sim_S$, namely the congruence of categories associated to the functor Sub_E.

The objects of E are unchanged, but we identify two parallel maps $f, g \colon A \to B$ of E when they have the same modular connection $(f_*, f^*) = (g_*, g^*) \colon \mathrm{Sub}A \to \mathrm{Sub}B$

$$\Pr E = E/\!\sim_S,$$
$$(f \sim_S g) \;\Leftrightarrow\; (f_* = g_*) \;\Leftrightarrow\; (f^* = g^*). \tag{6.41}$$

One easily verifies that the category $\Pr E$ is p-exact, with the same zero object, the same kernel and cokernel objects and the induced kernel and cokernel morphisms

$$\ker[f] = [\ker_E f] \colon \mathrm{Ker}\, f \rightarrowtail A,$$
$$\mathrm{cok}[f] = [\mathrm{cok}_E f] \colon A \twoheadrightarrow \mathrm{Cok}\, f. \tag{6.42}$$

Therefore, the projection functor P preserves and reflects exactness (including mono-, epi- and isomorphisms):

$$P \colon E \to \Pr E, \qquad A \mapsto A, \quad f \mapsto [f]. \tag{6.43}$$

The projective functor of E obviously factorises through P, yielding a faithful functor $S \colon \Pr E \to \mathsf{Mlc}$

$$\mathrm{Sub}_E = SP \colon E \to \Pr E \to \mathsf{Mlc},$$
$$S(A) = \mathrm{Sub}(A), \qquad S[f] = \mathrm{Sub}(f) = (f_*, f^*). \tag{6.44}$$

The functor S is isomorphic to the projective functor $\mathsf{Sub}_{\mathsf{E}'}$ of $\mathsf{E}' = \mathsf{Pr}\,\mathsf{E}$, through the natural isomorphism $\sigma\colon S \to \mathsf{Sub}_{\mathsf{E}'}$

$$\sigma A\colon S(A) \to \mathsf{Sub}_{\mathsf{E}'}(A), \qquad \sigma A(m) = [m]. \tag{6.45}$$

This shows that $\mathsf{Pr}\,\mathsf{E}$ is projective and $\mathsf{Pr}(\mathsf{Pr}\,\mathsf{E})$ can be identified with $\mathsf{Pr}\,\mathsf{E}$; it will be called the *projective* p-exact category associated to E.

$\mathsf{Pr}\,\mathsf{E}$ is determined up to isomorphism of categories by an obvious *universal property*: every exact functor $F\colon \mathsf{E} \to \mathsf{E}'$ with values in a projective p-exact category factorises in a unique way as $F = GP$, where $G\colon \mathsf{Pr}\,\mathsf{E} \to \mathsf{E}'$ is a functor (necessarily exact).

6.5.2 *Proposition* (The projective congruence of vector spaces)

In the abelian category $\mathsf{E} = K\,\mathsf{Vct}$ *of vector spaces over the (commutative) field* K, *the projective congruence* (6.41) *is characterised as:*

$$(f \sim_S g) \quad \Leftrightarrow \quad (\exists \lambda \in K^*\colon\ f = \lambda g). \tag{6.46}$$

Therefore the category $K\,\mathsf{Vct}$ *is projective if and only if* K *is a two-element field (isomorphic to* $\mathbb{Z}/2$).

Note. Property (6.46) is essentially well known, in different forms.

Proof The right-hand condition in (6.46) obviously implies that $f_* = g_*$. Conversely, let us suppose that $f_* = g_*$ (and $f^* = g^*$). Let $m = f_*(1) = g_*(1)$ and $p = \mathrm{cok}\,(f^*(0)) = \mathrm{cok}\,(g^*(0))$. There are unique isomorphisms u, v such that $f = mup$, $g = mvq$ and

$$u_* = (m^*m_*)u_*(p_*p^*) = m^*f_*p^* = m^*g_*p^* = v_*.$$

Therefore we only need to prove our property for two isomorphisms u, v: $A \to B$ with $u_* = v_*$. For every $x \in A$, $x \neq 0$, the mappings u_* and v_* coincide on the subspace $\langle x \rangle \subset A$ spanned by x, hence there is a unique scalar $\lambda_x \in K$ such that $u(x) = \lambda_x.v(x)$, and $\lambda_x \neq 0$.

Let $y \in A$, $y \neq 0$. If $y = \lambda x$ for some non-zero scalar λ, we have

$$\lambda_y.v(y) = u(y) = u(\lambda x) = \lambda\lambda_x.v(x) = \lambda_x.v(y)$$

and $\lambda_y = \lambda_x$. Otherwise, x and y are linearly independent and

$$\lambda_x.v(x) + \lambda_y.v(y) = u(x) + u(y) = u(x + y) = \lambda_{x+y}.v(x) + \lambda_{x+y}.v(y).$$

But $v(x)$ and $v(y)$ are also linearly independent, whence

$$\lambda_x = \lambda_{x+y} = \lambda_y.$$

$\square$

6.5.3 Projective spaces and projective maps

The associated projective category of $K\,\mathsf{Vct}$

$$K\,\mathsf{Prj} = \mathrm{Pr}(K\,\mathsf{Vct}) = K\,\mathsf{Vct}/\!\sim_S, \tag{6.47}$$

will be called the category *of projective spaces and projective linear maps.*

It is p-exact, with a projection functor $P\colon K\,\mathsf{Vct} \to K\,\mathsf{Prj}$ that preserves and reflects exactness (hence also the isomorphisms). We show below (in Theorem 6.5.5) that $K\,\mathsf{Prj}$ is not abelian, except in 'one' case, i.e. if K is the two-element field, when obviously $K\,\mathsf{Prj} = K\,\mathsf{Vct}$ 'is' the abelian category of abelian groups satisfying the axiom $2x = 0$, for all x.

Our quotient $K\,\mathsf{Prj}$ admits various concrete representations, in terms of faithful functors, that allow us to recognise the 'usual projective spaces' and their morphisms. The third, below, is the classical one.

(a) *Projective spaces as lattices of linear subspaces of vector spaces.* We use the composite

$$F_1 = US\colon K\,\mathsf{Prj} \to \mathsf{Mlc} \to \mathsf{Set},$$
$$F_1(X) = |\mathrm{Sub}_{\mathsf{E}}(X)|, \qquad F_1[f] = f_*, \tag{6.48}$$

of the transfer functor S, in (6.44), and the (covariant) forgetful functor U of Mlc, in (6.11).

(b) *Projective spaces as* pointed *sets, quotients of vector spaces.* We use the functor

$$F_2\colon K\,\mathsf{Prj} \to \mathsf{Set}_{\bullet},$$
$$F_2(X) = (|X|/\!\sim, [0]), \qquad (F_2[f])[x] = [f(x)], \tag{6.49}$$

where $F_2(X)$ is the quotient of the underlying set $|X|$ modulo the usual equivalence relation

$$x \sim y \iff (\exists \lambda \in K^*\colon x = \lambda y), \tag{6.50}$$

pointed at the singleton class $[0] = \{0\}$. F_2 is faithful because every linear subspace is the join of its 1-dimensional subspaces, and direct images preserve joins.

(c) *Projective spaces as* unpointed *sets.* We use the composed functor

$$F_3 = RF_2\colon K\,\mathsf{Prj} \to \mathsf{Set}_{\bullet} \to \mathcal{S}, \qquad F_3(X) = (|X| \setminus \{0\})/\!\sim, \tag{6.51}$$

where

$$R\colon \mathsf{Set}_{\bullet} \to \mathcal{S}, \qquad R(S, x_0) = S \setminus \{x_0\},$$

is the equivalence of categories between $\mathsf{Set}_{\bullet}$ and the category $\mathcal{S}$ of sets and partial mappings (see (1.69)).

$K\,\mathsf{Prj}$ is thus isomorphic to the category $K\,\mathsf{Prj}'$ whose objects are the K-vector spaces, while a morphism $\varphi\colon X \to Y$ is a *partial mapping* of sets

$$F_3(f)\colon (|X| \setminus \{0\})/\!\!\sim \ \to \ (|Y| \setminus \{0\})/\!\!\sim, \qquad (6.52)$$

induced by some linear mapping $f\colon X \to Y$, and defined on the set-theoretic complement of $\mathrm{Ker}\,f$ (see the remarks below).

6.5.4 *Exercises and complements*

(a) The reader can check that the realisation $K\,\mathsf{Prj}'$ of $K\,\mathsf{Prj}$ considered above amounts to the construction of 'projective linear mappings' as partial mappings, in Bourbaki [Bor3], Section 9, n. 10.

(b) Other authors use the term 'projective linear transformation' to mean an *isomorphism* of $K\,\mathsf{Prj}'$; this is a bijection

$$(|X| \setminus \{0\})/\!\!\sim \ \to \ (|Y| \setminus \{0\})/\!\!\sim,$$

induced by some isomorphism $X \to Y$ of vector spaces. (Such isomorphisms form a disconnected groupoid, with poor categorical properties.)

(c) For an arbitrary category E, the group $\pi = \mathrm{Aut}(\mathrm{id}\mathsf{E})$ of natural automorphisms of the identity functor defines a congruence: two morphisms $f, g\colon A \to B$ are π-equivalent if

$$(f \sim_\pi g) \ \Leftrightarrow \ (\exists \lambda \in \pi\colon \ f = g.\lambda_A = \lambda_B.g). \qquad (6.53)$$

If E is p-exact, this congruence plainly implies the projective one. One can prove that the quotient $\mathsf{E}/\!\!\sim_\pi$ is p-exact, the projection $\mathsf{E} \to \mathsf{E}/\!\!\sim_\pi$ is an exact functor, and the projection $\mathsf{E} \to \mathrm{Pr}\,\mathsf{E}$ factorises through the former.

If $\mathsf{E} = K\,\mathsf{Vct}$, π can be identified with the multiplicative group K^* and the congruence (6.53) coincides with the projective one, by 6.5.2; thus, $\mathsf{E}/\!\!\sim_\pi = \mathrm{Pr}\,\mathsf{E}$.

(d) On the other hand, if $\mathsf{E} = \mathsf{Ab} = \mathbb{Z}\,\mathsf{Mod}$, π 'is' the multiplicative group of invertible integers, $\{-1, 1\}$. The reader can check that the five homomorphisms $\mathbb{Z} \to \mathbb{Z}_5$ give *three* equivalence classes with respect to $\sim_\pi$ and *two* classes with respect to $\sim_S$; therefore, $\mathsf{Ab}/\!\!\sim_\pi$ is not projective.

6.5.5 *Theorem* (The lack of products)

Let the field K have more than two elements.

(a) The p-exact category $K\,\mathsf{Prj}$ lacks binary products and sums, and cannot have an exact embedding, or even a faithful half exact functor (see 6.3.4), with values in an abelian category.

(b) The biproduct $A \oplus B$ of K Vct does not induce a two-variable functor of projective spaces.

(c) Similarly, the p-exact category Mlc cannot have an exact embedding, or even a faithful half exact functor, with values in an abelian category.

Proof (a) The argument is based on the fact that the canonical injections

$$u_i \colon K \to K^2 \qquad (i = 1, 2),$$

give two projective linear maps $[u_i]$ that are *not* jointly epi in K Prj.

Indeed, let us choose some $\lambda \in K$ different from 0 and 1. The linear automorphisms

$$f = \mathrm{id} K^2, \qquad g \colon K^2 \to K^2, \quad g(x, y) = (x, \lambda y),$$

are not related by the projective congruence $\sim_S$ (because of Proposition 6.5.2); therefore $[f] \neq [g]$, but $[fu_i] = [gu_i]$, for $i = 1, 2$.

Suppose now that we have a half exact functor $F \colon K$ Prj $\to$ A with values in an abelian category. Then the composite $FP \colon K$ Vct $\to K$ Prj $\to$ A is a half exact functor between abelian categories and preserves biproducts (by 6.4.7), whence the pair $F[u_i]$ is jointly epi in A. Since a faithful functor reflects such pairs, F cannot be faithful.

As a consequence, K Prj is not abelian, and therefore cannot have binary products or sums.

(b) This statement also follows from the fact that the linear mapping $g \colon K^2 \to K^2$ sending (x, y) to $(x, \lambda y)$ is not equivalent to the identity. Therefore, the biproduct $A \oplus B$ of K Vct does not even induce a 'functor in each variable', on projective spaces.

Let us note that the butterfly diagram (6.34) with exact diagonals subsists in K Prj, but – without a $\mathbb{Z}$-linear structure – does not give a biproduct.

(c) Finally, it suffices to recall that the projective functor Sub: K Prj $\to$ Mlc is exact and faithful.

$$\square$$

6.5.6 The exponential law of projective spaces

The tensor product and the Hom functor of vector spaces

$$\otimes_K \colon \ K \text{ Vct} \times K \text{ Vct} \ \to \ K \text{ Vct},$$

$$\mathrm{Hom}_K \colon \ K \text{ Vct}^{\mathrm{op}} \times K \text{ Vct} \ \to \ K \text{ Vct}, \tag{6.54}$$

have been considered in 3.4.6, more generally for modules.

Both functors are easily seen to be consistent with the projective congruence, characterised in 6.5.2: if $f = \lambda f'$ and $g = \mu g'$, then

$$f \otimes g = \lambda\mu.f' \otimes g', \qquad \mathrm{Hom}(f, g) = \lambda\mu.\mathrm{Hom}(f', g').$$

Therefore, there are induced functors for $\mathsf{E} = K\,\mathsf{Prj}$ (that act on the objects as the previous ones):

$$\otimes_K : \mathsf{E} \times \mathsf{E} \to \mathsf{E}, \qquad\qquad [f] \otimes [g] = [f \otimes g],$$
$$\mathrm{Hom}_K : \mathsf{E}^{\mathrm{op}} \times \mathsf{E} \to \mathsf{E}, \qquad \mathrm{Hom}_K([f], [g]) = [g]. - .[f]. \tag{6.55}$$

The original adjunction of vector spaces

$$\varphi_{XY} : \ K\,\mathsf{Vct}(X \otimes_K A, Y) \ \to \ K\,\mathsf{Vct}(X, \mathrm{Hom}_K(A, Y)),$$
$$\varphi_{XY}(u)(x) = u(x \otimes -) : A \to Y, \tag{6.56}$$

(for $u : X \otimes_K A \to Y$ and $x \in X$) is also consistent with the projective congruence: if $u = \lambda v$ then $\varphi(u) = \lambda.\varphi(v)$.

We thus have an induced adjunction $F \dashv G$ between endofunctors of $K\,\mathsf{Prj}$

$$F(X) = X \otimes A, \qquad G(Y) = \mathrm{Hom}_K(A, Y) = Y^A, \tag{6.57}$$

where $\mathrm{Hom}_K(A, Y)$ *is still the original vector space*.

This gives a symmetric monoidal closed structure on $K\,\mathsf{Prj}$. Its identity is K, i.e. the one-point projective space in the classical representation 6.5.3(c).

6.6 Relations and induction for p-exact categories

A consistent part of Homological Algebra, from the homology of chain complexes to the theory of spectral sequences, is about *subquotients* and *induced morphisms* between them; the latter are defined through the categories of relations, since a subquotient in $R\,\mathsf{Mod}$ is simply a subobject in $\mathrm{Rel}\,(R\,\mathsf{Mod})$ (as shown in 6.6.2(f)).

The problem arises of the coherence of induction with composition, as already noted in Mac Lane's book 'Homology' ([M2], Section II.6). The problem can be apparently complicated, when we are working in complex systems. Yet, it has a simple solution, in the domain of p-exact categories and their relations.

The extension of relations from abelian to p-exact categories was introduced by Tsalenko [T1, T2], then studied by Brinkmann and Puppe [Bri, BriP]. Here we briefly recall this construction. (The interested reader will find the proofs and further information in these references, or in [G9].)

We end by considering subquotients and induced relations, with some hints at our Coherence Theorem for induction.

6.6.1 Relations of modules revisited

The category Rel (R Mod) of relations between modules has been reviewed in 1.8.5; we give now a more detailed analysis to prepare the construction of relations for all p-exact categories.

A relation $u \colon A \to B$ is a submodule of the direct sum $A \oplus B$. It can be viewed as a 'partially defined, multi-valued homomorphism', that sends an element $x \in A$ to the subset $\{y \in B \mid (x, y) \in u\}$ of B

The composite of u with $v \colon B \to C$ is

$$vu = \{(x, z) \in A \oplus C \mid \exists y \in B \colon (x, y) \in u \text{ and } (y, z) \in v\}.$$

Rel (R Mod) is an involutive ordered category. For two parallel relations $u, u' \colon A \to B$, $u \leqslant u'$ means that $u \subset u'$ as submodules of $A \oplus B$. The *opposite relation* of $u \colon A \to B$ is obtained by reversing pairs, and written as $u^\sharp \colon B \to A$. The involution $u \mapsto u^\sharp$ is *regular* in the sense of von Neumann, i.e. $uu^\sharp u = u$ for all relations u.

As a consequence, a *monorelation*, i.e. a monomorphism in the category Rel (R Mod), is characterised by the condition $u^\sharp u = 1$ (and is a split mono); an *epirelation* by the dual condition $uu^\sharp = 1$.

The category R Mod is embedded in the category of relations Rel (R Mod), identifying a homomorphism $f \colon A \to B$ with its graph

$$\{(x, y) \in A \oplus B \mid y = f(x)\}.$$

A relation $u \colon A \to B$ determines two submodules of A, called *definition* and *annihilator*

$$\operatorname{Def} u = \{x \in A \mid \exists y \in B \colon (x, y) \in u\},$$
$$\operatorname{Ann} u = \{x \in A \mid (x, 0) \in u\}, \tag{6.58}$$

and two submodules of B, called *values* and *indeterminacy*

$$\operatorname{Val} u = \{y \in B \mid \exists x \in A \colon (x, y) \in u\} = \operatorname{Def} u^\sharp,$$
$$\operatorname{Ind} u = \{y \in B \mid (0, y) \in u\} = \operatorname{Ann} u^\sharp. \tag{6.59}$$

6.6.2 Exercises and complements

(a) A relation $u \colon A \to B$ is a homomorphism if and only if it is *everywhere defined* (i.e. $\operatorname{Def} u = A$) and *single-valued* (i.e. $\operatorname{Ind} u = 0$), in which case annihilator and values coincide with kernel and image, respectively.

(b) The relation u is a monorelation if and only if $\operatorname{Def} u = A$ and $\operatorname{Ann} u = 0$.

(b*) The relation u is an epirelation if and only if $\operatorname{Val} u = B$ and $\operatorname{Ind} u = 0$.

(c) The relation u is invertible in Rel (R Mod) if and only if it is invertible in $R\,\mathsf{Mod}$, i.e. an isomorphism of modules, in which case $u^{\sharp}$ is the inverse isomorphism.

(d) Every monomorphism (resp. epimorphism) in $R\,\mathsf{Mod}$ is a monorelation (resp. epirelation).

(e) The relation u has a canonical factorisation $u = (nq^{\sharp}).i.(pm^{\sharp})$ (and a dual factorisation, by dashed arrows)

$$(6.60)$$

where m, n are embeddings of submodules, p, q are projections on quotients and i is an isomorphism; notice that the latter links a *subquotient* of A with a subquotient of B (see 6.3.9).

This factorisation consists of an epirelation $(ipm^{\sharp})$ followed by a monorelation $(s = nq^{\sharp})$: it is a factorisation epi-mono in $\mathsf{Rel\,Ab}$, and therefore *the image of u in this category should be defined as the subquotient* $\mathrm{Val}\,u/\mathrm{Ind}\,u$ (rather than the subobject $\mathrm{Val}\,u$).

(f) Following our analysis in 6.3.9, a subquotient $S = M/N$ of a module A is presented by the following bicartesian square (where $N \subset M$)

$$
\begin{array}{ccc}
M & \overset{m}{\rightarrowtail} & A \\
p\downarrow & \nearrow{}_{s} & \downarrow q \\
S & \underset{n}{\rightarrowtail} & A/N
\end{array}
\qquad (6.61)
$$

This bicartesian square determines one monorelation

$$s = mp^{\sharp} = q^{\sharp}n \colon S \to A,$$

that sends the class $[x] \in M/N$ to all the elements of the lateral $[x] = x + N \subset A$. Because of the factorisation (6.60) of a relation, every monorelation $s \colon S' \to A$ is of this type, up to isomorphism.

The *subquotients* of the module A amount thus to the *subobjects of A in* $\mathsf{Rel}\,(R\,\mathsf{Mod})$. This will be used in 6.6.6–6.6.8.

(g) All idempotents split (see 2.8.2).

6.6.3 Relations for p-exact categories

Every p-exact category E has an involutive ordered category of relations Rel E, which we now *describe*, without proofs. Let us recall that E has *pullbacks of monos* and *pushouts of epis* along arbitrary morphisms (see 6.3.6), and is assumed to be well powered (see 6.3.2).

A relation $u: A \to B$ has a *quaternary factorisation* $u = nq^{\sharp}pm^{\sharp}$, and a *coquaternary factorisation* $u = q'^{\sharp}n'm'^{\sharp}p'$, each of them determined up to three coherent isomorphisms; they form two *bicartesian* squares (pullback and pushout, see 6.3.6(b))

$$\tag{6.62}$$

The composed relation vu is computed, on *quaternary* factorisations, by *pullbacks of monos* (along arbitrary arrows) and *pushout of epis along epis*, as in the following diagram; the relation $u: A \to B$ is drawn in the upper row, the relation $v: B \to C$ in the right-hand column and $vu: A \to C$ along the dotted diagonal

$$\tag{6.63}$$

The square (x) is a pullback, (y) and (y') are commutative (by epi-mono factorisation), (z) and (z') are pullbacks and (u) is a pushout.

Dually, it can be computed on *coquaternary* factorisations, by means of *pushouts of epis* along arbitrary arrows and *pullbacks of monos along monos*.

The involution $u \mapsto u^{\sharp}$ is obvious; the order relation $u \leqslant v$ is expressed by the existence of a commutative diagram in E linking their quaternary factorisations (or equivalently the coquaternary ones)

$$A \xleftarrow{\ m\ } \bullet \xrightarrow{\ p\ } \bullet \xleftarrow{\ q\ } \bullet \xrightarrow{\ n\ } B \qquad (6.64)$$

In the construction of Rel E, a relation $A \to B$ is defined as an equivalence class $[m, p, q, n]$ of quaternary diagrams of E as above, up to three coherent *iso* morphisms u, v, w (forming a commutative diagram as above); the composition is defined as in diagram (6.63); the proof of associativity is not easy (see one of the references [T2, BriP, G9]).

The category Rel Gp can be constructed by 'quaternary classes', as shown in [BriP] (and not in the dual way). The category Cor Set mentioned in 1.8.9 can be constructed by 'coquaternary classes', as essentially shown in [Da1]. This explains why, in both cases, the involution is regular.

6.6.4 Exact squares

A commutative square of E is said to be *exact* if it is *bicommutative* in Rel E, which means that it stays commutative when one reverses two parallel edges of the square. To characterise this fact, first note that (by computing compositions in Rel E):

$$(6.65)$$

(i) a commutative square of monos (a) is exact if and only if it is a pullback,

(i*) a commutative square of epis (b) is exact if and only if it is a pushout,

(ii) a commutative *mixed* square (c) is exact if and only if it is a pullback, if and only if it is a pushout.

More generally, every preimage of a mono (by pullback) produces a bicommutative square, as well as every direct image of an epi (by pushout). Finally, a commutative square in E becomes bicommutative in Rel E if and only if its canonical factorisation is composed of four bicommutative

squares of the previous types

$$(6.66)$$

Every p-exact functor $F\colon \mathsf{E} \to \mathsf{E}'$ preserves preimages of monos and direct images of epis; therefore it has a unique extension to a functor that preserves involution and ordering

$$\mathrm{Rel}\,F\colon \mathrm{Rel}\,\mathsf{E} \to \mathrm{Rel}\,\mathsf{E}'. \qquad (6.67)$$

*Every natural transformation $\varphi\colon F \to G\colon \mathsf{E} \to \mathsf{E}'$ determines a *lax transformation of order-preserving functors*

$$\mathrm{Rel}\,\varphi\colon \mathrm{Rel}\,F \to \mathrm{Rel}\,G\colon \mathrm{Rel}\,\mathsf{E} \to \mathrm{Rel}\,\mathsf{E}', \qquad (6.68)$$

whose components are precisely those of φ (and belong to E').

For this notion the interested reader is referred to [G9]: essentially, a relation $u\colon A \to B$ in $\mathrm{Rel}\,\mathsf{E}$ produces a 'lax square' in the ordered category $\mathrm{Rel}\,\mathsf{E}'$, namely $\varphi B.(\mathrm{Rel}\,F)(u) \leqslant (\mathrm{Rel}\,G)(u).\varphi A$ (see (7.67)). *

6.6.5 The abelian case

Restricting everything to abelian categories gives little advantages in this domain, except for the existence of 'two-arrow factorisations' of relations, and the fact that the proof of associativity of composition can be given in the easier form of 4.5.4, since every abelian category is regular. (Yet, one should not forget the fact that (co)quaternary factorisations are adapted to the study of subquotients, as we have already seen in 6.6.2.)

In fact, every relation has now *binary factorisations* $u = gf^{\sharp}\colon A \to B$. Such a factorisation will be said to be *strong* if the pair (f, g) is jointly mono (corresponding to a monomorphism with values in $A \oplus B$); this can be obtained from a quaternary factorisation of u, by the pullback of the central epimorphisms:

$$(6.69)$$

Dually, there is a *strong cobinary factorisation* $u = f'^{\sharp}g'$, where (f', g') is jointly epi. Two cobinary factorisations yield the same relation u if and only if they have the same pullback, which is then a *strong* binary factorisation of u.

6.6.6 Induced relations and canonical isomorphisms

We end with some points on the study of induced relations and canonical isomorphisms between subquotients, at an introductory level.

Let us begin by working in a category of modules. We have seen in 6.6.2 that a subquotient can be represented as a subobject in Rel $(R \, \mathsf{Mod})$.

Given two subquotients $s \colon S \to A$ and $t \colon T \to B$, a relation $a \colon A \to B$ has an *induced relation*

$$a' = t^{\sharp} a s \colon S \to T. \tag{6.70}$$

Generally, this procedure does not give a commutative diagram, and is *not* preserved by composition.

More particularly, for $a = \mathrm{id} A$, two subquotients $s \colon S \to A$ and $t \colon T \to A$ have an *induced relation* $t^{\sharp} s \colon S \to T$. If it is an isomorphism, it is called the *canonical isomorphism* from S to T.

The reader is warned that, as already remarked in Mac Lane [M2], *this does not give a transitive relation*, in general: if we have a third subquotient $u \colon U \to A$ and two isomorphisms

$$t^{\sharp} s \colon S \to T, \qquad u^{\sharp} t \colon T \to U,$$

the composed isomorphism need not be the induced relation $u^{\sharp} s \colon S \to U$.

The term 'canonically isomorphic subquotients' should be reserved to situations where the corresponding relation is indeed an equivalence relation, as discussed below.

6.6.7 A case of incoherence.

The following examples show some instances of *incoherence of induction on subquotients*:

- first, canonical isomorphisms need not be closed under composition,

- second, if we extend them in this sense the result need not be uniquely determined.

In other words, canonical isomorphisms – in general – do not form nor generate a coherent system of isomorphisms (as defined in 1.4.9(b)).

As in Mac Lane's book [M2], our examples of inconsistency are based on the lattice Sub$A$ of subgroups of the abelian group $A = \mathbb{Z} \oplus \mathbb{Z}$, and more particularly on the (*non-distributive*) triple formed of the diagonal Δ and two of its complements, the subgroups A_1 and A_2

$$A_1 = \mathbb{Z} \oplus 0, \qquad A_2 = 0 \oplus \mathbb{Z},$$
$$A_i \vee \Delta = A, \qquad A_i \wedge \Delta = 0. \tag{6.71}$$

We consider the subquotients $m_i \colon A_i \to A$ and $s = p^\sharp \colon A/\Delta \to A$, in Rel Ab.

(a) The identity of A induces two canonical isomorphisms $u_i = pm_i \colon A_i \to A/\Delta$ (that are Noether isomorphisms), and a canonical isomorphism $u_2^{-1} \colon A/\Delta \to A_2$.

Then, the composed isomorphism $w = u_2^{-1}u_1 \colon A_1 \to A_2$ *is not canonical*. Indeed:

$$w \colon A_1 \to A/\Delta \to A_2,$$
$$(x, 0) \mapsto [(x, 0)] = [(0, -x)] \mapsto (0, -x), \tag{6.72}$$

while the canonical relation $m_2^\sharp.m_1 \colon A_1 \to A_2$ has graph $\{(0,0)\}$ (and is not even a homomorphism).

(b) Using the subgroup $\Delta' = \{(x, -x) \mid x \in \mathbb{Z}\}$ instead of the diagonal Δ, we get the *opposite* composed isomorphism from A_1 to A_2

$$A_1 \to A/\Delta' \to A_2 \qquad (x, 0) \mapsto [(x, 0)] = [(0, x)] \mapsto (0, x). \tag{6.73}$$

This shows that a composite $A_1 \to A_2$ of canonical isomorphisms between subquotients of $\mathbb{Z} \oplus \mathbb{Z}$ *is not determined*.

Now, a change of sign can be quite important, in Homological Algebra and Algebraic Topology. For instance, it is the case in the usual argument proving that 'even-dimensional spheres cannot be combed': if the sphere $\mathbb{S}^n$ has a non-null vector field, then its antipodal map $t \colon \mathbb{S}^n \to \mathbb{S}^n$ is homotopic to the identity, and the degree $(-1)^{n+1}$ of t must be 1, whence n must be odd. The conclusion cannot be obtained if we only know the induced homomorphism $t_{*n} \colon H_n(\mathbb{S}^n) \to H_n(\mathbb{S}^n)$ up to sign change.

(c) We are lead to think that these cases of incoherence might depend on the non-distributivity of the lattice of subobjects involved in our subquotients. We show now that it is indeed the case.

6.6.8 *Coherence and distributivity*

We come back to the framework of p-exact categories, where all the previous arguments work: even the abelian setting would now be insufficient. These informal comments are taken from the Introduction of [G9]; we refer to this book for the theory hinted at here.

Let us recall that a p-exact category E is said to be *distributive* if all its lattices of subobjects are distributive, or in other words if its transfer functor Sub: E → Mlc takes values in the full (p-exact) subcategory Dlc of distributive lattices and modular connections (see 6.3.3(c)). A nontrivial abelian category cannot be distributive: the lattice $Sub(A \oplus A)$ is not distributive, as soon as A is not a zero object. But there are distributive p-exact categories, like Dlc and other examples in 6.3.3.

Now, the coherence problem mentioned above is solved by Theorem 2.7.6 of [G9], whose basic part says the following.

Coherence Theorem of Homological Algebra. *In a p-exact category* E, *induced isomorphisms between subquotients are preserved by composition (and form a coherent system) if and only if* E *is distributive, if and only if its category of relations* Rel E *is orthodox (i.e. its idempotent endomorphisms are closed under composition).*

As a consequence, in a distributive p-exact category, all the diagrams of canonical isomorphisms between subquotients (induced by the identity of a given object) commute.

Now, every p-exact category (including the abelian ones!) has an associated distributive p-exact category, the distributive expansion Dst E (mentioned in 6.3.3(f) and constructed in [G9], 2.8.4.)

Combining the coherence theorem with the distributive expansion, it follows that the subquotients M/N of a given object A (in a p-exact, or abelian category) have a coherent system of isomorphisms (well-determined and closed under composition) *when we restrict to those subquotients whose numerator* (M) *and denominator* (N) *belong to a distributive sublattice X of the (modular) lattice of all subobjects.*

The crucial fact is that many important homological systems, and in particular most of those giving rise to spectral sequences (like the filtered complex, the double complex, Massey's exact couple and Eilenberg's exact system), actually 'live' in the distributive expansion of an abelian category, like Dst Ab, and can therefore be studied within *distributive p-exact categories.*

On the other hand, systems that lead to *non-*distributive sublattices of subobjects, like a *tri*filtered object or a *bi*filtered chain complex, *should be handled with great care,* to avoid the risk of using 'canonical isomorphisms' in an illegitimate way. Examples of such problems can be found in [G9], 1.4.5.

7

Hints at higher dimensional category theory

We sketch here the main two-dimensional categorical structures, namely 2-categories and double categories, together with their weak versions, namely bicategories and weak double categories.

The theory of 2-categories and bicategories (or weak 2-categories), briefly presented in Section 7.1, is the best known part of higher dimensional category theory. A general item is a 2-*cell* $\varphi\colon u \to v\colon X \to X'$, as in the left diagram below

$$X \xrightarrow[v]{\overset{u}{\underset{\downarrow\varphi}{\Rightarrow}}} Y \qquad\qquad \begin{array}{ccc} X & \xrightarrow{f} & X' \\ u\downarrow & \varphi & \downarrow v \\ Y & \xrightarrow{g} & Y' \end{array} \qquad\qquad \begin{array}{c} X \\ u\Big\downarrow \overset{\varphi}{\underset{\to}{}} \Big\downarrow v \\ Y \end{array} \qquad (7.1)$$

Strict and weak double categories (denoted by letters $\mathbb{A}, \mathbb{B}, \ldots$) are an extension of the former, presented in Section 7.2 and used in Sections 7.4, 7.5 for applications in Homological Algebra.

Now a general item is a *double cell* $\varphi\colon (u \,{}^{f}_{g}\, v)$ among horizontal and vertical arrows, as in the central diagram above. The previous case is obtained by reducing the horizontal arrows to vertical identities, as in the right diagram above. (We use the *horizontal* direction as the main one, that remains strict in the weak case; this choice, and its influence on terminology, will be discussed in 7.2.5.)

One of our main motivations for studying double categories, briefly recalled in 7.5.3, is that the general theory of two-dimensional adjunctions can only be formulated in this framework.

In fact, in the general case, *the left adjoint* $F\colon \mathbb{C} \to \mathbb{D}$ (between weak double categories, or *also* between bicategories) *is colax while the right adjoint* $G\colon \mathbb{D} \to \mathbb{C}$ *is lax*; they should not be composed (this would destroy their comparisons), but viewed as *vertical* and *horizontal* arrows, respectively, in a suitable double category Dbl introduced in [GP2]. The unit and counit of the adjunction are double cells of the latter

$$
\begin{array}{ccc}
\mathbb{C} =\!\!\!= \mathbb{C} & & \mathbb{D} \xrightarrow{G} \mathbb{C} \\
F \downarrow \quad \eta \quad \| & & \| \quad \varepsilon \quad \downarrow F \\
\mathbb{D} \xrightarrow{G} \mathbb{C} & & \mathbb{D} =\!\!\!= \mathbb{D}
\end{array}
\tag{7.2}
$$

and their coherence conditions are based – separately – on the comparisons of F and G.

Within 2-categories or bicategories, one can only formulate two particular cases: a colax-pseudo adjunction and a pseudo-lax adjunction, which do not cover the simple colax-lax adjunction presented in 7.5.3; furthermore, adjunctions of these different kinds can only be composed in the general setting.

Another motivation is that many 2-categories and bicategories, starting from Rel Ab and the bicategories of spans or cospans, have poor limits, while the corresponding (strict or weak) double categories have all horizontal limits, studied in [GP1].

Higher dimensional category theory naturally extends to finite and infinite dimension

- from (strict and weak) 2-categories to n-categories and ω-categories,

- from (strict and weak) double categories to cubical categories, or more generally to multiple categories.

The interested reader will look under these headings.

7.1 From categories to 2-categories and bicategories

The prime example of a 2-category is Cat, when we add to its objects and arrows the natural transformations $\varphi \colon F \to G \colon \mathsf{C} \to \mathsf{D}$, viewed as *2-cells*, or two-dimensional arrows between the functors – as in the left diagram (7.1), above.

We briefly describe the strict notion of 2-category (from [Eh2]) and the corresponding weak notion of bicategory [Be2]. The elementary cases of a locally ordered 2-category and a locally preordered bicategory have been presented in 4.5.1 and 4.5.5, respectively, while dealing with two approaches to relations on regular categories.

7.1.1 Sesquicategories

It is convenient to start from a preliminary notion, introduced in [St2]. A *sesquicategory* is a category C equipped with:

(a) for each pair of parallel morphisms $f, g\colon X \to Y$, a set of *2-cells*, or *homotopies*, $C_2(f, g)$ whose elements are written as $\varphi\colon f \to g$ (or $\varphi\colon f \to g\colon X \to Y$), so that each map f has an *identity endocell* $\mathrm{id}f\colon f \to f$,

(b) a *concatenation*, or *vertical composition* of 2-cells, written as $\psi\varphi$ or $\psi.\varphi$

$$X \quad \overset{\overset{f}{\longrightarrow}}{\underset{\underset{h}{\longrightarrow}}{\underset{\downarrow\psi}{\overset{\downarrow\varphi}{\longrightarrow}}}} \quad Y \qquad \psi\varphi\colon f \to h\colon X \to Y. \qquad (7.3)$$

(c) a *whisker composition* for 2-cells and maps, or *reduced horizontal composition*, written as $k\varphi h$ (or $k \circ \varphi \circ h$, when useful to distinguish it)

$$X' \overset{h}{\longrightarrow} X \overset{\overset{f}{\longrightarrow}}{\underset{g}{\underset{\downarrow\varphi}{\longrightarrow}}} Y \overset{k}{\longrightarrow} Y' \qquad (7.4)$$

$$k\varphi h\colon kfh \to kgh\colon X' \to Y'.$$

These data must satisfy the following axioms (for *associativities*, *identities* and *distributivity of the whisker composition*):

$$\chi(\psi\varphi) = (\chi\psi)\varphi, \qquad k'(k\varphi h)h' = (k'k)\varphi(hh'),$$

$$\varphi.\mathrm{id}f = \varphi = \mathrm{id}g.\varphi, \quad 1_Y \circ \varphi \circ 1_X = \varphi, \quad k \circ \mathrm{id}f \circ h = \mathrm{id}(kfh), \qquad (7.5)$$

$$k(\psi\varphi)h = (k\psi h).(k\varphi h).$$

Note that each set $C_2(f, g)$ is a category, under vertical composition.

The category Top equipped with homotopies $\varphi\colon f \to g\colon X \to Y$ does *not* become a sesquicategory, because the vertical composition of homotopies is not associative and has no identities: we have a 2-dimensional categorical structure of a more complicated nature: see [G8] and its references.

On the other hand it is easy to see that chain complexes, chain morphisms and homotopies, over an additive category A, form a sesquicategory $Ch_+(A)$.

7.1.2 Two-categories

A *2-category* can be defined as a sesquicategory C which satisfies the following *reduced interchange property*:

$$X \; \xrightarrow[g]{\overset{f}{\to}} \!\!{\scriptstyle\downarrow\varphi}\; Y \; \xrightarrow[k]{\overset{h}{\to}} \!\!{\scriptstyle\downarrow\psi}\; Z \qquad\qquad (\psi g).(h\varphi) = (k\varphi).(\psi f). \qquad (7.6)$$

To recover the usual definition one defines the *horizontal composition* of 2-cells φ, ψ which are *horizontally consecutive*, as in diagram (7.6), using the previous identity:

$$\psi\circ\varphi = (\psi g).(h\varphi) = (k\varphi).(\psi f)\colon hf \to kg\colon X \to Z. \qquad (7.7)$$

(The whisker composition is now a particular case: $h\varphi = \mathrm{id}h\circ\varphi$ and $\psi f = \psi\circ\mathrm{id}f$.)

The horizontal composition of 2-cells is then proved to be associative, to have identities (namely the identity cells of identity arrows) and to satisfy the *middle-four interchange property* with vertical composition (an extension of the reduced interchange property (7.6)):

$$X \; \xrightarrow{\overset{\downarrow\varphi}{\underset{\downarrow\psi}{}}} \; Y \; \xrightarrow{\overset{\downarrow\sigma}{\underset{\downarrow\tau}{}}} \; Z \qquad\qquad (\tau.\sigma)\circ(\psi.\varphi) = (\tau\circ\psi).(\sigma\circ\varphi). \qquad (7.8)$$

As a prime example of such a structure, Cat will also denote the *2-category* of small categories, their functors *and* their natural transformations, with the usual operations, defined in 1.5.3; we have seen in 1.5.4 that the axioms are satisfied.

We already noted that the usual definition of a 2-category is based on the complete horizontal composition, even though one usually works with the reduced one. Moreover, there are important sesquicategories where the reduced interchange property does not hold (and one does not define a full horizontal composition): for instance, the sesquicategories of chain complexes recalled above.

Two-dimensional limits are studied in [Kl3, St2, Gra].

Adjunctions, monads, equivalences and adjoint equivalences can be defined *inside* any 2-category. (For an adjunction one should use the 'algebraic' form (iii) of definition 3.1.2.)

An ordered category C, as defined in 1.8.2, is the same as a 2-category where each category $C_2(f,g)$ is an ordered set; it is also called a *locally ordered 2-category*. Typical examples are Ord, AdjOrd and Rel Set (see 4.5.1).

7.1.3 Bicategories

A *bicategory* is a weak version of a 2-category, introduced by Bénabou [Be2] in 1967.

The general framework of objects, arrows, 2-cells and their operations is the same as in a 2-category. The vertical composition of 2-cells is still categorical, and the interchange property still holds. But the horizontal composition $f \otimes g$ of arrows (now written in diagrammatic order) is associative and has units *up to comparisons cells* (for $f : A \to B$, $g : B \to C$, $h : C \to D$)

$$\kappa(f, g, h) : f \otimes (g \otimes h) \to (f \otimes g) \otimes h \qquad (associator),$$
$$\lambda(f) : \mathrm{id}A \otimes f \to f, \qquad \rho(f) : f \otimes \mathrm{id}B \to f \qquad (unitors), \tag{7.9}$$

that are invertible under vertical composition and natural with respect to 2-cells.

In Section 7.2 we shall see the more general structure of a weak double category, from which the interested reader can deduce the naturality and coherence axioms of the previous comparisons, in a bicategory. Typical examples, the bicategories of spans and cospans, will be seen in the extended form of weak double categories, in 7.2.6.

A monoidal (resp. strict monoidal) category C is the same as a bicategory (resp. 2-category) on one formal object $*$. The arrows $A : * \to *$ are the objects of C, the cells $f : A \to B$ are the maps of C, the vertical composition is that of C, the horizontal one is the tensor product $A \otimes B$.

A bicategory C is said to be *locally preordered* when each category $\mathsf{C}_2(f, g)$ is a preordered set, as in the structure $\mathrm{Rel}'(\mathsf{C})$ of 'jointly monic relations' on a regular category, described in 4.5.5.

In this case the comparison cells (7.9) are uniquely determined, as instances of the equivalence relation associated to the preorder

$$f \otimes (g \otimes h) \sim (f \otimes g) \otimes h, \qquad \mathrm{id}A \otimes f \sim f \sim f \otimes \mathrm{id}B,$$

and their coherence is automatically satisfied. A locally preorered bicategory C has an associated locally ordered 2-category $\mathsf{C}/{\sim}$.

The coherence theorem for bicategories can be found in [MaP]. For higher dimensional extensions of bicategories one can see Leinster [Le] and references therein.

7.1.4 Two-dimensional functors and transformations

A *2-functor* $F : \mathsf{C} \to \mathsf{D}$ between 2-categories sends objects to objects, arrows to arrows and cells to cells, strictly preserving the whole structure: (co)domains, units and compositions. (Lax versions can be found in [Be2, KlS]; here they will be defined for weak double categories, in 7.2.4.)

A 2-*natural transformation* $h\colon F \to G\colon \mathsf{C} \to \mathsf{D}$ between 2-functors is a natural transformation that behaves in a natural way also on every 2-cell $\varphi\colon f \to g\colon X \to Y$ in C

$$
\begin{array}{ccc}
FX & \xrightarrow{hX} & GX \\
{\scriptstyle F\varphi}\downarrow\downarrow & & \downarrow\downarrow{\scriptstyle G\varphi} \\
FY & \xrightarrow[hY]{} & GY
\end{array}
\qquad hY.F\varphi = G\varphi.hX. \qquad (7.10)
$$

It is now easy to define 2-adjunctions and 2-monads, by 2-functors and 2-natural transformations satisfying the same identities as in the ordinary one-dimensional case. In Section 7.6 we shall use 2-monads in Cat.

7.1.5 Two-dimensional universal arrows

Universal arrows of functors (defined in 2.7.1) have a strict and a weak extension to the two-dimensional case.

A 2-*universal arrow* from an object X of X to the 2-functor $U\colon \mathsf{A} \to \mathsf{X}$ is a pair $(A_0, h\colon X \to UA_0)$ which gives an isomorphism of categories (of arrows and cells, with vertical composition):

$$
\mathsf{A}(A_0, A) \to \mathsf{X}(X, UA), \qquad g \mapsto Ug.h. \qquad (7.11)
$$

This amounts to saying that the functor (7.11) is bijective on objects, full and faithful, i.e.

(i) for every A in A and every $f\colon X \to UA$ in X there exists precisely one $g\colon A_0 \to A$ in A such that $f = Ug.h$,

(ii) for every pair of arrows $g, g'\colon A_0 \to A$ in A and every cell $\varphi\colon Ug.h \to Ug'.h$ in X, there is precisely one cell $\psi\colon g \to g'$ in A such that $\varphi = U\psi.h$.

(Equivalently, one can use a global universal property: for every cell $\varphi\colon f \to f'\colon X \to UA$ in X, there is precisely one cell $\psi\colon g \to g'$ in A such that $\varphi = U\psi.h$. This implies that $f = Ug.h$ and $f' = Ug'.h$.)

More generally, a *biuniversal arrow* from X to $U\colon \mathsf{A} \to \mathsf{X}$ is a pair $(A_0, h\colon X \to UA_0)$ so that the functor (7.11) is an *equivalence* of categories. This can be rephrased saying that the functor (7.11) is essentially surjective on objects, full and faithful (see 1.5.5). In other words, we replace property (i) with a weaker version (and keep (ii) as it is):

(i') for every A in A and every $f\colon X \to UA$ in X there exists *some* $g\colon A_0 \to A$ in A such that $f \cong Ug.h$ (*isomorphic objects* in the category $\mathsf{X}(X, UA)$).

In concrete situations one can often replace (i) with the property of *surjectivity* on objects, which is intermediate between (i) and (i′):

(i″) for every A in A and every $f: X \to UA$ in X there exists *some* $g: A_0 \to A$ such that $f = Ug.h$.

Of course, the solution of a 2-universal problem is *determined up to isomorphism*, while the solution of a biuniversal one is *determined up to equivalence* (in a 2-category).

The *2-limit* (resp. *bilimit*) of a 2-functor is defined by a 2-universal (resp. biuniversal) property, extending 2.7.2. The interested reader can easily verify that Cat has (small) 2-limits. Further information on 2-limits and related matter can be found in [St2].

Note that the term 'biproduct' (in the present sense of product *and* coproduct, as in 2.1.9 and Section 6.4) clashes with the (more recent) 'bilimit'. The clash is limited by the fact that 2-products seem to be more important than the corresponding bilimits – essentially because they are 'flexible limits', see [BKPS].

7.1.6 Natural transformations and mates

Let us suppose that we have two adjunctions (between ordinary categories)

$$F: X \rightleftarrows Y : G, \qquad \eta: 1 \to GF, \qquad \varepsilon: FG \to 1,$$
$$F': X' \rightleftarrows Y' : G', \qquad \eta': 1 \to G'F', \qquad \varepsilon': F'G' \to 1, \qquad (7.12)$$

and two functors $H: X \to X'$, $K: Y \to Y'$.

As shown in [KlS], Section 2.2, there is a bijection between sets of natural transformations:

$$\mathrm{Nat}(F'H, KF) \to \mathrm{Nat}(HG, G'K), \qquad \varphi \leftrightarrow \psi,$$

$$\psi = G'K\varepsilon.G'\varphi G.\eta'HG: HG \to G'F'HG \to G'KFG \to G'K, \qquad (7.13)$$

$$\varphi = \varepsilon'KF.F'\psi F.F'H\eta: F'H \to F'HGF \to F'G'KF \to KF,$$

The natural transformations φ, ψ are said to be *mates under the given data*; or *under the given adjunctions* when H and K are identities,

Actually, this holds true for *internal* adjunctions in any 2-category.

7.1.7 The 2-category of adjoint functors

The category AdjCat of small categories and adjunctions

$$(F, G, \eta, \varepsilon) \colon \mathsf{C} \to \mathsf{D},$$

has been introduced in 3.2.1.

We show now that it has a natural 2-categorical structure, derived from that of Cat and extending the 2-categorical structure of AdjOrd given by its order (see 7.1.2).

Given two adjunctions between the same categories

$$(F, G, \eta, \varepsilon) \colon \mathsf{C} \to \mathsf{D}, \qquad (F', G', \eta', \varepsilon') \colon \mathsf{C} \to \mathsf{D}, \tag{7.14}$$

a 2-cell of AdjCat

$$(\varphi, \psi) \colon (F, G, \eta, \varepsilon) \to (F', G', \eta', \varepsilon') \colon \mathsf{C} \to \mathsf{D}, \tag{7.15}$$

is defined as a pair of mate natural transformations (under the given adjunctions), each of them determining the other

$$\varphi \colon F \to F' \colon \mathsf{C} \to \mathsf{D}, \qquad \psi \colon G' \to G \colon \mathsf{D} \to \mathsf{C},$$

$$\psi = (G\varepsilon'.G\varphi G'.\eta G' \colon G' \to GFG' \to GF'G' \to G), \tag{7.16}$$

$$\varphi = (\varepsilon F'.F\psi F'.F\eta' \colon F \to FG'F' \to FGF' \to F').$$

The vertical composition of 2-cells in AdjCat comes from the vertical composition of natural transformations

$$(\varphi', \psi').(\varphi, \psi) = (\varphi'\varphi, \psi\psi'). \tag{7.17}$$

The whisker composition of (φ, ψ) with two adjunctions

$$(H, K, \rho, \sigma) \colon \mathsf{C}' \to \mathsf{C}, \qquad (H', K', \rho', \sigma') \colon \mathsf{D} \to \mathsf{D}',$$

is given by two whisker compositions of natural transformations

$$\begin{aligned} H'\varphi H \colon H'FH \to H'F'H \colon \mathsf{C}' \to \mathsf{D}', \\ K\psi K' \colon KG'K' \to KGK' \colon \mathsf{D}' \to \mathsf{C}'. \end{aligned} \tag{7.18}$$

Finally the (contravariant) involution of (3.15) can be extended (covariantly) to 2-cells

$$(-)^{\mathrm{op}} \colon \mathsf{AdjCat} \to \mathsf{AdjCat}, \qquad \mathsf{C} \mapsto \mathsf{C}^{\mathrm{op}},$$

$$((F, G, \eta, \varepsilon) \colon \mathsf{C} \to \mathsf{D}) \mapsto ((G^{\mathrm{op}}, F^{\mathrm{op}}, \varepsilon^{\mathrm{op}}, \eta^{\mathrm{op}}) \colon \mathsf{D}^{\mathrm{op}} \to \mathsf{C}^{\mathrm{op}}),$$

$$((\varphi, \psi) \colon (F, G, \eta, \varepsilon) \to (F', G', \eta', \varepsilon')) \mapsto$$

$$((\psi^{\mathrm{op}}, \varphi^{\mathrm{op}}) \colon (G^{\mathrm{op}}, F^{\mathrm{op}}, \varepsilon^{\mathrm{op}}, \eta^{\mathrm{op}}) \to (G'^{\mathrm{op}}, F'^{\mathrm{op}}, \varepsilon'^{\mathrm{op}}, \eta'^{\mathrm{op}})). \tag{7.19}$$

In the same way (except for the involution) one can construct a 2-category AdjC of *internal adjunctions* in an arbitrary 2-category C.

7.2 Double categories

Strict double categories were introduced and studied by C. Ehresmann and Andrée C. Ehresmann [Eh1, Eh2, Eh3, Eh4, BasE, EE] since 1962; the weak notion much later, in [GP1]–[GP4]; many other references can be found in these articles.

The strict case extends the more usual notion of 2-category, while the weak one extends bicategories. The extension is made clear in 7.2.5.

Applications of double categories in Computer Science can be found, for instance, in [BruMM]. Applications in Homological Algebra will be presented in Sections 7.4 and 7.5.

7.2.1 Definition

A (strict) *double category* $\mathbb{A}$ consists of the following structure.

(a) A set $Ob\mathbb{A}$ of *objects* of $\mathbb{A}$.

(b) *Horizontal morphisms* $f: X \to X'$ between the previous objects; they form the category $Hor_0\mathbb{A}$ *of the objects and horizontal maps* of $\mathbb{A}$, with composition written as gf and identities $1_X: X \to X$.

(c) *Vertical morphisms* $u: X \to Y$ (often denoted by a dot-marked arrow) between the same objects; they form the category $Ver_0\mathbb{A}$ *of the objects and vertical maps* of $\mathbb{A}$, with composition written as $v \bullet u$ (or $u \otimes v$, in diagrammatic order) and identities written as $e_X: X \to X$ or $1_X^\bullet$.

(d) (*Double*) *cells* $a: (u\ {}^f_g\ v)$ with a *boundary* formed of two horizontal arrows f, g and two vertical arrows u, v (often marked with a dot)

$$
\begin{array}{ccc}
X & \xrightarrow{\ f\ } & X' \\
u \downarrow & a & \downarrow v \\
Y & \xrightarrow[\ g\ ]{} & Y'
\end{array}
\qquad (7.20)
$$

Writing $a: (X\ {}^X_g\ v)$ or $a: (e\ {}^1_g\ v)$ we mean that $f = 1_X$ and $u = e_X$. The cell a is also written as $a: u \to v$ (with respect to its *horizontal* domain and codomain, which are *vertical* arrows) or as $a: f \to g$ (with respect to its *vertical* domain and codomain, which are *horizontal* arrows).

We refer now to the following diagrams of cells, where the first is called a *consistent matrix* $\left(\begin{smallmatrix} a & b \\ c & d \end{smallmatrix}\right)$ of cells

$$
\begin{array}{ccc}
X \xrightarrow{\ f\ } X' \xrightarrow{\ f'\ } X'' & & \\
\Big\downarrow u \quad a \quad \Big\downarrow v \quad b \quad \Big\downarrow w & X \xrightarrow{\ 1\ } X & X \xrightarrow{\ f\ } X' \\
Y \xrightarrow{-g>} Y' \xrightarrow{-g'>} Y'' & \Big\downarrow u \quad 1_u \quad \Big\downarrow u & \Big\downarrow e \quad e_f \quad \Big\downarrow e \qquad (7.21) \\
\Big\downarrow u' \quad c \quad \Big\downarrow v' \quad d \quad \Big\downarrow w' & Y \xrightarrow{\ 1\ } Y & X \xrightarrow{\ f\ } X' \\
Z \xrightarrow{\ h\ } Z' \xrightarrow{\ h'\ } Z'' & &
\end{array}
$$

(e) Cells have a *horizontal composition*, consistent with the horizontal composition of arrows and written as

$$(a \mid b): (u \; {}^{f'f}_{g'g} \; w),$$

or $a|b$; this composition gives the category $\mathrm{Hor}_1\mathbb{A}$ *of vertical arrows and cells* $a: u \to v$ of $\mathbb{A}$, with identities $1_u: (u \; {}^1_1 \; u)$.

(f) Cells have also a *vertical composition*, consistent with the vertical composition of arrows and written as

$$\left(\frac{a}{c}\right): (u' \bullet u \; {}^f_h \; v' \bullet v),$$

or $\frac{a}{c}$, or $a \otimes c$; this composition gives the category $\mathrm{Ver}_1\mathbb{A}$ *of horizontal arrows and cells* $a: f \to g$ of $\mathbb{A}$, with identities $e_f = 1^{\bullet}_f: (e \; {}^f_f \; e)$.

(g) The two compositions satisfy the *interchange laws* (for binary and zeroary compositions), which means that we have, in diagram (7.21):

$$
\left(\frac{a \mid b}{c \mid d}\right) = \left(\frac{a}{c} \middle| \frac{b}{d}\right), \qquad \left(\frac{1_u}{1_{u'}}\right) = 1_{u' \bullet u}, \tag{7.22}
$$

$$(e_f \mid e_{f'}) = e_{f'f}, \qquad\qquad 1_{e_X} = e_{1_X}.$$

The first condition says that a consistent matrix $\left(\begin{smallmatrix} a & b \\ c & d \end{smallmatrix}\right)$ has a precise *pasting*; the last says that an object X has an *identity cell*

$$\square_X = 1_{e_X} = e_{1_X}.$$

The expressions $(a \mid f')$ and $(f \mid b)$ will stand for $(a \mid e_{f'})$ and $(e_f \mid b)$, when this makes sense.

The double category $\mathbb{A}$ is said to be *flat* if every double cell $a: (u \; {}^f_g \; v)$ is determined by its boundary – namely the arrows f, g, u, v. A standard example is the double category $\mathbb{R}\mathsf{elSet}$ of sets, mappings and relations, described below in 7.2.7.

7.2.2 Weak double categories

More generally, in a *weak double category* $\mathbb{A}$ the horizontal composition behaves categorically (and we still have ordinary categories $\mathrm{Hor}_0\mathbb{A}$ and $\mathrm{Hor}_1\mathbb{A}$), while the composition of vertical arrows is categorical up to *comparison cells*; the latter are *special* (which means that their horizontal arrows are identities) and *horizontally invertible*.

- For $u\colon X \to Y$ we have a *left unitor* and a *right unitor*

$$\lambda u\colon e_X\otimes u \to u, \qquad\qquad \rho u\colon u\otimes e_Y \to u, \qquad\qquad (7.23)$$

which are natural for a cell $a\colon (u\,{}^f_g\,v)$ in the sense that

$$(e_f\otimes a \mid \lambda v) = (\lambda u \mid a), \qquad (a\otimes e_g \mid \rho v) = (\rho u \mid a).$$

- For three consecutive vertical arrows $u\colon X \to Y$, $v\colon Y \to Z$ and $w\colon Z \to T$ we have an *associator*

$$\kappa(u,v,w)\colon u\otimes(v\otimes w) \to (u\otimes v)\otimes w, \qquad\qquad (7.24)$$

natural for cells $a\colon (u\,{}^f_g\,u')$, $b\colon (v\,{}^g_h\,v')$, $c\colon (w\,{}^h_k\,w')$, in the sense that

$$(a\otimes(b\otimes c) \mid \kappa(u',v',w')) = (\kappa(u,v,w) \mid (a\otimes b)\otimes c).$$

Interchange between horizontal and vertical composition of cells holds strictly, as above. *Finally, the comparison cells* λ, ρ, κ *are assumed to satisfy coherence axioms*, listed below.

$\mathbb{A}$ is said to be *unitary* if the unitors are identities, so that the vertical identities behave as strict units: a constraint which in concrete cases can often be easily met.

The terminology of the strict case is extended to the weak one, as far as possible.

7.2.3 Coherence conditions

In a weak double category the following diagrams of cells must commute under horizontal composition.

(a) *Coherence pentagon of the associator* κ, on four consecutive vertical arrows x, y, z, u

$$x\otimes(y\otimes(z\otimes u)) \qquad\qquad ((x\otimes y)\otimes z)\otimes u \qquad (7.25)$$

(b) *Coherence conditions for the unitors*, for $x\colon A \to B$ and $y\colon B \to C$

$$x\otimes(e_B\otimes y) \xrightarrow{\ \kappa\ } (x\otimes e_B)\otimes y$$
$$1\otimes\lambda \searrow \quad \swarrow \rho\otimes1$$
$$x\otimes y$$
$$(7.26)$$

$$e_A\otimes(x\otimes y) \xrightarrow{\ \kappa\ } (e_A\otimes x)\otimes y$$
$$\lambda \searrow \quad \swarrow \lambda\otimes1$$
$$x\otimes y$$
$$(7.27)$$

$$x\otimes(y\otimes e_C) \xrightarrow{\ \kappa\ } (x\otimes y)\otimes e_C$$
$$1\otimes\rho \searrow \quad \swarrow \rho$$
$$x\otimes y$$
$$(7.28)$$

$$\lambda(eA) = \rho(eA)\colon eA\otimes eA \to eA. \tag{7.29}$$

We write the conditions (b) in the form used by Mac Lane in his classical paper on coherence of monoidal categories [M3]. As proved by Kelly [Kl1], this set of axioms is redundant: properties (7.25) and (7.26) imply the other three.

This is also true in the present more general case. Listing all the previous conditions has nevertheless some interest: there are defective structures where the unitors are *not* invertible and the redundancy recalled above disappears.

7.2.4 Lax functors and horizontal transformations

A *lax (double) functor* $F\colon \mathbb{X} \to \mathbb{A}$ between weak double categories amounts to assigning:

(a) two functors $\mathrm{Hor}_0 F$ and $\mathrm{Hor}_1 F$, consistent with domain and codomain

$$
\begin{array}{ccc}
\mathrm{Hor}_1\mathbb{X} \xrightarrow{\mathrm{Hor}_1 F} \mathrm{Hor}_1\mathbb{A} \\
\mathrm{Dom}\downarrow \qquad \downarrow\mathrm{Dom} \\
\mathrm{Hor}_0\mathbb{X} \xrightarrow[\mathrm{Hor}_0 F]{} \mathrm{Hor}_0\mathbb{A}
\end{array}
\qquad
\begin{array}{ccc}
\mathrm{Hor}_1\mathbb{X} \xrightarrow{\mathrm{Hor}_1 F} \mathrm{Hor}_1\mathbb{A} \\
\mathrm{Cod}\downarrow \qquad \downarrow\mathrm{Cod} \\
\mathrm{Hor}_0\mathbb{X} \xrightarrow[\mathrm{Hor}_0 F]{} \mathrm{Hor}_0\mathbb{A}
\end{array}
\tag{7.30}
$$

(b) for any object X in $\mathbb{X}$ a special cell, the *identity comparison* of F

$$\underline{F}(X)\colon e_{FX} \to Fe_X\colon FX \to FX,$$

(c) for any vertical composite $u \otimes v \colon X \to Y \to Z$ in $\mathbb{X}$ a special cell, the *composition comparison*

$$\underline{F}(u, v) \colon Fu \otimes Fv \to F(u \otimes v) \colon FX \to FZ.$$

Again, these comparisons must satisfy axioms of naturality and coherence with the comparisons of $\mathbb{X}$ and $\mathbb{A}$ (cf. [GP2]).

A *colax (double) functor* has comparisons in the opposite direction. A *pseudo (double) functor* is a lax functor whose comparison cells are horizontally invertible; it is made colax by the inverse cells.

A *horizontal transformation of lax functors* $h \colon F \to G \colon \mathbb{X} \to \mathbb{A}$ has the following components:

(a) a horizontal map $hX \colon FX \to GX$ in $\mathbb{A}$, for each object X in $\mathbb{X}$,

(b) a cell $hu \colon (Fu \begin{smallmatrix} hX \\ hY \end{smallmatrix} Gu)$ for each vertical map $u \colon X \to Y$ in $\mathbb{X}$,

under conditions of naturality and coherence with the comparisons of F and G (cf. [GP2]).

7.2.5 Dualities

A weak double category has a *horizontal opposite* $\mathbb{A}^h$ (reversing the horizontal direction) and a *vertical opposite* $\mathbb{A}^v$ (reversing the vertical direction). A strict structure also has a *transpose* $\mathbb{A}^t$ (interchanging the horizontal and vertical issues).

The prefix 'co', as in *colimit, coequaliser* or *colax double functor*, refers to horizontal duality, the main one.

Let us note that a weak double category whose horizontal arrows are identities is the same as a *bicategory written in vertical* (as in the right diagram of (7.1)), with arrows and weak composition in the vertical direction and strict composition of cells in the horizontal one. This is why *the op*lax *functors of bicategories correspond here to* co*lax double functors.*

(Transposing the theory of double categories, as is done in some papers, would avoid this conflict of terminology, but would produce other conflicts at a more basic level: for instance, colimits in Set would become 'op-limits' in $\mathbb{R}$elSet and $\mathbb{S}$panSet.)

7.2.6 Spans and cospans

For a category C with (a fixed choice of) pullbacks there is a weak double category $\mathbb{S}$pan(C) *of spans* over C.

Objects, horizontal arrows and their composition come from C, so that $\mathsf{Hor}_0(\mathbb{S}\mathsf{pan}\mathsf{C}) = \mathsf{C}$.

A vertical arrow $u\colon X \to Y$ is a span $u = (u', u'')$, i.e. a diagram

$$X \leftarrow U \to Y$$

in C, or equivalently a functor $u\colon \mathsf{V} \to \mathsf{C}$ defined on the formal-span category $\bullet \leftarrow \bullet \to \bullet$. A vertical identity is a pair $e_X = (1_X, 1_X)$.

A cell $\sigma\colon (u \,{}^f_g\, v)$ is a natural transformation $u \to v$ of such functors and amounts to the commutative left diagram below

$$
\begin{array}{ccc}
X \xrightarrow{\ f\ } X' & \qquad X \longrightarrow X' \longrightarrow X'' \\
\end{array}
$$

$$\text{(7.31)}$$

We say that the cell σ is *represented* by its middle arrow $m\sigma\colon U \to V$, which determines it together with the boundary (the present structure is not flat).

The horizontal composition $\sigma \mid \tau$ of σ with a second cell $\tau\colon v \to w$ is a composition of natural transformations as in the right diagram above; it gives the category $\mathsf{Hor}_1(\mathbb{S}\mathsf{pan}\mathsf{C}) = \mathsf{Cat}(\mathsf{V}, \mathsf{C})$.

The vertical composition $u \otimes v$ of spans is computed by (chosen) pullbacks in C

$$W = U \times_Y V. \qquad \text{(7.32)}$$

This composition is extended to double cells, in the obvious way. For the sake of simplicity *we make* $\mathbb{S}\mathsf{pan}(\mathsf{C})$ *unitary*, by adopting the 'unit constraint' for pullbacks (as in 2.4.6): the chosen pullback of an identity along any morphism is an identity. The associator κ is determined by the universal property of pullbacks.

Dually, for a category C with (a fixed choice of) pushouts there is a unitary weak double category $\mathbb{C}\mathsf{osp}(\mathsf{C})$ *of cospans* over C, that is horizontally dual to $\mathbb{S}\mathsf{pan}(\mathsf{C}^{\mathrm{op}})$. We have now

$$\mathsf{Hor}_0(\mathbb{C}\mathsf{osp}\mathsf{C}) = \mathsf{C}, \qquad \mathsf{Hor}_1(\mathbb{C}\mathsf{osp}\mathsf{C}) = \mathsf{Cat}(\wedge, \mathsf{C}), \qquad \text{(7.33)}$$

where $\wedge = \mathsf{V}^{\mathrm{op}}$ is the formal-cospan category $\bullet \to \bullet \leftarrow \bullet$.

A vertical arrow $u = (u', u'') \colon \wedge \to \mathsf{C}$ is a cospan, i.e. a diagram $X \to U \leftarrow Y$ in C, and a cell $\sigma \colon u \to v$ is a natural transformation of such functors. Their vertical composition is computed with pushouts in C; again, we generally follow the 'unit constraint' for pushouts.

We write as $\mathrm{Span}(\mathsf{C})$ and $\mathrm{Cosp}(\mathsf{C})$ the bicategories obtained from the previous weak double categories, by restricting horizontal arrows to identities.

7.2.7 Other examples

We briefly recall other strict or weak double categories. The interested reader will find the details in [GP1].

(a) The ordered category Rel Set of sets and relations has been introduced in 1.8.3. We now form a flat double category $\mathbb{A} = \mathbb{R}\mathrm{elSet}$, with $\mathrm{Hor}_0\mathbb{A} = \mathrm{Set}$, $\mathrm{Ver}_0\mathbb{A} = \mathrm{Rel\,Set}$ and double cells of the following kind

$$
\begin{array}{ccc}
X & \xrightarrow{\ f\ } & X' \\
u \downarrow & \leqslant & \downarrow v \\
Y & \xrightarrow[g]{} & Y'
\end{array}
\qquad gu \leqslant vf \quad (\text{in Rel Set}). \qquad (7.34)
$$

Double cells are determined by their boundary; their horizontal and vertical composition amount to composing mappings or relations, respectively.

Similarly, one can form a double category $\mathbb{R}\mathrm{el}(\mathsf{C})$ over any regular category, 'pasting' C with the ordered category $\mathrm{Rel}(\mathsf{C})$ constructed in Section 4.5. Or over any p-exact category, using the ordered category constructed in Section 6.6; this structure will be used in Section 7.5.

(b) The ordered category AdjOrd of ordered sets and Galois connections has been introduced in 1.7.4. We now form a flat double category $\mathbb{A} = \mathrm{AdjOrd}$, with $\mathrm{Hor}_0\mathbb{A} = \mathrm{Ord}$, $\mathrm{Ver}_0\mathbb{A} = \mathrm{AdjOrd}$ and double cells of the following kind

$$
\begin{array}{ccc}
X & \xrightarrow{\ f\ } & X' \\
u \downarrow & \downarrow & \downarrow v \\
Y & \xrightarrow[g]{} & Y'
\end{array}
\qquad
\begin{aligned}
v_\bullet f &\leqslant g u_\bullet \colon X \to Y' \quad (\text{in Ord}), \\[4pt]
f u^\bullet &\leqslant v^\bullet g \colon Y \to X' \quad (\text{in Ord}).
\end{aligned}
\qquad (7.35)
$$

(The two conditions are easily seen to be equivalent.) Again, the horizontal and vertical composition amount to composing horizontal or vertical arrows.

We are also interested in the double subcategory $\mathrm{Adj}_0\mathrm{Ord}$ where we restrict the double cells to the *bicommutative* ones, which here means that $v_\bullet f = g u_\bullet$ *and* $f u^\bullet = v^\bullet g$.

(c) More generally there is a double category AdjCat of (small) *categories, functors and adjunctions.*

Here $\mathrm{Hor}_0(\mathsf{AdjCat}) = \mathsf{Cat}$ and $\mathrm{Ver}_0(\mathsf{AdjCat}) = \mathsf{AdjCat}$, the category of adjunctions introduced in 3.2.1. A vertical arrow is thus an adjunction, conventionally directed as the *left* adjoint

$$u = (u_\bullet, u^\bullet, \eta, \varepsilon)\colon X \to Y, \qquad (u_\bullet\colon X \to Y) \dashv (u^\bullet\colon Y \to X),$$
$$\eta\colon 1_X \to u^\bullet u_\bullet, \qquad\qquad\qquad \varepsilon\colon u_\bullet u^\bullet \to 1_Y. \tag{7.36}$$

A double cell $a = (a_\bullet, a^\bullet)$ is a pair of *mate* natural transformations (cf. 7.1.6)

$$
\begin{array}{ccc}
X & \xrightarrow{\;f\;} & X' \\
u \downarrow & \downarrow a & \downarrow v \\
Y & \xrightarrow{\;g\;} & Y'
\end{array}
\qquad
\begin{array}{l}
a_\bullet\colon v_\bullet f \to g u_\bullet\colon X \to Y', \\[2mm]
a^\bullet\colon f u^\bullet \to v^\bullet g\colon Y \to X',
\end{array}
\tag{7.37}
$$

Each of them determines the other via the units and counits of the adjunctions u and v

$$a^\bullet = (f u^\bullet \to v^\bullet v_\bullet f u^\bullet \to v^\bullet g u_\bullet u^\bullet \to v^\bullet g),$$
$$a_\bullet = (v_\bullet f \to v_\bullet f u^\bullet u_\bullet \to v_\bullet v^\bullet g u_\bullet \to g u_\bullet). \tag{7.38}$$

(d) This can be further extended to a double category AdjC, where C is any 2-category.

(e) Profunctors, or distributors, form a bicategory [Be3, Lw1]. The weak double category Cat *of categories, functors and profunctors* was introduced in [GP1], Section 3.1. Objects are small categories, a horizontal arrow is a functor and a vertical arrow is a *profunctor* $u\colon X \to Y$, defined as a functor $u\colon X^{\mathrm{op}} \times Y \to \mathsf{Set}$. A cell $a\colon (u\, {}^f_g\, v)$ is a natural transformation $a\colon u \to v(f^{\mathrm{op}} \times g)\colon X^{\mathrm{op}} \times Y \to \mathsf{Set}$.

Compositions and comparisons can be seen *loc. cit.*

7.2.8 Vertical involutions

As we have hinted at in 1.8.1, the (covariant) reversor of Ord and the (contravariant) involution of AdjOrd have the same action $X \mapsto X^{\mathrm{op}}$ on the objects, and can be combined to form a *vertical involution* of the double category AdjOrd: this is horizontally covariant (on increasing mappings) and vertically contravariant (on adjunctions).

More precisely, we have an involutive double functor defined on the vertical dual $\mathbb{AdjOrd}^v$

$$(-)^{\mathrm{op}} \colon \mathbb{AdjOrd}^v \to \mathbb{AdjOrd}, \qquad X \mapsto X^{\mathrm{op}}, \qquad (7.39)$$

which acts as follows on the arrows and the (flat) double cells

$$
\begin{array}{ccc}
\begin{array}{ccc}
X & \xrightarrow{f} & X' \\
u\downarrow & \downarrow & \downarrow v \\
Y & \xrightarrow{g} & Y'
\end{array}
& \mapsto &
\begin{array}{ccc}
Y^{\mathrm{op}} & \xrightarrow{f^{\mathrm{op}}} & Y'^{\mathrm{op}} \\
u^{\mathrm{op}}\downarrow & \downarrow & \downarrow v^{\mathrm{op}} \\
X^{\mathrm{op}} & \xrightarrow{g^{\mathrm{op}}} & X'^{\mathrm{op}}
\end{array}
\end{array}
\qquad (7.40)
$$

On a strict double category $\mathbb{A}$, a *vertical involution* will thus be an involutive double functor $\mathbb{A}^v \to \mathbb{A}$.

This also exists in $\mathbb{RelSet}$. More generally, we have a vertical involution in $\mathbb{AdjC}$, where C is any 2-category, and in $\mathbb{Rel}(\mathsf{C})$, where C is a regular or p-exact category.

For a weak double category $\mathbb{A}$, we must relax the previous procedure: one generally encounters a *weak vertical involution*, defined as an involutive *pseudo* double functor $\mathbb{A}^v \to \mathbb{A}$.

In fact, let us consider the weak double category $\mathbb{Span}(\mathsf{C})$ of 7.2.6, where vertical arrows are spans composed with a choice of pullbacks. We cannot assume that this choice is symmetric, as already remarked in 2.3.2(c). Therefore, the procedure of reversing a span is only consistent with composition *up to isomorphism of spans* (contravariantly, of course).

Similarly, the weak double category $\mathbb{Cat}$ of categories, functors and profunctors has a weak vertical involution $\mathsf{C} \mapsto \mathsf{C}^{\mathrm{op}}$.

7.3 Enriched, internal and ordered categories

The classical references for enriched categories are the article [EiK], by Eilenberg and Kelly, and Kelly's book [Kl2]. Categories internal to a category were introduced by C. Ehresmann [Eh3]; see also [St3].

7.3.1 Enriched categories

In an additive category C, like $R\,\mathsf{Mod}$ and Ban, or more generally in a preadditive one, each hom-set $\mathsf{C}(X, Y)$ is equipped with a structure of abelian group, so that composition is bilinear over $\mathbb{Z}$.

We show now that, if C has small hom-sets, this amounts to saying that C is 'enriched' over Ab, with respect to the symmetric monoidal structure of its tensor product.

In general, let $V = (V, \otimes, E)$ be a symmetric monoidal category, whose comparisons will be left understood. A V-*enriched category*, or V-*category*, C consists of:

(a) a set ObC, of objects,

(b) for every pair X, Y in ObC, a V-object of *morphisms* $C(X, Y)$,

(c) for every $X \in$ ObC, a V-morphism of *identity* $\mathrm{id}_X : E \to C(X, X)$.

(d) for every triple X, Y, Z in ObC, a V-morphism of *composition*

$$k : C(X, Y) \otimes C(Y, Z) \to C(X, Z).$$

These data are to satisfy three axioms of unitarity and associativity. Namely, the following diagrams must commute, for all X, Y, Z, U in ObC

$$
\begin{array}{ccc}
E \otimes C(X, Y) & \xrightarrow{\mathrm{id} \otimes 1} & C(X, X) \otimes C(X, Y) \\
 & \searrow{\scriptstyle 1} & \downarrow{\scriptstyle k} \\
 & & C(X, Y)
\end{array}
\tag{7.41}
$$

$$
\begin{array}{ccc}
C(X, Y) \otimes E & \xrightarrow{1 \otimes \mathrm{id}} & C(X, Y) \otimes C(Y, Y) \\
 & \searrow{\scriptstyle 1} & \downarrow{\scriptstyle k} \\
 & & C(X, Y)
\end{array}
\tag{7.42}
$$

$$
\begin{array}{ccc}
C(X, Y) \otimes C(Y, Z) \otimes C(Z, U) & \xrightarrow{1 \otimes k} & C(X, Y) \otimes C(Y, U) \\
{\scriptstyle k \otimes 1} \downarrow & & \downarrow{\scriptstyle k} \\
C(X, Z) \otimes C(Z, U) & \xrightarrow{k} & C(X, U)
\end{array}
\tag{7.43}
$$

An enriched category has an *underlying category* $|C|$, with the same objects and hom-sets constructed with the canonical forgetful functor $U = V(E, -) : V \to$ Set (not necessarily faithful, see (3.32))

$$|C|(X, Y) = U(C(X, Y)) = V(E, C(X, Y)). \tag{7.44}$$

The identity of X in $|C|$ is precisely $\mathrm{id}_X : E \to C(X, X)$. Given two consecutive morphisms $f \in |C|(X, Y)$, $g \in |C|(Y, Z)$ of the underlying category, their composite gf is defined by the canonical mapping φ of (3.33)

$$gf = (Uk).\varphi(f, g), \tag{7.45}$$

$$gf : |C|(X, Y) \times |C|(Y, Z) \to U(C(X, Y) \otimes C(Y, Z)) \to |C|(X, Z).$$

7.3.2 Enriched functors and transformations

A V-*functor* $F: C \to D$ between V-categories consists of:

(a) a mapping $F_0: ObC \to ObD$ whose action is written as $X \mapsto F(X)$,

(b) for every pair X, Y in ObC, a V-morphism

$$F_{XY}: C(X,Y) \to D(FX, FY)$$

(which will be written as $f \mapsto F(f)$ *when* V *is a category of structured sets*).

The following diagrams are assumed to commute, for all X, Y, Z in ObC

$$
\begin{array}{ccc}
E & \xrightarrow{\ \ \mathrm{id}\ \ } & C(X,X) \\
& \searrow{\scriptstyle \mathrm{id}} & \downarrow{\scriptstyle F} \\
& & D(FX, FX)
\end{array}
\qquad (7.46)
$$

$$
\begin{array}{ccc}
C(X,Y) \otimes C(Y,Z) & \xrightarrow{\ k\ } & C(X,Z) \\
{\scriptstyle F \otimes F}\downarrow & & \downarrow{\scriptstyle F} \\
D(FX, FY) \otimes D(FY, FZ) & \xrightarrow[k]{} & D(FX, FZ)
\end{array}
\qquad (7.47)
$$

A V-*transformation* $\varphi: F \to G: C \to D$ between V-functors consists of a family of V-morphisms

$$\varphi X: E \to D(FX, GX) \qquad\qquad (X \in ObC),$$

making the following diagrams commutative

$$
\begin{array}{ccccc}
E \otimes C(X,Y) & \xrightarrow{\varphi X \otimes G} & D(FX, GX) \otimes D(GX, GY) & \xrightarrow{\ k\ } & D(FX, GY) \\
\| & & & & \| \\
C(X,Y) \otimes E & \xrightarrow[F \otimes \varphi Y]{} & D(FX, FY) \otimes D(FY, GY) & \xrightarrow[k]{} & D(FX, GY)
\end{array}
$$
$$(7.48)$$

Appropriate compositions give the 2-category V-Cat of V-categories (see [Kl2]).

Again, a V-functor has an underlying functor and a V-transformation has an underlying natural transformation, so that we have a 2-functor

$$|-|: \text{V-Cat} \to \text{Cat}.$$

7.3.3 Exercises and complements

The reader should examine and complete the following examples. They are based on the cartesian structure of the categories Set, Cat, Ord, **2**, and the

symmetric monoidal structure of the categories Ab and Abm. In each case
the enriched functors and their transformations are the obvious ones.

(a) A Set-enriched category is the same as an ordinary category with small
hom-sets.

(b) A Cat-enriched category is the same as a 2-category with small hom-
categories.

(c) An Ord-enriched category is an ordered category, as defined in 1.8.2,
with small hom-sets. (A different notion of 'ordered category' is recalled
in 7.3.6.)

(d) The arrow category **2** is equivalent to the subcategory of Set consisting
of all sets of cardinality $\leqslant 1$. It has all limits, and an obvious cartesian
structure. A **2**-enriched category is a preordered set, possibly large.

(e) Let C be a *preadditive category* (see 6.4.1) with small hom-sets: as
we have already recalled at the beginning of this section, this amounts to
saying that C is enriched over Ab.

Completing some remarks in 1.2.1, every object X of C has a (possibly
large, unitary) ring $\mathrm{End}(X)$. A ring is essentially the same as a preadditive
category on a single object, while a preadditive category can be thought to
be a 'multi-object generalisation' of a ring.

(f) Similarly an N-linear category (see 6.4.1) with small hom-sets is enriched
over Abm, the symmetric monoidal category of abelian monoids, with tensor
product over N defined 'as' in Ab. An N-linear category is a 'multi-object
generalisation' of a semiring.

(g) A symmetric monoidal closed category C is canonically enriched on
itself, by its internal hom $\mathrm{Hom}(X, Y)$, with $\mathrm{id}_X \colon E \to \mathrm{Hom}(X, X)$ corresponding to the unitor $\lambda X \colon E \otimes X \to X$.

Examining under this aspect the symmetric monoidal closed structure of
Ban_1 reviewed in 3.4.7(h), it is important to note that the enrichment of
Ban_1 on itself does not mean to provide the set $\mathrm{Ban}_1(X, Y)$ with a structure
of Banach space (!), *but to introduce a Banach space* $\mathrm{Hom}(X, Y)$ *whose unit
ball is that set.* The 'naive view' of enriching a given ordinary category by
providing *its hom-sets* of a V-structure is only adequate when the canonical
forgetful functor of V does give (what we view as) the underlying set of any
V-object. The next example is even farer from this naive view.

7.3.4 *Lawvere metric spaces*

The strict monoidal category $\mathbb{R}_+$ was introduced by Lawvere [Lw1] to formalise 'generalised metric spaces' as enriched categories. An object of $\mathbb{R}_+$

is an 'extended' positive real number $\lambda \in [0, \infty]$, an arrow $\lambda \to \mu$ is given by the order relation $\lambda \geqslant \mu$; the tensor product is given by the ordinary sum $\lambda + \lambda'$.

A category enriched over $\mathbb{R}_+$ is a set X with *hom-values*

$$\delta(x, y) = X(x, y),$$

satisfying the axioms:

$$\delta(x, y) \in [0, \infty],$$
$$\delta(x, x) = 0, \qquad \delta(x, y) + \delta(y, z) \geqslant \delta(x, z). \tag{7.49}$$

Note that 'composition' becomes the triangular inequality, and symmetry is not required (while it was for an extended semimetric, in 5.1.7).

We call this structure X a *directed metric space*, or a *δ-metric space*.

A morphism $f \colon X \to Y$ is a weak contraction, satisfying

$$\delta(x, x') \geqslant \delta(f(x), f(x')),$$

for all $x, x' \in X$. This forms the category $\delta\mathsf{Mtr} = \mathbb{R}_+\mathsf{Cat}$.

(All this leads to the important notion of Cauchy-completeness of enriched categories, see [Lw1].)

In the context of Directed Algebraic Topology, $\delta\mathsf{Mtr}$ is studied in [G8] as a category of 'directed spaces'. The reversor $R \colon \delta\mathsf{Mtr} \to \delta\mathsf{Mtr}$ sends the space X to the *opposite δ-metric space* $R(X) = X^{\mathrm{op}}$, with opposite δ-metric, $\delta^{\mathrm{op}}(x, y) = \delta(y, x)$.

The reader is invited to compute the category underlying a δ-metric space X.

7.3.5 Internal categories

(a) Let us begin by remarking that a small category C can be described as a diagram in Set having the form of a 3-truncated simplicial set (see Section 5.3)

$$C_0 \;\underset{e_0}{\overset{\partial_i}{\rightleftarrows}}\; C_1 \;\underset{e_i}{\overset{\partial_i}{\rightleftarrows}}\; C_2 \;\underset{e_i}{\overset{\partial_i}{\rightleftarrows}}\; C_3 \tag{7.50}$$

Here $C_0 = \mathrm{Ob}\mathsf{C}$ is the (small) set of objects of C, $C_1 = \mathrm{Mor}\mathsf{C}$ is the set of morphisms, C_2 (resp. C_3) is the set of consecutive pairs (resp. triples) of morphisms.

The faces ∂_i and the degeneracies e_i are defined as follows (on an object

$X \in C_0$, an arrow $f \in C_1$, a pair $(f, g) \in C_2$ and a triple $(f, g, h) \in C_3$):

$$e_0(X) = \mathrm{id}_X, \qquad \partial_0(f) = \mathrm{Dom}\,(f), \qquad \partial_1(f) = \mathrm{Cod}\,(f),$$

$$e_0(f) = (f, 1), \qquad e_1(f) = (1, f).$$

$$\partial_0(f, g) = f, \qquad \partial_1(f, g) = gf, \qquad \partial_2(f, g) = g,$$

$$e_0(f, g) = (f, g, 1), \qquad e_1(f, g) = (f, 1, g), \qquad e_2(f, g) = (1, f, g),$$

$$\partial_0(f, g, h) = (f, g), \qquad \partial_1(f, g, h) = (f, hg), \qquad \partial_2(f, g, h) = (gf, h),$$

$$\partial_3(f, g, h) = (g, h).$$

The simplicial identities hold, so that (7.50) is indeed a 3-truncated simplicial set:

$$\partial_i \partial_j = \partial_{j-1} \partial_i \ \text{ for } i < j, \qquad e_j e_i = e_i e_{j-1} \ \text{ for } i < j,$$

$$\partial_i e_j = e_{j-1} \partial_i \ \text{ for } i < j, \qquad \partial_i e_j = e_j \partial_{i-1} \ \text{ for } i > j+1, \qquad (7.51)$$

$$\partial_i e_j = \mathrm{id} \ \text{ for } i = j, j+1.$$

Finally, the following two squares are pullbacks (of sets)

$$
\begin{array}{ccc}
C_2 \xrightarrow{\ \partial_0\ } C_1 & \qquad & C_3 \xrightarrow{\ \partial_0\ } C_2 \\
{\scriptstyle \partial_2}\downarrow \qquad \downarrow{\scriptstyle \partial_1} & & {\scriptstyle \partial_3}\downarrow \qquad \downarrow{\scriptstyle \partial_2} \\
C_1 \xrightarrow[\ \partial_0\ ]{} C_0 & & C_2 \xrightarrow[\ \partial_0\ ]{} C_1
\end{array}
\qquad (7.52)
$$

The morphism $m = \partial_1 : C_2 \to C_1$ is called the *composition morphism* of C, or *partial multiplication*.

(b) Replacing Set with an arbitrary category X, an *internal category* C in X, or a *category object* in C, consists of a diagram (7.50) in X where $C_0 = \mathrm{Ob}\,C$ is called the *object of objects* of C, $C_1 = \mathrm{Mor}\,C$ is called the *object of morphisms* and C_2 is called the *object of consecutive pairs of morphisms*.

The following (redundant) axioms must be satisfied:

(i) the simplicial identities (7.51) hold, so that (7.50) is a 3-truncated simplicial object in X,

(ii) the square diagrams of (7.52) are pullbacks in X.

(Note that the category X is not required to have all pullbacks: this would exclude cases of interest.)

In a category with finite limits, an internal monoid with respect to the cartesian product, as defined in 3.4.8, is the same as an internal category where $\mathrm{Ob}\,C$ is the terminal object.

(c) An *internal functor* $F : C \to D$ between internal categories in X is defined as a morphism of 3-simplicial objects. In other words we have four arrows in X

$$F_i \colon C_i \to D_i \qquad (i = 0, ..., 3), \qquad (7.53)$$

that commute with faces and degeneracies. The components F_2, F_3 are determined by the pullback condition, so that for $X = \mathsf{Set}$ we simply have

$$F_2(f, g) = (F_1 f, F_1 g), \qquad F_3(f, g, h) = (F_1 f, F_1 g, F_1 h).$$

The composition of internal functors is obvious.

(d) An *internal transformation* $\varphi \colon F \to G \colon \mathsf{C} \to \mathsf{D}$ between internal functors in X is given (or represented) by an X-morphism $\hat{\varphi}$ satisfying the following conditions

$$\hat{\varphi} \colon C_0 \to D_1, \qquad \partial_0.\hat{\varphi} = F_0, \qquad \partial_1.\hat{\varphi} = G_0,$$
$$m.\langle F_1, \hat{\varphi}\partial_1 \rangle = m.\langle \hat{\varphi}\partial_0, G_1 \rangle. \qquad (7.54)$$

For the last condition (of *naturality*), $m = \partial_1 \colon D_2 \to D_1$ is the partial multiplication of D. The morphism $\langle F_1, \hat{\varphi}\partial_1 \rangle \colon C_1 \to D_2$ with values in the pullback D_2 is well defined because

$$\partial_1 F_1 = F_0 \partial_1 = \partial_0 \, \hat{\varphi} \, \partial_1.$$

Similarly $\langle \hat{\varphi}\partial_0, G_1 \rangle \colon C_1 \to D_2$ is legitimate because

$$\partial_1 \, \hat{\varphi} \, \partial_0 = G_0 \partial_0 = \partial_0 G_1.$$

The interested reader can verify that there is a 2-category $\mathsf{Cat}(X)$ of internal categories, internal functors and internal transformations in X. The vertical composition of internal transformations and their whisker composition with internal functors are defined as follows

$$(\psi.\varphi)\hat{} = m.\langle \hat{\varphi}, \hat{\psi} \rangle \colon C_0 \to D_1, \qquad \text{for } \psi \colon G \to H \colon \mathsf{C} \to \mathsf{D},$$
$$(M\varphi L)\hat{} = M_1.\hat{\varphi}.L_0 \colon C_0' \to D_1', \quad \text{for } L \colon \mathsf{C}' \to \mathsf{C}, \ M \colon \mathsf{D} \to \mathsf{D}. \qquad (7.55)$$

7.3.6 Exercises and complements

(a) We have already seen that an internal category in Set is a small category.

(b) An internal category in Cat is a small double category, as defined in 7.2.1. Internal functors are the double functors; internal transformations are the horizontal ones (or the vertical ones, according to the way we are 'presenting' the double category).

*(c) We say that a category C is *internally ordered* when the sets $\mathrm{Ob}\mathsf{C}$ and $\mathrm{Mor}\mathsf{C}$ are equipped with consistent order relations (respected by domain,

codomain, identity and composition). If C is small, this the same as an internal category in Ord. (An internally ordered category with a discrete order on the set of objects is the same as an ordered category in the present enriched sense: see 1.8.2.)

This notion was extensively used by C. Ehresmann for his theory of 'pseudogroups' in Differential Geometry. But let us recall that an 'ordered category' in the sense of Ehresmann is also assumed to satisfy the following condition: the order induced on each hom-set $C(X, Y)$ is discrete (see [Eh4], p. 711).

The main example of such a structure is the category Set ordered by (weak) inclusion, of objects ($X \subset X'$) and morphisms: the mapping $f\colon X \to Y$ being *included* in $f'\colon X' \to Y'$ if

$$X \subset X', \quad Y \subset Y', \qquad f(x) = f'(x) \quad \text{for } x \in X.$$

(Indeed, if this holds with $X = X'$ and $Y = Y'$, then $f = f'$).

This example, as many others used by Ehresmann, rests heavily on set-theory. One can replace this internally ordered category by the category $\mathcal{S}$ of sets and partial mappings (see 1.8.6), ordered in the present sense, where inclusion is only used between subsets of given sets.

Similarly, the ordered category $\mathcal{C}$ of partial continuous mappings defined on open subspaces (introduced in 1.8.8) can be used in the theory of topological manifolds, while its differential analogues can be used for differentiable manifolds. The interested reader is referred to [G2].

7.4 *Double categories of lattices in Homological Algebra

Continuing Section 6.2, we show, *in an informal way*, how double categories of lattices can be of help in Homological Algebra.

For the sake of simplicity, we still work in categories of modules, but everything can be extended to Puppe-exact categories, which include the abelian ones.

7.4.1 A double category of modular lattices

The transfer functor for subobjects of $R\,\mathsf{Mod}$

$$\mathrm{Sub}_R\colon R\,\mathsf{Mod} \to \mathsf{Mlc}, \tag{7.56}$$

has been introduced in 6.2.1, and shown to be exact in 6.2.6.

Now let $F\colon R\,\mathsf{Mod} \to S\,\mathsf{Mod}$ be an exact functor. We have seen (in 6.3.8) that it also preserves subobjects, their meets and joins. For every

R-module A we have thus a *homomorphism of lattices*, which belongs to the category Mlh of 1.2.6 (*not* to Mlc)

$$(\mathrm{Sub}_F)_A \colon \mathrm{Sub}_R(A) \to \mathrm{Sub}_S(FA), \qquad X \mapsto F(X). \tag{7.57}$$

Furthermore F preserves direct and inverse images along morphisms: for a homomorphism $f \colon A \to B$ in $R\,\mathrm{Mod}$ we have two commutative squares (of increasing mappings)

$$
\begin{array}{ccc}
\mathrm{Sub}_R(A) & \xrightarrow{\;(\mathrm{Sub}_F)_A\;} & \mathrm{Sub}_S(A) \\[2pt]
f_* \downarrow \,\uparrow f^* & & Ff_* \downarrow \,\uparrow Ff^* \\[2pt]
\mathrm{Sub}_R(B) & \xrightarrow[\;(\mathrm{Sub}_F)_B\;]{} & \mathrm{Sub}_S(B)
\end{array}
\tag{7.58}
$$

$$F(f_* X) = (Ff)_*(FX),$$
$$F(f^* Y) = (Ff)^*(FY),$$

because F preserves epi-mono factorisations and pullbacks). This diagram is a cell in the double category $\mathrm{Adj}_0\,\mathrm{Ord}$ of ordered sets, increasing mappings (as horizontal arrows), Galois connections (as vertical arrows) and bicommutative cells, defined in 7.2.7.

To express this interaction between lattice-homomorphisms and modular connections we shall use – more precisely – a double subcategory Mlhc $\subset$ $\mathrm{Adj}_0\,\mathrm{Ord}$ introduced below (in 7.4.2) and consisting of: modular lattices, homomorphisms, modular connections and bicommutative cells.

The transfer functor (7.56) can then be seen as a *vertical functor*

$$\mathrm{Sub}_R \colon R\,\mathrm{Mod} \to \mathbb{Mlhc}, \tag{7.59}$$

i.e. a double functor defined on the obvious 'vertical' double category $\mathbb{V}(R\,\mathrm{Mod})$ of modules, identities and homomorphisms.

For every exact functor $F \colon R\,\mathrm{Mod} \to S\,\mathrm{Mod}$ we have a *horizontal transformation of vertical functors*

$$\mathrm{Sub}_F \colon \mathrm{Sub}_R \to \mathrm{Sub}_S . F \colon R\,\mathrm{Mod} \to \mathbb{Mlhc}, \tag{7.60}$$

where the vertical functor $\mathrm{Sub}_S . F$ is – more precisely – the composite

$$\mathrm{Sub}_S . \mathbb{V}F \colon \ \mathbb{V}(R\,\mathrm{Mod}) \to \mathbb{V}(S\,\mathrm{Mod}) \to \mathbb{Mlhc}.$$

7.4.2 Double categories of lattices

The double categories $\mathbb{AdjOrd}$ and $\mathbb{Adj}_0\,\mathrm{Ord}$ introduced in 7.2.7 have double subcategories that are of interest here.

(a) First we have two *cellwise-full* double subcategories of AdjOrd:

- AdjLt, the double category of *lattices, homomorphisms and adjunctions*, which amalgamates the category Lth of lattices and homomorphisms with the category Ltc of lattices and Galois connections (see 6.4.5(c)),

- AdjMl $\subset$ AdjLt, the double category of *modular lattices, homomorphisms and modular connections*, which amalgamates the category Mlh of modular lattices and homomorphisms with the category Mlc of modular lattices and modular connections.

Both have double cells as in (7.35)

$$
\begin{array}{ccc}
X & \xrightarrow{f} & X' \\
u\downarrow & \downarrow & \downarrow v \\
Y & \xrightarrow{g} & Y'
\end{array}
\qquad
\begin{array}{l}
v_\bullet f \leqslant gu_\bullet \quad \text{(in Ord)}, \\
\\
(\Leftrightarrow \quad fu^\bullet \leqslant v^\bullet g).
\end{array}
\qquad (7.61)
$$

(b) But we are more interested in their double subcategories contained in $\mathrm{Adj}_0\mathrm{Ord}$ (see 7.2.7), namely

$$
\begin{aligned}
\mathbb{L}\mathrm{thc} &= \mathrm{Adj}_0\mathrm{Lt} \subset \mathrm{AdjLt}, \\
\mathbb{M}\mathrm{lhc} &= \mathrm{Adj}_0\mathrm{Ml} \subset \mathrm{AdjMl}.
\end{aligned}
\qquad (7.62)
$$

These have the same objects and arrows as in (a), but only bicommutative cells:

$$
v_\bullet f = gu_\bullet, \qquad\qquad fu^\bullet = v^\bullet g. \qquad (7.63)
$$

We have already seen above, in 7.4.1, that $\mathbb{M}$lhc comes out naturally when we want to formalise direct and inverse images of subobjects for modules (or abelian groups).

The double category $\mathbb{L}$thc plays a similar role in the much more general context of 'semiexact' and 'homological' categories, as one can see in [G10].

7.4.3 Vertical limits and colimits for adjoints

It is interesting to note (or it may be, for some reader) that the kernels and cokernels of $\mathrm{Mlc} = \mathrm{Ver}_0\mathbb{M}$lhc become vertical double (co)limits in $\mathbb{M}$lhc. These vertical (co)limits also exist in $\mathbb{L}$thc $= \mathrm{Adj}_0\mathrm{Lt}$.

To begin with, the zero-object $\{*\}$ of the category Mlc (and Ltc) becomes a vertical double limit and colimit in $\mathbb{M}$lhc (and $\mathbb{L}$thc), with uniquely determined bicommutative cells $f \to 1_{\{*\}} \to f$

$$
\begin{array}{ccc}
X & \xrightarrow{\ f\ } & X' \\
{\scriptstyle t}\downarrow & & \downarrow{\scriptstyle t} \\
\{*\} & \longrightarrow & \{*\} \\
{\scriptstyle s}\downarrow & & \downarrow{\scriptstyle s} \\
X & \xrightarrow[\ f\ ]{} & X'
\end{array}
\qquad
\begin{array}{l}
t^{\bullet}(*) = 1, \\[2mm]
s_{\bullet}(*) = 0.
\end{array}
\tag{7.64}
$$

We have also seen that the kernel and cokernel of a covariant connection $u = (u_{\bullet}, u^{\bullet})\colon X \to Y$ in the category Mlc (and Ltc), are computed as follows

$$
\begin{aligned}
m\colon &\ \downarrow(u^{\bullet}0) \to X, & m_{\bullet}(x) &= x, & m^{\bullet}(x) &= x \wedge u^{\bullet}0, \\
p\colon &\ Y \to \uparrow(u_{\bullet}1), & p_{\bullet}(y) &= y \vee u_{\bullet}1, & p^{\bullet}(y) &= y.
\end{aligned}
\tag{7.65}
$$

These become a vertical double kernel and a vertical double cokernel in Mlhc (or Lthc). Indeed, given the central bicommutative cell $a\colon (u \overset{f}{\underset{g}{\,}} v)$ below (with $gu_{\bullet} = v_{\bullet}f$, $fu^{\bullet} = v^{\bullet}g$), the vertical kernels $m = \ker(u)$ and $n = \ker(v)$ determine the upper bicommutative cell of the following diagram, where the homomorphism f' is the restriction of f (and preserves the new maxima because $f(u^{\bullet}0) = v^{\bullet}g(0) = v^{\bullet}(0)$)

$$
\begin{array}{ccc}
\downarrow(u^{\bullet}0) & \xrightarrow{\ f'\ } & \downarrow(v^{\bullet}0) \\
{\scriptstyle m}\downarrow & & \downarrow{\scriptstyle n} \\
X & \;-\,f\to\; & X' \\
{\scriptstyle u}\downarrow & & \downarrow{\scriptstyle v} \\
Y & \;-\,g\to\; & Y' \\
{\scriptstyle p}\downarrow & & \downarrow{\scriptstyle q} \\
\uparrow(u_{\bullet}1) & \xrightarrow[\ g'\ ]{} & \uparrow(v_{\bullet}1)
\end{array}
\tag{7.66}
$$

This cell is the kernel of a in $\mathsf{Ver}_1\mathsf{Mlhc}$ (or $\mathsf{Ver}_1\mathsf{Lthc}$). Dually the cokernel of a is the lower cell above, where $p = \mathrm{cok}(u)$, $q = \mathrm{cok}(v)$ and the homomorphism g' is the restriction of g.

The double category Mlhc is *vertically Puppe-exact*, in an obvious sense:

(i) the category $\mathsf{Ver}_0\mathsf{Mlhc} = \mathsf{Mlc}$ is p-exact,

(ii) the category $\mathsf{Ver}_1\mathsf{Mlhc}$ is p-exact ,

(iii) the faces and degeneracy functors between them are exact.

In fact we already know (i) and have now verified that $Ver_1\mathbb{M}lhc$ has kernels and cokernels, preserved by faces and degeneracy. Property (ii) follows now easily: the induced (bicommutative) cell from $Cok(\ker(a))$ to $Ker(\cok(a))$ is vertically invertible because its vertical arrows are.

Finally we recall that the category Ltc has a semiadditive structure, with finite biproducts (see 6.4.5). These 'limits-colimits' become vertical biproducts in $\mathbb{L}thc$. (One can prove that the double category $\mathbb{L}thc$ is *vertically homological*, extending as above the notion of homological category defined in [G10].)

7.5 *Double categories of relations in Homological Algebra

Double categories of relations are also of interest. In particular, an adjunction between abelian (or p-exact) categories has a natural extension to a colax/lax adjunction between their double categories of relations.

Double adjunctions are studied in [GP2], where the interested reader can find the complete definitions; the crucial point – as recalled in the Introduction to this chapter – is that colax and lax double functors must be kept 'in separate directions', say vertical and horizontal, because composing them would destroy their comparisons.

E, E' are always p-exact categories. This is the natural framework for the present issues; but the reader can think of abelian categories, or even of categories of modules.

7.5.1 *Double categories of relations on p-exact categories*

We have seen in 6.6.3 that the p-exact category E can be embedded in the involutive ordered category $Rel\,E$ of its relations, constructed by quaternary factorisations.

Proceeding as in 7.2.7(a) for $\mathbb{R}elSet$, we now form a flat double category $\mathbb{A} = \mathbb{R}el E$ with $Hor_0\mathbb{A} = E$, $Ver_0\mathbb{A} = Rel\,E$ and double cells of the following kind

$$
\begin{array}{ccc}
A & \xrightarrow{\;f\;} & B \\
u\downarrow & \leqslant & \downarrow v \\
A' & \xrightarrow[g]{} & B'
\end{array}
\qquad gu \leqslant vf \quad (\Leftrightarrow fu^\sharp \leqslant v^\sharp g). \qquad (7.67)
$$

This amounts to a commutative diagram of E based on quaternary factorisations, where the three dashed arrows are uniquely determined, by

induction on subobjects and quotients

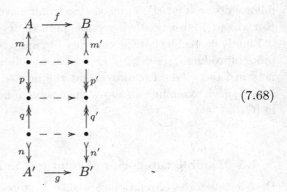

$$(7.68)$$

Equivalently, one can use a dual diagram for coquaternary factorisations.

An exact functor $F \colon \mathsf{E} \to \mathsf{E}'$ can be uniquely extended to a double functor $\mathrm{Rel}(F) \colon \mathbb{Rel}\mathsf{E} \to \mathbb{Rel}\mathsf{E}'$ that preserves involution and order. But we are interested in more general situations.

7.5.2 Left exact functors

Let us take a *left exact* functor $S \colon \mathsf{E} \to \mathsf{E}'$, i.e. a functor which preserves kernels (whence the zero-object, monomorphisms and their pullbacks).

Since the functor S preserves preimages of monomorphisms, which become bicommutative squares in $\mathbb{Rel}\mathsf{E}'$ (by 6.6.4), we can extend it to relations using – equivalently – the quaternary or coquaternary factorisation:

$$S'(nq^\sharp pm^\sharp) = S(n)(Sq)^\sharp (Sp)(Sm^\sharp),$$
$$S'(q'^\sharp n'm'^\sharp p') = (Sq')^\sharp (Sn')(Sm')^\sharp (Sp'). \qquad (7.69)$$

Note that S need *not* preserve epis, so that the right-hand parts of these formulas are *not* (co)quaternary factorisations, generally.

We obtain thus a *lax* double functor $S' = \mathbb{Rel}S \colon \mathbb{Rel}\mathsf{E} \to \mathbb{Rel}\mathsf{E}'$ (see 7.2.4). The fact that S' preserves cells follows trivially from the fact that the cell (7.67) amounts to a commutative diagram (7.68) in E, and factorises in four cells of $\mathbb{Rel}\mathsf{E}$, whose images are cells of $\mathbb{Rel}\mathsf{E}'$.

The comparison special cells for vertical composition $S'v.S'u \leqslant S'(v.u)$

are an easy consequence of the composition $v.u$ as computed in (6.63)

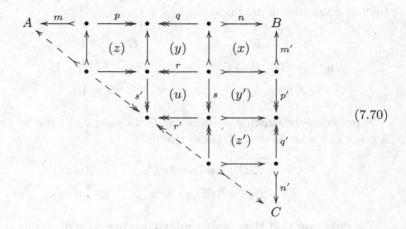

$$(7.70)$$

Indeed the functor S preserves the pullbacks of monos $(x), (z), (z')$, that are bicommutative in the category of relations, and the commutative squares $(y), (y')$. Furthermore it carries the pushout (u) to a commutative square of morphisms $Ss'.Sr = Sr'.Ss$ in E'; this gives

$$(Ss)(Sr)^\sharp \leqslant (Sr')^\sharp(Ss),$$

in $\mathrm{Rel}\,\mathsf{E}'$. It follows immediately that $S'v.S'u \leqslant S'(v.u)$.

Again, a natural transformation $\varphi\colon S \to T\colon \mathsf{E} \to \mathsf{E}'$ of left exact functors gives a horizontal transformation of lax double functors

$$\mathrm{Rel}\varphi\colon \mathrm{Rel}S \to \mathrm{Rel}T\colon \mathbb{R}\mathrm{el}\mathsf{E} \to \mathbb{R}\mathrm{el}\mathsf{E}',$$

with the same components as φ (the coherence with the comparison cells of $\mathbb{R}\mathrm{el}S$ and $\mathbb{R}\mathrm{el}T$ is here automatic, since our double categories are flat).

Dually, a *right exact* functor $S\colon \mathsf{E} \to \mathsf{E}'$ (which preserves cokernels) can be extended to a *colax* double functor $S' = \mathbb{R}\mathrm{el}S\colon \mathbb{R}\mathrm{el}\mathsf{E}' \to \mathbb{R}\mathrm{el}\mathsf{E}$, by the same definition as above, in (7.69).

Of course if S is exact we get the double functor already considered in 7.5.1.

7.5.3 Adjoints between categories of relations

Let us start from an arbitrary adjunction $F \dashv G$ between p-exact categories (or just categories of modules)

$$
\begin{aligned}
F \colon \mathsf{E} \to \mathsf{E}', && G \colon \mathsf{E}' \to \mathsf{E}, && \\
\eta \colon 1 \to GF \colon \mathsf{E} \to \mathsf{E}, && \varepsilon \colon FG \to 1 \colon \mathsf{E}' \to \mathsf{E}', && (7.71) \\
\varepsilon F.F\eta = 1_F, && G\varepsilon.\eta G = 1_G. &&
\end{aligned}
$$

F preserves the existing colimits and G the existing limits; we have thus a colax and a lax extension, respectively

$$
\begin{aligned}
F' = \mathbb{R}\mathrm{el}F \colon \mathbb{R}\mathrm{el}\mathsf{E} \to \mathbb{R}\mathrm{el}\mathsf{E}' && (colax), \\
G' = \mathbb{R}\mathrm{el}G \colon \mathbb{R}\mathrm{el}\mathsf{E}' \to \mathbb{R}\mathrm{el}\mathsf{E} && (lax).
\end{aligned} \qquad (7.72)
$$

We show now that they form a colax-lax adjunction of double categories (as defined in [GP2], Section 3).

As remarked in the Introduction to this chapter, this is not a trivial extension: the composites $G'F'$ and $F'G'$ are neither lax nor colax, generally.

The extended adjunction consists of two double cells η, ε in a suitable double category $\mathbb{D}\mathrm{bl}$ of double categories, with lax and colax double functors as horizontal or vertical arrows, respectively

$$
(7.73)
$$

that satisfy the *triangle identities* (here automatically, by flatness)

$$
\eta \otimes \varepsilon = 1_{F'}, \qquad \varepsilon \mid \eta = 1^{\bullet}_{G'}.
$$

The unit has components

$$
\eta A \colon A \to GFA, \qquad \eta u \colon u \to G'F'u,
$$

where ηA is the original one while ηu is defined below, for every vertical arrow u of $\mathbb{R}\mathrm{el}\mathsf{E}$. Their coherence conditions (cf. [GP2], Section 2) are based, separately, on the comparisons of F' and G' (but here, again, they hold automatically).

The double cell ηu of $\mathbb{R}\mathsf{el}E$ is constructed as follows

$$
\begin{array}{ccccc}
A & = & A & \xrightarrow{\ \eta A\ } & GFA & = & GFA \\
& & \uparrow{\scriptstyle m} & & \uparrow{\scriptstyle GFm} & & \\
& & \bullet \longrightarrow \bullet & & & & \\
& & \downarrow{\scriptstyle p} & & \downarrow{\scriptstyle GFp} & & \\
u\ \Big| & = & \bullet \longrightarrow \bullet & & \leqslant & G'F'u & \\
& & \uparrow{\scriptstyle q} & & \uparrow{\scriptstyle GFq} & & \\
& & \bullet \longrightarrow \bullet & & & & \\
& & \downarrow{\scriptstyle n} & & \downarrow{\scriptstyle GFn} & & \\
B & = & B & \xrightarrow{\ \eta B\ } & GFB & = & GFB
\end{array}
\tag{7.74}
$$

The right-hand inequality comes from the colax-property of G' and the definition of F'

$$
\begin{aligned}
(GFn)(GFq)^{\sharp}(GFp)(GFm^{\sharp}) &\leqslant G'((Fn)(Fq)^{\sharp}(Fp)(Fm^{\sharp})) \\
&= G'F'u.
\end{aligned}
\tag{7.75}
$$

The components of the counit are defined dually.

7.5.4 The abelian case

A left exact functor S between abelian categories preserves arbitrary pullbacks; its extension $S' = \mathbb{R}\mathsf{el}S$ can be equivalently computed over quaternary factorisations, or coquaternary factorisations, or *strong* binary factorisations, or *arbitrary* cobinary factorisations; again, *we get a* lax *double functor*, as in the more general p-exact case considered above. Dually for right exact functors.

As an example, starting from the following adjunction $F \dashv G$ of abelian groups

$$
F = - \otimes A \colon \mathsf{Ab} \to \mathsf{Ab}, \qquad G = \mathrm{Hom}(A, -) \colon \mathsf{Ab} \to \mathsf{Ab},
\tag{7.76}
$$

with $A = \mathbb{Z}/2$, both the extended functors fail to be strict (because F and G are not exact).

To show that the lax functor $G' = \mathbb{R}\mathsf{el}(G)$ is not strict we take the canonical projection $p \colon \mathbb{Z} \to \mathbb{Z}/2$; then $pp^{\sharp} = 1 \colon \mathbb{Z}/2 \to \mathbb{Z}/2$ but $G(p) = 0 \colon 0 \to \mathbb{Z}/2$, whence $(Gp)(Gp)^{\sharp} \neq G'(pp^{\sharp}) = 1$. This also shows that we cannot compute G' over the (non-strong) binary factorisation $1 = pp^{\sharp}$.

Similarly, the colax extension $F' = \mathbb{R}\mathsf{el}(F)$ is not strict. The monomorphism $m = 2.- \colon \mathbb{Z} \to \mathbb{Z}$ gives $m^{\sharp}m = 1_{\mathbb{Z}}$, but $F(m) = 0 \colon \mathbb{Z}/2 \to \mathbb{Z}/2$, whence $(Fm)^{\sharp}(Fm) \neq F'(m^{\sharp}m) = 1$.

7.6 *Factorisation systems as pseudo algebras for 2-monads

Factorisation systems, both the general ones and the proper ones (see Section 2.5), have an interesting, non obvious relationship with two-dimensional monads and their pseudo algebras (or weak algebras).

We only give some hints to these notions and this relationship; the interested reader can see the papers cited below. General pseudo algebras can be found in Street [St1].

7.6.1 A biuniversal property

The starting point is the fact that the category C^2, equipped with the factorisation system (E, M) introduced in 2.5.1(b), is the *weakly free category with factorisation system* generated by C.

This means that, with respect to the diagonal embedding

$$\eta_C \colon C \to C^2,$$
$$\eta_C(X) = 1_X, \qquad \eta_C(f \colon X \to Y) = (f, f) \colon 1_X \to 1_Y, \tag{7.77}$$

every functor $F \colon C \to D$ with values in a category with an assigned factorisation system (E', M') can be extended to a functor $G \colon C^2 \to D$ that preserves the systems (i.e. $G(E) \subset E'$, $G(M) \subset M'$); the latter is determined up to a unique functorial isomorphism.

More precisely, (C^2, η_C) is a biuniversal arrow from C to the forgetful 2-functor fsCat $\to$ Cat, defined on the obvious 2-category of categories equipped with a factorisation system (see 7.1.5).

7.6.2 A two-dimensional monad

We have seen, in 3.4.9, the structure of the arrow category **2** as an internal comonoid in Cat

$$\mathbf{1} \xleftarrow{\ e\ } \mathbf{2} \xrightarrow{\ d\ } \mathbf{2} \times \mathbf{2} \qquad d = (1,1), \tag{7.78}$$

with the diagonal d as comultiplication. By contravariant exponentials, the category Cat inherits a monad, with unit $\eta_C \colon C \to C^2$ as above

$$P \colon \mathsf{Cat} \to \mathsf{Cat}, \qquad P(C) = C^2,$$
$$\eta_C \colon C \to C^2, \qquad \mu_C \colon C^{2 \times 2} \to C^2, \tag{7.79}$$
$$\mu_C(q \colon \mathbf{2} \times \mathbf{2} \to C) = qd \colon \mathbf{2} \to C.$$

The multiplication simply turns a commutative square q of C into its diagonal qd.

(This monad is often called 'the squaring monad' on Cat, which is somewhat misleading: of course C^2 is not the cartesian power C^2. A better name might be 'the path monad', viewing 2 as the standard directed interval of Cat, and C^2 as the category of directed paths in C, as in [G8].)

(P, η, μ) is actually a 2-monad (as defined in 7.1.4) for the 2-category structure of Cat, and allows us to consider its 'pseudo algebras' (satisfying the axioms of algebras up to coherent invertible cells). It is known, since some hints in Coppey [Cop] and a full proof in Korostenski–Tholen [KrT], that the factorisation systems of C are in bijective correspondence with the pseudo-isomorphism classes of unitary pseudo algebras of P.

The idea of the proof can be described as follows: a pseudo algebra $(C, F \colon C^2 \to C)$ assigns to each morphism $f \colon X \to Y$ an object $F(f)$ (to be viewed as the *image* of f), and the factorisation of $\eta_C(f) = (f, f) \colon 1_X \to 1_Y$ in the factorisation system of C^2 (see 2.5.1(b))

$$
\begin{array}{ccccc}
X & \xrightarrow{\;1\;} & X & \xrightarrow{\;f\;} & Y \\
{\scriptstyle 1}\downarrow & & {\scriptstyle f}\downarrow & & \downarrow{\scriptstyle 1} \\
X & \xrightarrow[\;f\;]{} & Y & \xrightarrow[\;1\;]{} & Y
\end{array}
\qquad\qquad (7.80)
$$

yields a factorisation $X \to F(f) \to Y$ of f in C.

7.6.3 The case of proper factorisation systems

Proper factorisation systems on C can also be viewed as pseudo algebras for a 2-monad.

One starts from an interesting, unexpected quotient $\mathrm{Fr}(C) = C^2/R$, introduced by Freyd [Fr2] for a particular (and very interesting) goal: to embed the stable homotopy category of spaces in an abelian category.

Namely, two parallel morphisms $f = (f', f'') \colon x \to y$ and $g = (g', g'') \colon x \to y$ of C^2 (i.e. commutative squares of C, written as in 2.5.1(b)) are R-equivalent whenever their diagonals $\overline{f}, \overline{g}$ coincide

$$
vf' = f''u = vg' = g''u. \qquad\qquad (7.81)
$$

This quotient was later studied by the present author, in a general context, to study weak limits and proper factorisation systems [G4, G6].

The first paper shows that $\mathrm{Fr}(C)$ has an induced proper factorisation system (easily described in the strict form of 2.5.3(f)), and is the *weakly free category with proper factorisation system* generated by C. The second studies the induced 2-monad on the 2-functor $\mathrm{Fr} \colon \mathsf{Cat} \to \mathsf{Cat}$, and proves that there is a canonical bijection between proper factorisation systems

in C and pseudo-isomorphism classes of unitary pseudo algebras of this
2-monad.

Similar, simpler relations hold in the *strict* case: *strict factorisation systems are strictly monadic on* Cat, *as well as the proper strict factorisation
systems.*

References

[AHS] J. Adámek, H. Herrlich, G.E. Strecker, Abstract and concrete categories. The joy of cats, John Wiley & Sons, New York 1990.

[AnF] F.W. Anderson, K.R. Fuller, Rings and categories of modules, Springer-Verlag 1974.

[AT] H. Appelgate, M. Tierney, Categories with models, in: Seminar on triples and categorical homology theory, Lecture Notes in Math. Vol. 80, Springer-Verlag 1969, pp. 156–244.

[BaL] J.C. Baez, A.D. Lauda, A prehistory of n-categorical physics, in Deep Beauty: Mathematical Innovation and the Search for an Underlying Intelligibility of the Quantum World, ed. Hans Halvorson, Cambridge University Press, Cambridge 2011, pp. 13–128. Available at: https://arxiv.org/abs/0908.2469

[BaP] J.C. Baez, B.S. Pollard, A Compositional Framework for Reaction Networks, 2017. Available at: https://arxiv.org/abs/1704.02051

[Bar] M. Barr, Exact categories, in Lecture Notes in Math. Vol. 236, Springer-Verlag 1971, pp. 1–120.

[BarB] M. Barr, J. Beck, Homology and standard constructions, in Seminar on Triples and Categorical Homology Theory, Springer-Verlag 1969, pp. 245–335.

[BarW] M. Barr, C. Wells, Category Theory for Computing Science, Prentice Hall, New York 1990.

[BasE] A. Bastiani, C. Ehresmann, Multiple Functors I. Limits Relative to Double Categories, Cahiers Top. Géom. Diff. 15 (1974), 215–292.

[BBP] M.A. Bednarczyk, A.M. Borzyszkowski, W. Pawlowski, Generalized congruences-epimorphisms in Cat, Theory Appl. Categ. 5 (1999), No. 11, 266–280.

[Bc] J.M. Beck, Triples, algebras and cohomology, Repr. Theory Appl. Categ. 2 (2003).

[Be1] J. Bénabou, Catégories avec multiplication, C. R. Acad. Sci. Paris 256 (1963), 1887-1890.

[Be2] J. Bénabou, Introduction to bicategories, in: Reports of the Midwest Category Seminar, Lecture Notes in Math. Vol. 47, Springer-Verlag 1967, pp. 1–77.

[Be3] J. Bénabou, Les distributeurs, Inst. de Math. Pure et Appliquée, Université Catholique de Louvain, Rapport n. 33, 1973.

[Bi] G. Birkhoff, Lattice theory, 3rd ed., Amer. Math. Soc. Coll. Publ. 25, Providence 1973.

[BKPS] G.J. Bird, G.M. Kelly, A.J. Power, R.H. Street, Flexible limits for 2-categories, J. Pure Appl. Algebra 61 (1989), 1–27.

[Bo1] F. Borceux, Handbook of categorical algebra 1, Cambridge University Press, Cambridge 1994.

[Bo2] F. Borceux, Handbook of categorical algebra 2, Cambridge University Press, Cambridge 1994.

[Bo3] F. Borceux, Handbook of categorical algebra 3, Cambridge University Press, Cambridge 1994.

[Bo4] F. Borceux, A survey of semi-abelian categories, in: Galois theory, Hopf algebras, and semiabelian categories, Fields Inst. Commun. Vol. 43, Amer. Math. Soc., Providence 2004, p. 27–60.

[BoB] F. Borceux, D. Bourn, Mal'cev, protomodular, homological and semi-abelian categories, Kluwer Academic Publishers, Dordrecht 2004.

[BoJ] F. Borceux, G. Janelidze, Galois theories, Cambridge University Press, Cambridge 2001.

[Bor1] N. Bourbaki, Algèbre multilinéaire, Hermann, Paris 1948.

[Bor2] N. Bourbaki, Topologie Générale, Ch. X, Hermann, Paris 1961.

[Bor3] N. Bourbaki, Algèbre Linéaire, 3e édition, Hermann, Paris 1962.

[Bou1] D. Bourn, Normalization equivalence, kernel equivalence and affine categories, in: Category theory (Como, 1990), Lecture Notes in Math. Vol. 1488, Springer-Verlag 1991, pp. 43–62.

[Bou2] D. Bourn, Moore normalization and Dold–Kan theorem for semi-abelian categories, in: Categories in algebra, geometry and mathematical physics, 105–124, Amer. Math. Soc., Providence 2007.

[Bra] H. Brandt, Über eine Verallgemeinerung des Gruppenbegriffes, Math. Ann. 96 (1927), 360–366.

[Bri] H.B. Brinkmann, Relations for exact categories, J. Algebra 13 (1969), 465–480.

[BriP] H.B. Brinkmann, D. Puppe, Abelsche und exacte Kategorien, Korrespondenzen, Lecture Notes in Math. Vol. 96, Springer-Verlag 1969.

[Bro1] R. Brown, Elements of modern topology, McGraw–Hill, New York 1968.

[Bro2] R. Brown, Topology and groupoids, Third edition of Elements of modern topology, BookSurge, LLC, Charleston SC 2006.

[BroH1] R. Brown, P.J. Higgins, On the algebra of cubes, J. Pure Appl. Algebra 21 (1981), 233–260.

[BroH2] R. Brown, P.J. Higgins, Tensor products and homotopies for w-groupoids and crossed complexes, J. Pure Appl. Algebra 47 (1987), 1–33.

[BruMM] R. Bruni, J. Meseguer, U. Montanari, Symmetric monoidal and Cartesian double categories as a semantic framework for tile logic, Math. Structures Comput. Sci. 12 (2002), 53–90.

[Bu1] D.A. Buchsbaum, Exact categories and duality, Trans. Amer. Math. Soc. 80 (1955), 1–34.

[Bu2] D.A. Buchsbaum, Appendix, in: H. Cartan, S. Eilenberg, Homological algebra, Princeton University Press, Princeton 1956.

[BunGS] M.C. Bunge, F. Gago, A.M. San Luis, Synthetic Differential Topology, to appear.

[CaG] A. Carboni, M. Grandis, Categories of projective spaces, J. Pure Appl. Algebra 110 (1996) 241–258.

[CaJKP] A. Carboni, G. Janelidze, G.M. Kelly, R. Paré, On localization and stabilization for factorization system, Appl. Categ. Structures 5 (1997), 1–58.

[CE] H. Cartan, S. Eilenberg, Homological algebra, Princeton University Press,

Princeton 1956.

[ClP] A.H. Clifford, G.B. Preston, The algebraic theory of semigroups, Vol. 1, Math. Surveys of the Amer. Math. Soc. 7, Providence 1961.

[Coh] P.M. Cohn, Universal algebra, Harper & Row, New York 1965.

[Cop] L. Coppey, Algèbres de décompositions et précategories, Diagrammes 3 Suppl. (1980), 1–139.

[Cr] R.L. Crole, Categories for Types, Cambridge University Press, Cambridge 1993.

[Da1] G. Darbo, Aspetti algebrico-categoriali della teoria dei dispositivi, in: Symposia Mathematica, Vol. IV, INDAM, Roma 1968/69. Academic Press, London 1970, pp. 303–336.

[Da2] G. Darbo, Un'algebra locale reticolata che interviene nella teoria dei sistemi, Rend. Sem. Mat. Univ. Padova 60 (1978), 115–139.

[Dg] J. Dugundij, Topology, Allyn & Bacon, Boston 1966.

[Ds] J. Duskin, Variations on Beck's tripleability criterion, Lecture Notes in Math. Vol. 106, Springer-Verlag 1969, pp. 74–129.

[EE] A. Ehresmann, C. Ehresmann, Multiple functors II. The monoidal closed category of multiple categories, Cahiers Top. Géom. Diff. 19 (1978), 295–333.

[Eh1] C. Ehresmann, Catégories des foncteurs types, Rev. Un. Mat. Argentina 20 (1962), 194–209.

[Eh2] C. Ehresmann, Catégories structurées, Ann. Sci. Ecole Norm. Sup. 80 (1963), 349–425.

[Eh3] C. Ehresmann, Catégories et structures, Dunod, Paris 1965.

[Eh4] C. Ehresmann, Oeuvres complètes et commentées, Partie II.2, Editée et commentée par Andrée C. Ehresmann, Amiens 1982.

[EiM] S. Eilenberg, S. Mac Lane, General theory of natural equivalences, Trans. Amer. Math. Soc. 58 (1945), 231294.

[EiK] S. Eilenberg, G.M. Kelly, Closed categories, in Proc. Conf. Categorical Algebra, La Jolla 1965, Springer-Verlag 1966, pp. 421–562.

[EiS] S. Eilenberg, N. Steenrod, Foundations of algebraic topology, Princeton University Press, Princeton 1952.

[FGR1] L. Fajstrup, E. Goubault, M. Raussen, Detecting deadlocks in concurrent systems, in: CONCUR'98 (Nice), 332–347, Lecture Notes in Comput. Sci., 1466, Springer-Verlag 1998.

[FGR2] L. Fajstrup, E. Goubault, M. Raussen, Algebraic topology and concurrency, Theor. Comput. Sci. 357 (2006), 241–178. (Revised version of a preprint at Aalborg, 1999.)

[Fo] B. Fong, The Algebra of Open and Interconnected Systems, Ph.D. thesis, Department of Computer Science, University of Oxford, 2016.

[Fr1] P. Freyd, Abelian categories, An introduction to the theory of functors, Harper & Row, New York 1964. Republished in: Reprints Theory Appl. Categ. 3 (2003).

[Fr2] P. Freyd, Stable homotopy, in: Proc. Conf. Categ. Algebra, La Jolla, 1965, Springer-Verlag 1966, pp. 121–176.

[Fr3] P. Freyd, On the concreteness of certain categories, in: Symposia Mathematica, Vol. IV, INDAM, Roma 1968/69. Academic Press, London 1970, pp. 431–456.

[Fr4] P. Freyd, Homotopy is not concrete, in: The Steenrod algebra and its applications, Battelle Memorial Inst., Columbus, OH, 1970, Lecture Notes in Math. Vol. 168, Springer-Verlag 1970, pp. 25–34.

[FrK] P. Freyd, G.M. Kelly, Categories of continuous functors, J. Pure Appl. Algebra 2 (1972), 1–18.

[FrS] P.J. Freyd, A. Scedrov, Categories, allegories, North-Holland Publ. Co., Amsterdam 1990.

[Ga] R. Garner, A homotopy-theoretic universal property of Leinster's operad for weak w-categories, Math. Proc. Cambridge Philos. Soc. 147 (2009), 615–628.

[G1] M. Grandis, Transfer functors and projective spaces, Math. Nachr. 118 (1984), 147–165.

[G2] M. Grandis, Cohesive categories and manifolds, Ann. Mat. Pura Appl. 157 (1990), 199–244.

[G3] M. Grandis, Cubical monads and their symmetries, in: Proceedings of the Eleventh International Conference on Topology, Trieste 1993, Rend. Ist. Mat. Univ. Trieste 25 (1993), 223–262.

[G4] M. Grandis, Weak subobjects and the epi-monic completion of a category, J. Pure Appl. Algebra 154 (2000), 193–212.

[G5] M. Grandis, Finite sets and symmetric simplicial sets, Theory Appl. Categ. 8 (2001), No. 8, 244–252.

[G6] M. Grandis, On the monad of proper factorisation systems in categories, J. Pure Appl. Algebra 171 (2002) 17–26.

[G7] M. Grandis, The role of symmetries in cubical sets and cubical categories (On weak cubical categories, I), Cah. Topol. Géom. Différ. Catég. 50 (2009), 102–143.

[G8] M. Grandis, Directed Algebraic Topology, Models of non-reversible worlds, Cambridge University Press, Cambridge 2009. Available at: http://www.dima.unige.it/~grandis/BkDAT_page.html

[G9] M. Grandis, Homological Algebra, The interplay of homology with distributive lattices and orthodox semigroups, World Scientific Publishing Co., Hackensack NJ 2012.

[G10] M. Grandis, Homological Algebra in strongly non-abelian settings, World Scientific Publishing Co., Hackensack NJ 2013.

[GM] M. Grandis, L. Mauri, Cubical sets and their site, Theory Appl. Categ. 11 (2003), No. 8, 185–211.

[GP1] M. Grandis, R. Paré, Limits in double categories, Cah. Topol. Géom. Différ. Catég. 40 (1999), 162–220.

[GP2] M. Grandis, R. Paré, Adjoint for double categories, Cah. Topol. Géom. Différ. Catég. 45 (2004), 193–240.

[GP3] M. Grandis, R. Paré, Kan extensions in double categories (On weak double categories, III), Theory Appl. Categ. 20 (2008), No. 8, 152–185.

[GP4] M. Grandis, R. Paré, Lax Kan extensions for double categories (On weak double categories, Part IV), Cah. Topol. Géom. Différ. Catég. 48 (2007), 163–199.

[GT] M. Grandis, W. Tholen, Natural weak factorisation systems, Archivum Mathematicum (Brno) 42 (2006), 397–408.

[Gr1] G. Grätzer, Universal algebra, Van Nostrand Co., Princeton 1968.

[Gr2] G. Grätzer, General lattice theory, Academic Press, New York 1978.

[Gra] J.W. Gray, The existence and construction of lax limits, Cah. Topol. Géom. Différ. 21 (1980), 277–304.

[Grt] P.A. Grillet, Exact categories and categories of sheaves, in Lecture Notes in Math. Vol. 236, Springer-Verlag 1971, pp. 121222.

[Gt] A. Grothendieck, Sur quelques points d'algèbre homologique, Tôhoku Math. J. 9 (1957), 119– 221.

[Ha] A. Hatcher, Algebraic topology, Cambridge University Press, Cambridge 2002. Available at: https://www.math.cornell.edu/~hatcher/AT/ATpage.html

[HeS] H. Herrlich, G.E. Strecker, Category theory, an introduction, Allyn and Bacon, Boston 1973.

[Hi] P.J. Hilton, Correspondences and exact squares, in: Proc. of the Conf. on Categorical Algebra, La Jolla 1965, Springer-Verlag 1966, 255–271.

[HiW] P.J. Hilton, S. Wylie, Homology theory: An introduction to algebraic topology, Cambridge University Press, Cambridge 1960.

[Ho] J.M. Howie, An introduction to semigroup theory, Academic Press, London 1976.

[In] H. Inassaridze, Non-abelian homological algebra and its applications, Kluwer Academic Publishers, Dordrecht 1997.

[JaMT] G. Janelidze, L. Márki, W. Tholen, Semi-abelian categories, in: Category theory 1999 (Coimbra), J. Pure Appl. Algebra 168 (2002), 367–386.

[Je] T. Jech, Set theory, Academic Press, New York 1978.

[Jo1] P.T. Johnstone, The point of pointless topology, Bull. Amer. Math. Soc. 8 (1983), 4153.

[Jo2] P.T. Johnstone, The art of pointless thinking: a student's guide to the category of locales, in: Category theory at work (Bremen, 1990) Heldermann, Berlin 1991, pp. 85–107.

[Jo3] P.T. Johnstone, Sketches of an elephant: a topos theory compendium, Vol. 1, Oxford University Press, New York 2002.

[Jo4] P.T. Johnstone, Sketches of an elephant: a topos theory compendium, Vol. 2, Oxford University Press, New York 2002.

[K1] D.M. Kan, Abstract homotopy I, Proc. Nat. Acad. Sci. U.S.A. 41 (1955), 1092–1096.

[K2] D.M. Kan, Adjoint functors, Trans. Amer. Math. Soc. 87 (1958), 294–329.

[Ke] J.L. Kelley, General topology, Van Nostrand, New York 1955.

[Kl1] G.M. Kelly, On Mac Lane's conditions for coherence of natural associativities, commutativities, etc., J. Algebra 1 (1964), 397–402.

[Kl2] G.M. Kelly, Basic concepts of enriched category theory, Cambridge University Press, Cambridge 1982.

[Kl3] G.M. Kelly, Elementary observations on 2-categorical limits, Bull. Austral. Math. Soc. 39 (1989), 301–317.

[KlS] G.M. Kelly, R. Street, Review of the elements of 2-categories, in: Category Seminar, Sydney 1972/73, Lecture Notes in Math. Vol. 420, Springer-Verlag 1974, pp. 75–103.

[Ko] A. Kock, Synthetic differential geometry, Cambridge University Press, Cambridge 1981.

[KrT] M. Korostenski, W. Tholen, Factorization systems as Eilenberg–Moore algebras, J. Pure Appl. Algebra 85 (1993), 57–72.

[LaS] J. Lambek, P.J. Scott, Introduction to higher order categorical logic, Cambridge University Press, Cambridge 1986.

[Law] M.V. Lawson, Inverse semigroups, The theory of partial symmetries, World Scientific Publishing Co., River Edge 1998.

[Lw1] F.W. Lawvere, Metric spaces, generalized logic and closed categories, Rend. Sem. Mat. Fis. Univ. Milano 43 (1974), 135–166. Republished in: Reprints Theory Appl. Categ. 1 (2002).

[Lw2] F.W. Lawvere, Functorial semantics of algebraic theories and some algebraic problems in the context of functorial semantics of algebraic theories, Repr. Theory Appl. Categ. 5 (2004).

[LwR] F.W. Lawvere, R. Rosebrugh, Sets for mathematics, Cambridge University Press, Cambridge 2003.

[LwS] F.W. Lawvere, S.H. Schanuel, Conceptual mathematics. A first introduction to categories. Second edition. Cambridge University Press, Cambridge 2009.

[Le] T. Leinster, Higher operads, higher categories, Cambridge University Press, Cambridge 2004.

[M1] S. Mac Lane, An algebra of additive relations, Proc. Nat. Acad. Sci. USA 47 (1961), 1043–1051.

[M2] S. Mac Lane, Homology, Springer-Verlag 1963.

[M3] S. Mac Lane, Natural associativity and commutativity, Rice University Studies 49 (1963), 28–46.

[M4] S. Mac Lane, Categories for the working mathematician, Springer-Verlag 1971.

[MaM] S. Mac Lane, I. Moerdijk, Sheaves in geometry and logic. A first introduction to topos theory, Springer-Verlag 1994.

[MaP] S. Mac Lane, R. Paré, Coherence for bicategories and indexed categories, J. Pure Appl. Algebra 37 (1985), 59–80.

[MkR] M. Makkai, G.E. Reyes, First order categorical logic. Model-theoretical methods in the theory of topoi and related categories, Lecture Notes in Math. Vol. 611, Springer-Verlag 1977.

[Mas] W.S. Massey, Singular homology theory, Springer-Verlag 1980.

[Mat] M. Mather, Pull-backs in homotopy theory, Can. J. Math. 28 (1976), 225–263.

[Me] J. Meisen, Relations in regular categories, in: Localization in group theory and homotopy theory and related topics, Lecture Notes in Math. Vol. 418, Springer-Verlag 1974, pp. 96–102.

[Mic] E. Michael, Local compactness and Cartesian products of quotient maps and k-spaces, Ann. Inst. Fourier (Grenoble) 18 (1968), 281–286.

[Mit] B. Mitchell, Theory of categories, Academic Press, New York 1965.

[Mo] K. Morita, Duality for modules and its applications to the theory of rings with minimum condition, Sci. Rep. Tokyo Kyoiku Daigaku, Sect. A 6 (1958), 83–142.

[Pa] R. Paré, Simply connected limits, Canad. J. Math. 42 (1990), 731–746.

[Pu] D. Puppe, Korrespondenzen in abelschen Kategorien, Math. Ann. 148 (1962), 1–30.

[Qu] D. Quillen, Higher algebraic K-theory, I, in: Lecture Notes in Math. Vol. 341, Springer-Verlag, pp. 85–147.

[Se] Z. Semadeni, Banach spaces of continuous functions, Polish Sci. Publ., Warszawa 1971.

[St1] R. Street, Fibrations and Yoneda's lemma in a 2-category, in: G.M. Kelly Ed., Category Seminar, Sydney 1972/73, Lecture Notes in Math. Vol. 420, Springer-Verlag 1974, pp. 104–133.

[St2] R. Street, Limits indexed by category-valued 2-functors, J. Pure Appl. Alg. 8 (1976), 149–181.

[St3] R. Street, Cosmoi of internal categories, Trans. Amer. Math. Soc. 258 (1980), 271–318.

[T1] M.S. Tsalenko, Correspondences over a quasi exact category, Dokl. Akad. Nauk SSSR 155 (1964), 292–294.

[T2] M.S. Tsalenko, Correspondences over a quasi exact category, Mat. Sbornik 73 (1967), 564–584.

[Vi] J.W. Vick, Homology theory. An introduction to algebraic topology. Second edition, Springer-Verlag 1994.

Index

Printed in the United States
By Bookmasters